MICROSTRUCTURAL SCIENCE VOLUME 8

Sponsored by
The International Metallographic Society

MICROSTRUCTURAL SCIENCE VOLUME 8

Proceedings of the Twelfth Annual Technical Meeting of the International Metallographic Society

Edited by

Donald W. Stevens
Xerox Corporation

George F. Vander Voort
Bethlehem Steel Corporation

James L. McCall
Battelle — Columbus Laboratories

ELSEVIER
NEW YORK • OXFORD

Elsevier North Holland, Inc.
52 Vanderbilt Avenue, New York, New York 10017

Distributors outside the United States and Canada:

Elsevier/North Holland
335 Jan van Galenstraat, P.O. Box 211
Amsterdam, The Netherlands

LC 73-10895
ISBN 0-444-00561-7

Manufactured in the United States of America

CONTENTS

Contributors

Adler, T.A.
Virginia Polytechnic Institute
Blacksburg, Virginia USA

Augis, J.A.
Bell Telephone Laboratories
Columbus, Ohio USA

Bacon, R.J.
Systems Research Laboratories
Dayton, Ohio USA

Bagnall, C.
Westinghouse Electric Corporation
Madison, Pennsylvania USA

Banerji, S.K.
Foote Mineral Company
Exton, Pennsylvania USA

Bennett, J.E.
Bell Telephone Laboratories
Columbus, Ohio USA

Bernett, M.K.
Naval Research Laboratory,
Washington, D.C. USA

Biss, V.
Climax Molybdenum Company
of Michigan
Ann Arbor, Michigan USA

Blattner, R.J.
University of Illinois at Urbana
Urbana, Illinois USA

Blau, P.J.
The Ohio State University
Columbus, Ohio USA

Braun, J.
Monsanto Research Corporation
Miamisburg, Ohio USA

Briant, C.L.
General Electric Company
Schenectady, New York USA

Cameron, T.B.
University of Connecticut
Storrs, Connecticut USA

Canonico, D.A.
Oak Ridge National Laboratory
Oak Ridge, Tennessee USA

Clausing, R.E.
Oak Ridge National Laboratory
Oak Ridge, Tennessee USA

Conrad, H.
University of Kentucky
Lexington, Kentucky USA

Crouse, R.S.
Oak Ridge National Laboratory
Oak Ridge, Tennessee USA

DeAngelis, R.J.
University of Kentucky
Lexington, Kentucky USA

Eylon, D.
University of Cincinnati
Cinnati, Ohio USA

Eyre, T.S.
Brunel University
Uxbridge, Middlesex, UK

Froes, F.H.
Wright-Patterson Air Force Base
Dayton, Ohio USA

Ganesan, P.
University of Kentucky
Lexington, Kentucky USA

Golden, J.F.
E. Leitz, Incorporated
Rockleigh, New Jersey USA

Goldstein, J.I.
Lehigh University
Bethlehem, Pennsylvania USA

Gray, R.J.
Oak Ridge National Laboratory
Oak Ridge, Tennessee USA

Griess, J.C., Jr.
Oak Ridge National Laboratory
Oak Ridge, Tennessee USA

Henson, T.J.
Oak Ridge National Laboratory
Oak Ridge, Tennessee USA

Hoegfeldt, J.M.
Honeywell, Incorporated
Hopkins, Minnesota USA

Houck, C.W.
Oak Ridge National Laboratory
Oak Ridge, Tennessee USA

Hunn, W.
E. Leitz, Incorporated
Rockleigh, New Jersey USA

Irene, E.A.
IBM Thomas J. Watson Research Center
Yorktown Heights, New York USA

Jacobus, N.
EG&G ORTEC
Oak Ridge, Tennessee USA

Johnson, W.C.
Perkin-Elmer Corporation
Eden Prairie, Minnesota USA

Kenny, L.D.
Aluminum Company of Canada, Ltd.
Kingston, Ontario, Canada

Keshavan, M.K.
University of Kentucky
Lexington, Kentucky USA

Kim, Y.–W.
Wright-Patterson Air Force Base
Dayton, Ohio USA

Krishnadev, M.R.
Laval University
Quebec, Canada

Kuo, H.K.
University of Kentucky
Lexington, Kentucky USA

LeMay, I.
University of Saskatchewan
Saskatoon, Canada

Levy, S.A.
Reynolds Metals Company
Richmond, Virginia USA

Little, B.
National Space Technology Laboratory
Station, Mississippi USA

Louthan, M.R., Jr.
Virginia Polytechnic Institute
Blacksburg, Virginia USA

Lui, M.–W.
Meyer Aluminum Ltd.
Kowloon, Hong Kong

McCall, J.L.
Battelle-Columbus Laboratories
Columbus, Ohio USA

McNitt, R.P.
Virginia Polytechnic Institute
Blacksburg, Virginia USA

Morris, L.R.
Aluminum Co. of Canada, Ltd.
Kingston, Ontario, Canada

Nelson, J.A.
Buehler, Ltd.
Evanston, Illinois USA

Odom, T.R.
Paducah Gaseous Diffusion Plant
Paducah, Kentucky USA

Omlor, R.E.
Systems Research Laboratories
Dayton, Ohio USA

Pavlichko, W.R.
Consumers Power Company
Jackson, Michigan USA

Pinnel, M.R.
Bell Telephone Laboratories
Columbus, Ohio USA

Pirs, J.
Technical Faculty Rijeka
Rijeka, Yugoslavia

Ravner, H.
Naval Research Laboratory
Washington, D.C. USA

Rhines, F.N.
University of Florida
Gainesville, Florida USA

Rhude, H.V.
Argonne National Laboratory
Argonne, Illinois USA

Ryvola, M.
Aluminum Co. of Canada, Ltd.
Kingston, Ontario, Canada

Sagues,A.A.
Kentucky Center for Energy Research Laboratory
Lexington, Kentucky USA

Samuels, L.E.
Australian Dept. of Defense
Ascot Vale, Victoria, Australia

Sargent, C.M.
University of Saskatchewan
Saskatoon, Canada

Slaughter, G.M.
Oak Ridge National Laboratory
Oak Ridge, Tennessee

Sojka, G.J.
Universite Laval
Quebec, Canada

Solomon, A.
Consumers Power Company
Jackson, Michigan USA

Stevens, D.W.
Xerox Corporation
Webster, New York USA

Teeters, G.L.
Systems Research Laboratories
Dayton, Ohio USA

Tierney, E.
IBM Thomas J. Watson Research Center
Yorktown Heights, New York USA

Tiner, N.A.
Microprocess Laboratory
Laguna Beach, California USA

Tompkins, H.G.
Bell Telephone Laboratories
Columbus, Ohio USA

VanderVoort, G.F.
Bethlehem Steel Corporation
Bethlehem, Pennsylvania USA

Wheeler, B.D.
EG&G ORTEC
Oak Ridge, Tennessee USA

White, W.E.
University of Calgary
Calgary, Canada

Witkowski, R.E.
Westinghouse Electric Corporation
Madison, Pennsylvania USA

Wollenberg, R.L.
General Electric Company
Coshocton, Ohio USA

Zakraysek, L.
General Electric Company
Syracuse, New York USA

Preface

This is the Eighth Volume of the *Microstructural Science* series and contains most of the papers presented at the Twelfth Annual Technical Meeting of the International Metallographic Society (IMS) at Tamiment, Pennsylvania, USA, in the Pocono Mountains, July 10 and 11, 1979. The papers span a rich variety of materials structure investigations under the increasingly broad umbrella of the art and science of metallography. The subject matter includes grain boundary embrittlement, phase diagram determination, surface oxidation of thin films, fractography, failure analysis of engineering components, wear, erosion, ocean biofouling, solvent cleaning, oxygen absorption, sintering, graphitization, the theory and practice of image analysis, and advances in metallographic techniques. The principal metallographic techniques employed in the above investigations are classical (light optics) metallography, scanning and transmission electron microscopy (frequently with energy dispersive x-ray analysis to complement structure/morphology information with local chemistry), quantitative image analysis, x-ray diffraction, and surface chemistry analysis by Auger electron and x-ray photoelectron spectroscopy, etc. Nearly half the papers are focussed on the application of surface analysis techniques to relate surface chemistry to properties and behavior not otherwise understood sufficiently from structure and chemistry analysis of bulk material.

The numerous, diverse entries at the meeting's International Metallographic Exhibit (sponsored jointly by IMS and the American Society for Metals) comprised an enviable display of metallography at its finest; the Exhibit awards are announced in the appendix, along with other prestigious awards given by IMS. In addition to recognizing outstanding achievements and contributions by individuals, these awards reflect the Society's goal of continually upgrading the professional standards of metallographers in applying their knowledge and skill towards the understanding of engineering materials and their intelligent application to the needs of man.

Many persons contributed, either directly or indirectly, to the success of this meeting. To all of them go our warmest thanks. This meeting was held under the capable guidance of the recent Past-President of IMS, Robert J. Gray, and his Vice-President (now President), P. Michael French. The General Chairmanship was performed vigorously by George F. Vander Voort. The Technical Meeting Chairman and Co-Chairman were Donald W. Stevens and William E. White, respectively. C.K.H. "Bud" DuBose was Metallographic Exhibit Chairman, an arduous task well done. Again, James L. McCall was the spark plug and generous contributor in putting this volume together. Connie McCall rose once more to the occasion of transforming a diverse array of papers into a skillfully typed, homogeneous manuscript.

Donald W. Stevens
Xerox Corporation
Rochester, New York

George F. VanderVoort
Bethlehem Steel Corporation
Bethlehem, Pennsylvania

James L. McCall
Battelle-Columbus Laboratories
Columbus, Ohio

METALLOGRAPHIC CHARACTERIZATION OF SURFACES: I

Chairpersons: J.D. Braun and R.S. Crouse

Stevens, Vander Voort, McCall, eds. Microstructural Science, Volume 8

GRAIN BOUNDARY CHEMISTRY AND TEMPERED MARTENSITE EMBRITTLEMENT IN STEELS

S.K. Banerji* and C.L. Briant**

ABSTRACT

This paper demonstrates the use of present day analytical techniques to understand the impurity induced tempered martensite embrittlement of ultra-high strength steels. The most harmful elements in this embrittlement are P, S, and N. Auger electron spectroscopy provided the basic tool for the understanding of the role of P. Electron microprobe analysis explained the S effects. Transmission electron microscopy was necessary to understand the behavior of N. With the aid of such techniques it is now possible to understand this and other kinds of intergranular embrittlement in steels.

INTRODUCTION

Segregation of residual impurities to grain boundaries are known to cause severe intergranular embrittlement in low-alloy steels[1]. Tempered martensite embrittlement (TME) in commercial, ultra-high strength steels is an example of such impurity-induced embrittlement.[1-2] The phenomenology of this embrittlement is illustrated schematically in Fig. 1. When the quenched martensite is tempered for a short time, usually 1–2 hours, at relatively low temperatures (100–500°C), the hardness and strength of the steel decrease (Fig. 1a & b) with increasing tempering temperature. With this general behavior one expects an improvement in toughness. However, in the range of 250–400°C there is an anomalous decrease in notched-bar fracture energy (Fig. 1c). This loss of toughness, or embrittlement trough, is usually accompanied by the onset of brittle intergranular failure along prior austenite grain boundaries (Fig. 1d) and a peak in the ductile to brittle transition temperature (Fig. 1e). These observations are the hallmarks of tempered martensite embrittlement.

The microstructural changes associated with the tempering of martensite are illustrated in Fig. 1f. The onset of TME coincides with the transformation of ϵ-carbides to plate-like cementite.[3,4] Restoration of impact toughness at higher tempering temperatures is associated with coarsening and spheroidization of cementite. This suggests that the precipitation of thin plate-like cementite is a necessary criterion for the occurrence of this embrittlement. It does not, however, explain why the fracture mode accompanying this embrittlement is intergranular because precipitation of cementite is not localized at the grain boundaries. Nor does it explain why a high-purity steel is immune to this intergranular embrittlement (Fig. 1g).[2,5] Recent work[2] has shown clear evidence of P, S, and N on the intergranular fracture surfaces of embrittled, commercial 4340 steels. It is known that these impurities can cause a drastic reduction in the strength of the grain boundaries.[1] Therefore, their presence is also required for this embrittlement.

*Foote Mineral Co., Exton, Pennsylvania 19341 USA.
**Corporate Research & Development, General Electric Co., Schenectady, New York 12301 USA.

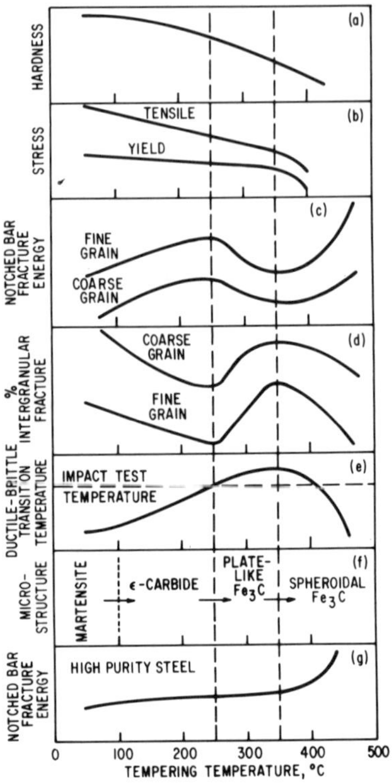

Fig. 1. Characteristics of tempered martensite embrittlement.

This paper summarizes some recent work to show how the modern analytical and metallographic techniques have allowed us to uncover the effects of residual impurities on this type of embrittlement. The individual role of P, S, and N in the embrittlement process was investigated by doping high-purity vacuum melted steels (0.3C, 3.5Ni, 1.7Cr)* with controlled amounts of the desired impurity. The embrittlement was characterized by the measurement of Charpy V-notch fracture energy.

EFFECT OF PHOSPHORUS

Three doping levels of P, 0.01, 0.03, and 0.06% were examined and the results were compared with a high-purity undoped steel (0.004S, 0.004P) of the same base composition, Fig. 2. Even 0.01%P in this steel causes severe embrittlement. Scanning electron microscopy of the fracture surfaces showed that the amount of intergranular fracture inversely correlated with the fracture energy. The embrittlement trough always coincided with a peak in the amount of intergranular fracture. This intercrystalline separation is very smooth and brittle.

Why, then, is the embrittlement intergranular and why is it enhanced by increased P concentration? Auger electron spectroscopy (AES) provided the explanation. An Auger spectrum yields a chemical analysis of the first monolayer of the analyzed surface, the subsequent layers contributing only a very small fraction to the signal. The

*All compositions are in weight percent.

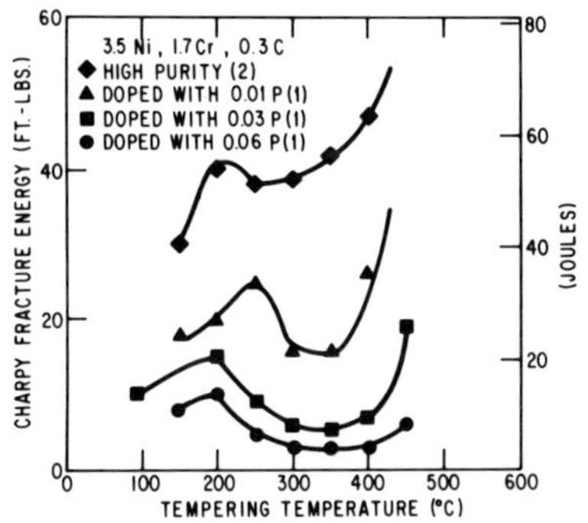

Fig. 2. Room temperature fracture energy plotted as a function of tempering temperature for the three P doped heats and a high purity heat. The numbers in parentheses are ASTM grain sizes.

use of a scanning Auger microprobe allowed unambiguous identification of the intergranular facets to be analyzed and a chemical analysis of each exposed facet. In this way the occurrence of preferential segregation at the grain boundaries was determined by comparing spectra from an intergranular facet and a transgranular area. All fractures and analyses were performed at 1×10^{-10} torr.

In these steels, segregation of phosphorus was observed on the prior austenite grain boundaries. The P to Fe peak-height ratio was used as a measure of the amount of segregation. Grain to grain analysis of the grain boundaries showed that not all of them contained the same amount of P. This was to be expected due to the differences in orientation and energy of various grain boundaries.[6] Therefore, a large number of grain boundaries were analyzed and averaged to represent the segregation in any given sample.

The most important result of this Auger investigation is that all the phosphorus segregation important in this embrittlement occurred during austenitization and not during tempering, Fig. 3. The amount of P that segregates to the grain boundaries in the austenite was found to depend directly on the bulk P concentration. This explains the severity of embrittlement with increasing bulk P content of the steel.

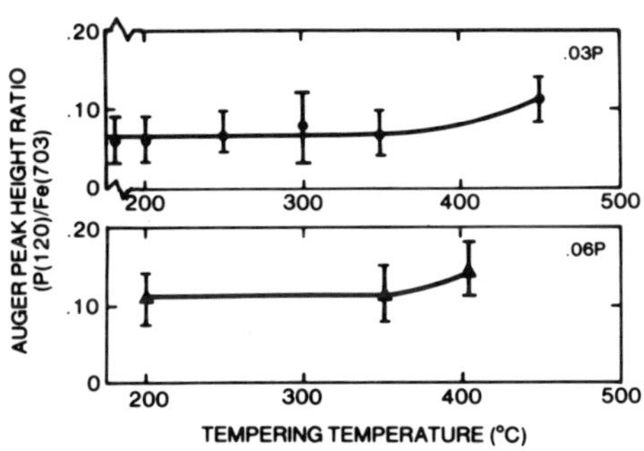

Fig. 3. Grain boundary P concentration plotted as a function of tempering temperature. Point to left of break in abscissa is for as quenched state. Samples had an ASTM grain size of 1.

If the segregation of P to the grain boundaries is inherited from the austenite, why does the embrittlement trough occur on tempering in the 250–400°C range? The impurity segregation in the austenite lowers the cohesive strength of the grain boundaries irrespective of the tempering conditions. When the steel is tempered in the 250–400°C range, the cementite platelets which precipitate on the already impurity-weakened grain boundaries provide a sufficiently strong slip-barrier to help initiate an intergranular crack along the weak grain boundaries.

In summary, there are two essential steps in this intergranular embrittlement process: (a) grain boundary segregation of phosphorus in the austenite which lowers grain boundary cohesion, and (b) precipitation of grain boundary cementite during tempering which aids intergranular cracking and the onset of the embrittlement.

EFFECT OF SULFUR

It was shown earlier that the high-purity steel with only 40 ppm each of S and P and a grain size of ASTM No. 2 (austenitized at 1000°C/1 hr) showed a very weak embrittlement trough, Fig. 2. Figure 4 shows that at a coarser grain size of ASTM No. 2 achieved by austenitizing at a high temperature (1200°C/3h) produced a rather large TME trough. Fractographic examination clearly showed that this embrittlement trough is associated with intergranular fracture.[7]

The Auger examination of the intergranular fracture surfaces showed S segregated to the boundaries, Fig. 5. As in the case of phosphorus, the average grain boundary sulfur concentration remained unchanged upon tempering below 400°C, so segregation must have occurred during austenitization.

However, sulfur has a higher affinity for compound formation with some of the alloying or residual elements in steels. Samples that were first austenitized at 1200°C for 3 hours and then cooled slowly to 850°C and held there for 1 hour before quenching, showed no embrittlement trough, although the fracture energy at all tempering temperatures was lowered, Fig. 6. The fracture surfaces of these specimens contained a large number of intergranular facets.

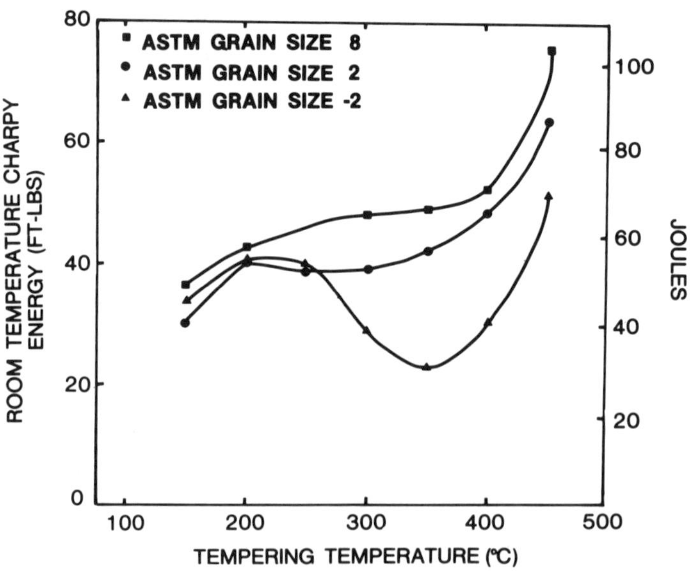

Fig. 4. Room temperature fracture plotted as a function of tempering temperature for the high purity heat.

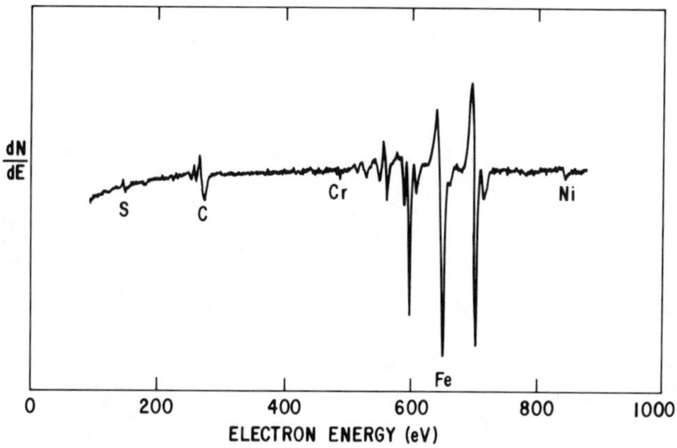

Fig. 5. An Auger spectrum from a grain boundary of the high purity heat.

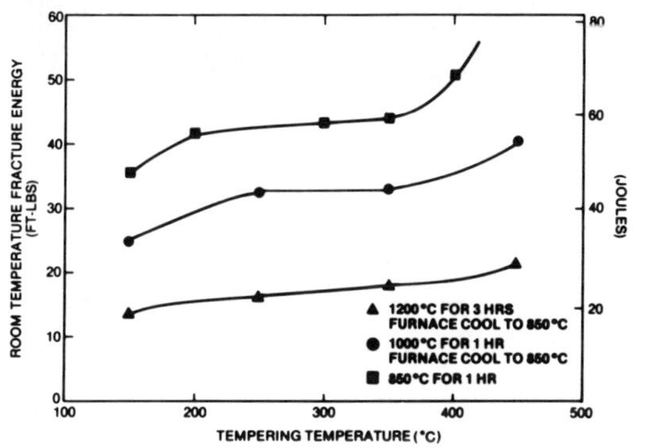

Fig. 6. Fracture energy of high purity samples slowly cooled through the austenite.

A closer examination of the fracture surface revealed that these intergranular facets were not smooth but ductile. The microvoids on the grain boundary surfaces contained small particles. Both electron microprobe and EDAX identified these as chromium sulfide, Fig. 7. Thus, at low austenitizing temperatures, no TME is observed because all the sulfur is still tied up with the undissolved chromium sulfides and cannot segregate to the austenite grain boundaries. With increasing austenitizing temperature, the sulfides dissolve and more free sulfur is available to segregate to the grain boundaries. This causes the appearance of the embrittlement trough, and the associated smooth intergranular fracture.

EFFECT OF NITROGEN

Nitrogen is present in most commercial steels at least in the 0.005–0.015% range, and previous workers[5,8] reported that N may well be the most deleterious element in TME. A 0.01% N-doped steel was used to study this problem. Figure 8 shows that when the grain size is coarse the typical embrittlement trough is observed. No embrit-

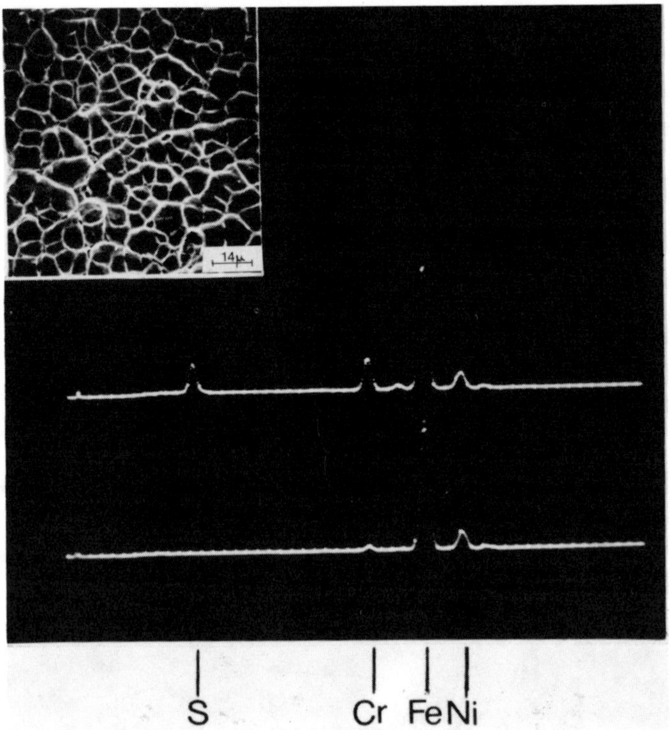

Fig. 7. Energy dispersive X-ray analysis of particles on grain boundary facet (upper curve) and matrix (lower curve) of high purity samples.

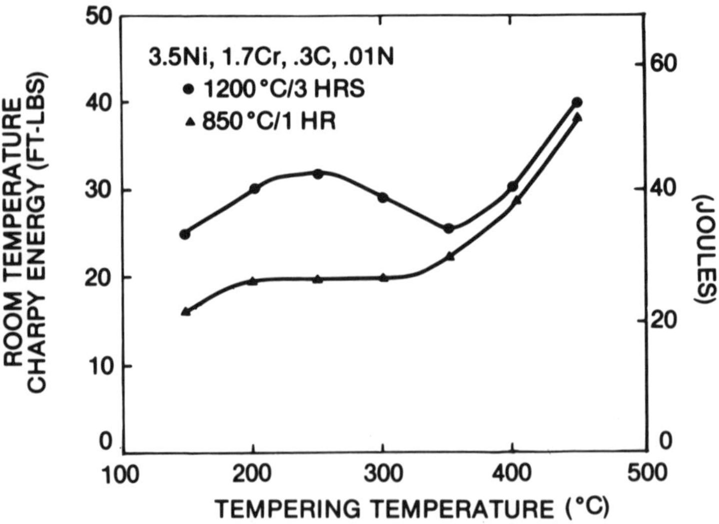

Fig. 8. Room temperature fracture of the nitrogen doped heat plotted as a function of tempering temperature. The two curves are for ASTM grain sizes 2 and 8 which were fixed by using the upper and lower austenitizing temperatures shown on the figure.

tlement is observed when the grain size is fine. Also, the coarse grain steel has higher fracture energies than the fine grain samples after all tempering temperatures up to 400°C. This is anomalous.

Fractographic examination showed another unexpected behavior. The TME trough observed in the large grain samples was not associated with any change in fracture mode. The fracture was microvoid coalescence (MVC) with only some of it occurring along the grain boundaries. This general behavior persisted at all tempering temperatures up to 450°C, Fig. 9a. The presence of very fine particles in the microvoids were noticed but could not be identified by electron microprobe due to intergranular MVC increased rapidly and some large particles could be observed inside the grain boundary microvoids. The electron microprobe analysis showed these particles to be rich in Cr. The fracture behavior of the fine grain samples was mainly intergranular MVC, Fig. 9b, and also did not change on tempering up to about 450°C. The discreet particles found inside the grain boundary microvoids were similar to the Cr-rich particles found in the large grain samples (aust. 1200°C/3h) tempered at 650°C.

Transmission electron microscopy was used to examine the microstructural changes associated with tempering. In early stages of tempering below 250°C before the onset of the embrittlement trough, the microstructure of the large grain samples consisted of a fine dispersion of Fe-Ni nitrides in the matrix as well as on the grain boundaries, Fig. 10a. In addition, ϵ-carbides resulting from the decomposition of martensite were also present in the microstructure. In the tempering temperature range of 250–450°C, the fine Fe-Ni nitrides transformed to larger disc-like iron-nitrides and some chromium-nitrides, Fig. 10b. The transformation of ϵ-carbides to thin plate-like cementite also occurred in this range of tempering temperature. After tempering above 450°C, the iron-nitrides transformed to the more stable chromium-nitrides which continued to grow at the grain boundaries, Fig. 10c. The fine grain samples austenitized at 850°C showed the presence of these grain boundary chromium-nitrides at all tempering temperatures up to 450°C in addition to ϵ-carbides and cementite at appropriate temperatures.

These observations can now rationalize the anomalous embrittlement behavior of the N-doped steel. The dependence of the embrittlement on austenitizing temperature is similar to the case of S-induced embrittlement because N is also a compound forming element. Chromium-nitride is the stable compound in this case. When the austenizing temperature is high; e.g., 1200°C, all the chromium nitride has dissolved and nitrogen is free to segregate to the austenite grain boundaries. Because of high diffusivity of nitrogen even at low tempering temperatures (<250°C), this quenched-in free nitrogen has sufficient mobility to form the kinetically favored Fe-Ni nitrides. These nitrides are too small to impair the impact strength of the steel, so the fracture energy increases. When the tempering temperature is between 250–450°C, the formation of larger disc-like iron-nitrides act as potential sites for micro-crack nucleation and cause a drop in the fracture energy. This produces the transgranular MVC. At still higher tempering temperatures; e.g., 650°C, the transformation of these iron-nitrides to chromium nitrides and their growth at the grain boundaries leads to increased intergranular MVC. This also inhibits the sharp increase in fracture energy normally expected at high tempering temperatures. These chromium nitrides are stable well into the austenite range. Therefore, when the austenitizing temperature is low; e.g., 850°C (fine grain size), there is no free nitrogen available in the quenched martensite. As a result, upon tempering there is no parallel sequence of precipitating nitrides as seen in the case of high austenitizing temperature and tempered martensite embrittlement is not observed.

CONCLUSIONS

In summary, this paper illustrates how modern analytical techniques have been used to characterize the chemical constituents of fracture surfaces. Particularly, the availa-

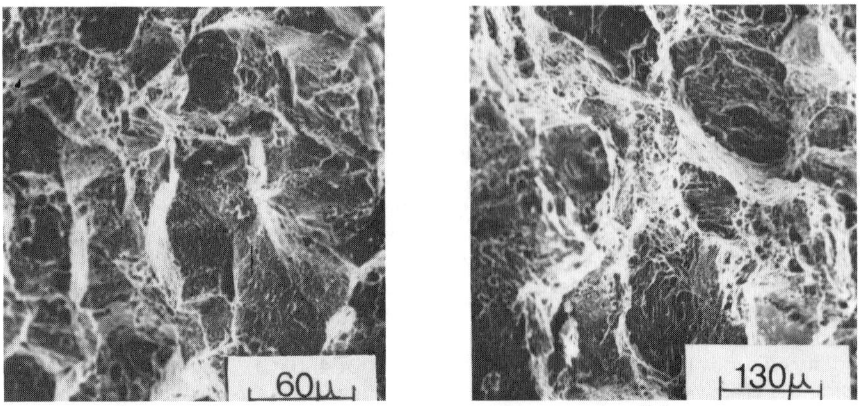

(a) (b)

Fig. 9. Fracture surfaces of the nitrogen doped steel. (a) large grain sample and (b) fine grain sample.

bility of surface sensitive techniques for the chemical analysis of uncontaminated fracture surfaces has provided a quantum jump in the understanding of fracture behavior in many kinds of intergranular embrittlement in steels. Hopefully, this understanding will lead to better alloy design to meet the increasingly severe demands on performance in future years of new emerging technologies.

ACKNOWLEDGMENT

The authors wish to thank Foote Mineral Company and General Electric Company for their permission to publish this work. Assistance of Ann Ritter with the transmission electron microscopy is gratefully acknowledged.

REFERENCES

1. C.L. Briant and S.K. Banerji, "Intergranular Failure in Steels: The Role of Grain-Boundary Composition", *International Metals Reviews,* Vol. 23, pp. 164–199, (1978).
2. S.K. Banerji, C.J. McMahon, Jr., and H.C. Feng, "Intergranular Fracture in 4340-Type Steels: Effects of Impurities and Hydrogen", *Metallurgical Trans. A,* Vol. 9A, pp. 237–247, (1978).
3. L.J. Klingler, W.J. Barnett, R.P. Frohmberg and A.R. Troiano, "The Embrittlement of Alloy Steel at High Strength Levels", *Transaction of the ASM,* Vol. 46, pp. 1557–1598, (1954).
4. B.S. Lement, B.L. Averbach and M. Cohen, "Microstructural Changes on Tempering Iron-Carbon Alloys", *Trans. of the ASM,* Vol. 46, pp. 851–881, (1954).
5. J.M. Capus and G. Mayer, "The Influence of Trace Elements on Embrittlement Phenomena in Low-Alloy Steels", *Metallurgia,* Vol. 62, pp. 133–138, (1960).
6. A. Joshi, "Segregation at Selective Grain Boundaries and Its Role in Temper Embrittlement of Low Alloy Steels", *Scripta Metallurgica,* Vol. 9, pp. 251–260, (1975).
7. C.L. Briant and S.K. Banerji, "Tempered Martensite Embrittlement in a High Purity Steel", to be published in *Metallurgical Trans. A,* (1979).
8. T. Inoue, "Grain Boundary Segregation of Impurity and Alloying Elements in Low Carbon Steels", in *Grain Boundaries in Engineering Materials* (4th Bolton Landing Conference), ed. J.L. Walter et al., Claitor's Publishing Div., Baton Rouge, LA, p. 553, (1975).

(a)

(b)

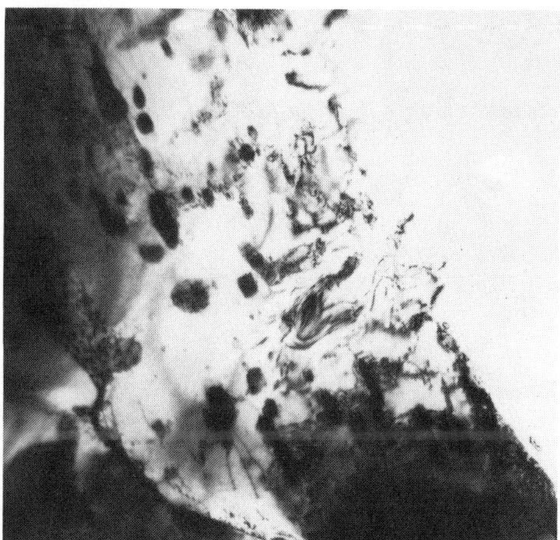

(c)

Fig. 10. Precipitates produced after tempering the nitrogen doped heat that had been austenitized at 1200°C. (a) is for a sample tempered at 150°C. (b) is for a sample tempered at 350°C. (c) is for a sample tempered at 650°C. Magnification: 20,000X.

Stevens, Vander Voort, McCall, eds. Microstructural Science, Volume 8

THE EFFECT OF GOLD FILM GRAIN SIZE ON COPPER DIFFUSION AND SURFACE ACCUMULATION BY OXIDATION

M.R. Pinnel*, H.G. Tompkins* and J.A. Augis*

ABSTRACT

For the processes of diffusion of bulk substrate copper through overplates of gold and of surface oxidation of the diffused copper, the influence of gold film grain size on process rates has been studied. Samples were prepared by sputtering 300 nm of gold onto NaCl substrates while controlling the substrate temperature to alter the fold film grain size. Films with average grain sizes of 8nm, 115nm and single crystals, as evaluated in a TEM, were prepared. Following the deposition of 5μm of copper on one surface of the gold, the specimens were aged at 100 and 150°C for times from 1 h to 56 days in air, and the surface accumulation of copper was measured using Auger spectroscopy and characterized in the SEM.

Results show that rapid transport of copper occurs along gold grain boundaries and that the initial rate of copper accumulation on the surface of gold is a function of the grain boundary density. The accumulation slows with time, as the surface sinking reaction of oxidation becomes rate controlling. The time required for leveling off in accumulation is a function of temperature and the gold film grain size. The results support the boundary diffusion-rate controlling surface reaction model.

INTRODUCTION

In previous work directed at the understanding of the low temperature ($< 150^\circ$C) diffusion [1-3] and oxidation [4-6] behavior in gold layer over base metal (copper, nickel) systems, several factors became rather clear. The base metal is transported to the gold surface via diffusion along the gold grain boundaries [1-3,7]. However, shortly after base metal reaches the gold/air interface, the surface reaction rate, such as oxidation, becomes rate controlling in most cases [8]. Thus, it is primarily the oxidation step which controls the rate of base metal film growth on the gold surface, thereby leading to resistive surface film failure modes in devices. Also, it was found that modest changes in atmospheric pollutant levels of elements such as chlorine can have a very significant effect on this film formation and growth process in copper [1, 4,6].

The model depicted in Fig. 1 is believed applicable to describing these processes in both the gold/copper substrate and gold/nickel substrate systems. The base metal is transported rapidly along the gold boundaries to the surface region until these paths are "saturated", ie, the pipes filled condition. It is emphasized that this "saturated" condition is not necessarily indicative of the gold boundaries being totally composed of base metal. The base metal at the surface concurrently reacts with the atmosphere at the locations where these paths intersect the surface. After nucleation of oxide caps over the boundaries, the boundary diffusion rate is decreased as it becomes limited by other factors which determine the rate at which base metal can be removed from the boundaries. One process is further thickening and lateral expansion of the oxide caps, which is controlled by diffusion through the oxide. Another

*Bell Telephone Laboratories, Columbus, Ohio 43213 USA.

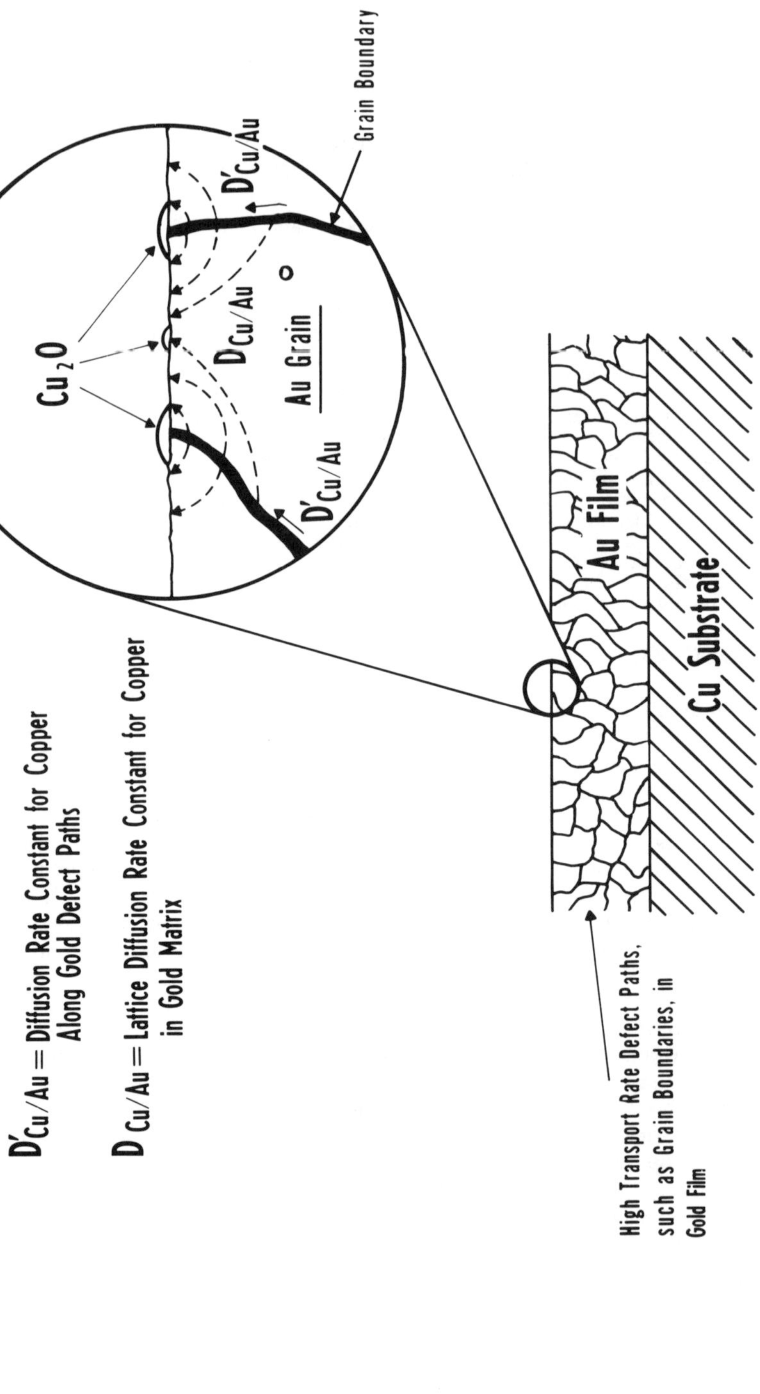

Fig. 1. Schematic depicting model for the diffusion-oxidation process in a gold film—copper substrate system.

process is removal of base metal atoms from the boundaries via diffusion through the grain interior, ie., the slower lattice diffusion process, to the surface and there to nucleate additional oxide patches. Simultaneous operation of both these processes can eventually produce complete coverage of the surface with resistive films and a consequent increase in contact resistance.

Several aspects of this model have received verification from studies on the gold-silver system [9,10]. Those results demonstrated the rapid transport of silver along the gold grain boundaries and the subsequent accumulation and spreading on the gold surface. The spreading occurred by surface diffusion, since the samples were heated and analyzed in vacuum. Auger scanning analysis demonstrated that the silver accumulated along and spread from the grain boundaries, but was constrained to within about 1μm of the boundaries which surrounded the approximately 10μm grains. This is analogous to that depicted in the model.

This present work was directed at similar goals for the copper-gold system in which the surface sinking reaction was the oxidation of copper. A further verification of the diffusion-oxidation model presented was sought, as well as quantitative results on the rate of surface accumulation of the resistive oxide film as a function of the grain size in the gold layer, when exposed in more common environments.

EXPERIMENTAL PROCEDURE

Preparation of Samples

It was found[11] that the mean grain size of gold thin film could be varied over many orders of magnitude in both thin electron transparent films (40 to 60 nm) and thicker films (300 nm) suitable for the diffusion-oxidation study by controlling the substrate temperature (T_D) during deposition. Diffusion samples were prepared by a two step deposition onto NaCl substrates at controlled substrate temperatures using magnetically enhanced sputtering. For each value of T_D, one NaCl wafer was first carefully dusted and installed in the bell jar of a conventionally pumped sputtering machine. The substrate was kept in close contact to a water cooled heating platen. A chromel-alumel thermocouple, clamped to the upper surface of the wafer, monitored the deposition temperature and controlled the power supply of the platen. Each substrate was first glow-discharge cleaned at an average current density of 1.2 mA/cm^2. Power was then applied to the platen to bring the upper surface of the substrate close to the desired T_D. The relatively low heat conduction of NaCl accounts for an additional temperature increment during deposition close to 25°C. Gold was sputtered onto the wafer from a 99.99% pure gold target under 8.5 millitorr of argon at a constant deposition rate of 5nm/sec as measured by a crystal monitor to a thickness of 300nm. Three values of T_D were used in sample preparation.

Following the gold deposition, the chamber was backfilled with pure dry nitrogen to install the gold film sample under a 99.99% pure copper target. After system evacuation, the gold covered substrates were once more glow cleaned. A large flow of water was forced through the platen to assure minimum heating of the gold during copper deposition. Copper was also sputtered under 8.5 millitorr of argon, but at a deposition rate of 3.2nm/sec to a thickness of 5μm. The substrate temperature (T_D) during copper deposition never exceeded 55°C.

Similar procedures were followed to obtain the samples for TEM examination with the exceptions that the thickness of the gold film was limited to about 40nm and no copper was deposited.

Diffusion-Oxidation Anneals

The gold/copper films were scribed into small specimen squares and floated off the NaCl substrates in a distilled water bath. Each specimen was clamped between a solid sheet of nickel and a sheet of molybdenum which contained a 2mm diameter hole. The specimen was placed in the holder such that the copper was against the nickel and the gold face was exposed to the environment through the hole. Annealing was carried out in glass tube furnaces,[4] which for most anneals were arranged to maintain an atmosphere of flowing zero-grade bottled air* at 50% relative humidity, but in a number of cases were also opened to typical laboratory air. Annealing temperatures of 100°C and 150°C and times ranging from one h to 56 days, with numerous times in between, were evaluated. Specimens of all three grain size samples were always simultaneously exposed in the same furnace to assure that any changes in temperature, atmosphere, or any other potentially variable annealing parameter would not alter the validity of the comparison and evaluation of gold grain size on diffusion-oxidation behavior.

Scanning Electron Microscopy Analysis

Following heat treatments, the specimens were transferred to the scanning electron microscope in the holders so to avoid touching or disturbing them. The surface was characterized in routine fashion at 20 KV using secondary electrons. The distribution of copper oxide nodules on the surface was analyzed and photographed. An indication of the composition of the nodules was determined using energy dispersive X-ray analysis (EDAX).

Auger Electron Spectroscopy Analysis

Following SEM analysis, the specimens were transferred to the Auger spectrometer in the same holders. Several Auger spectra were taken at various locations on the surface of each specimen to assure that the recorded peak heights for the various elements detected (Cu, Au, Cl, O, C) were representative. The specimens were then progressively sputtered for short time increments and a spectrum recorded at the same location after each sputtering interval. This procedure was continued until the copper signal diminished to a background level which represents either the amount of copper in the gold grain boundaries in steady state or the noise level. The sputtering ion beam current density used was $9.4 \mu A/cm^2$, which yields sputtering rates of about 3 nm/min on gold and 0.9nm/min on copper oxide for this system.[4] Due to the nonuniformity of the surface film, neither sputtering rate is precisely correct for converting sputtering time to layer thickness. However, in most cases a large gold signal appeared after very brief sputtering times, which primarily removed just the usual carbonaceous film. Therefore, the sputtering rate for gold was consistently used to convert sputtering times to film thickness; thus the results are at least self consistent.

Representative specimens of each grain size sample were heavily sputtered to near the midplane of the gold film. This was done to determine the background level of copper present in the grain boundaries in the "saturated" condition following extensive diffusion. In all cases, this level was below or barely at the detectability limit of the Auger technique using routine efforts.

RESULTS AND DISCUSSION

Transmission Electron Microscopy (TEM) – Grain Size Analysis

Shown in Figs. 2 – 4 are TEM bright field micrographs and corresponding Selected Area Diffraction (SAD) patterns for the gold foils deposited at substrate temperatures of 55°C, 225°C and 305°C, respectively. The lowest temperature or finest

*As specified by Matheson Gas Products, certified analysis less than 2.0 ppm total hydrocarbons as CH_4.

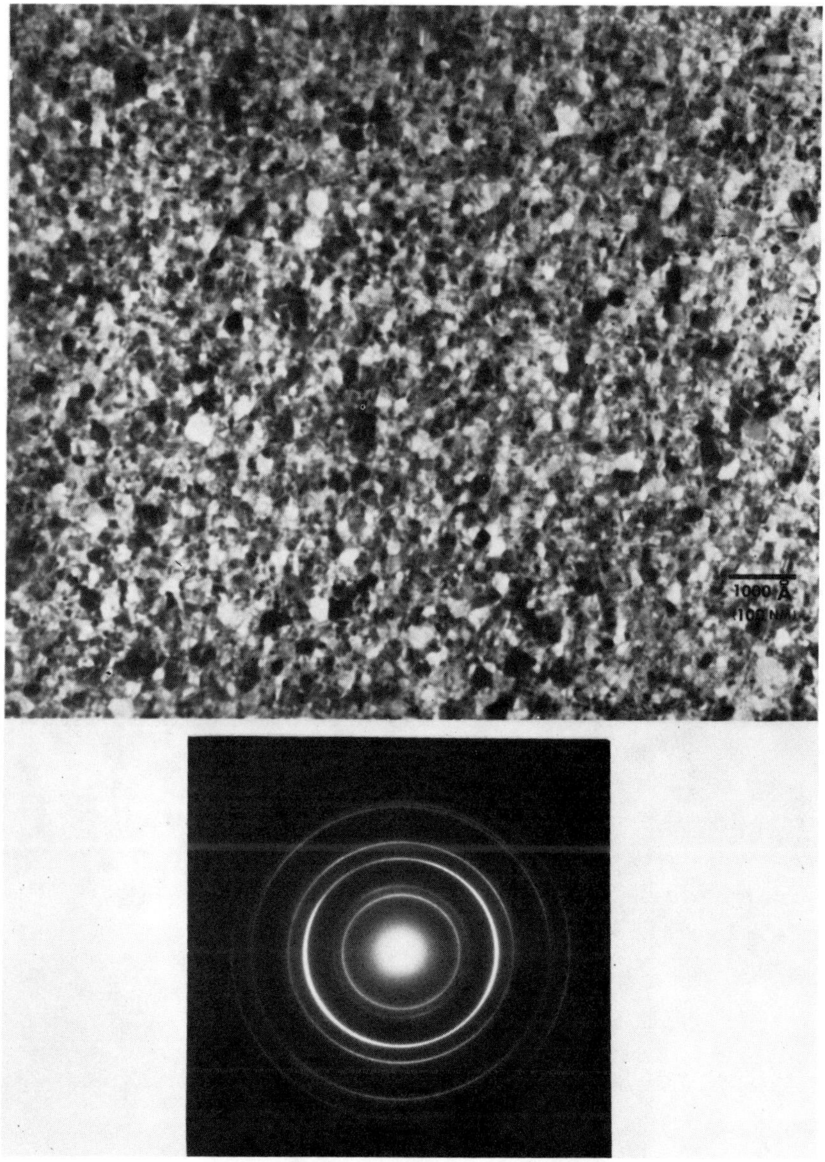

Fig. 2. Transmission electron micrograph and corresponding selected area diffraction pattern from a 40nm gold film sputtered onto a NaCl substrate at 55°C. Average grain size is approximately 8nm. Accelerating voltage is 100 KeV.

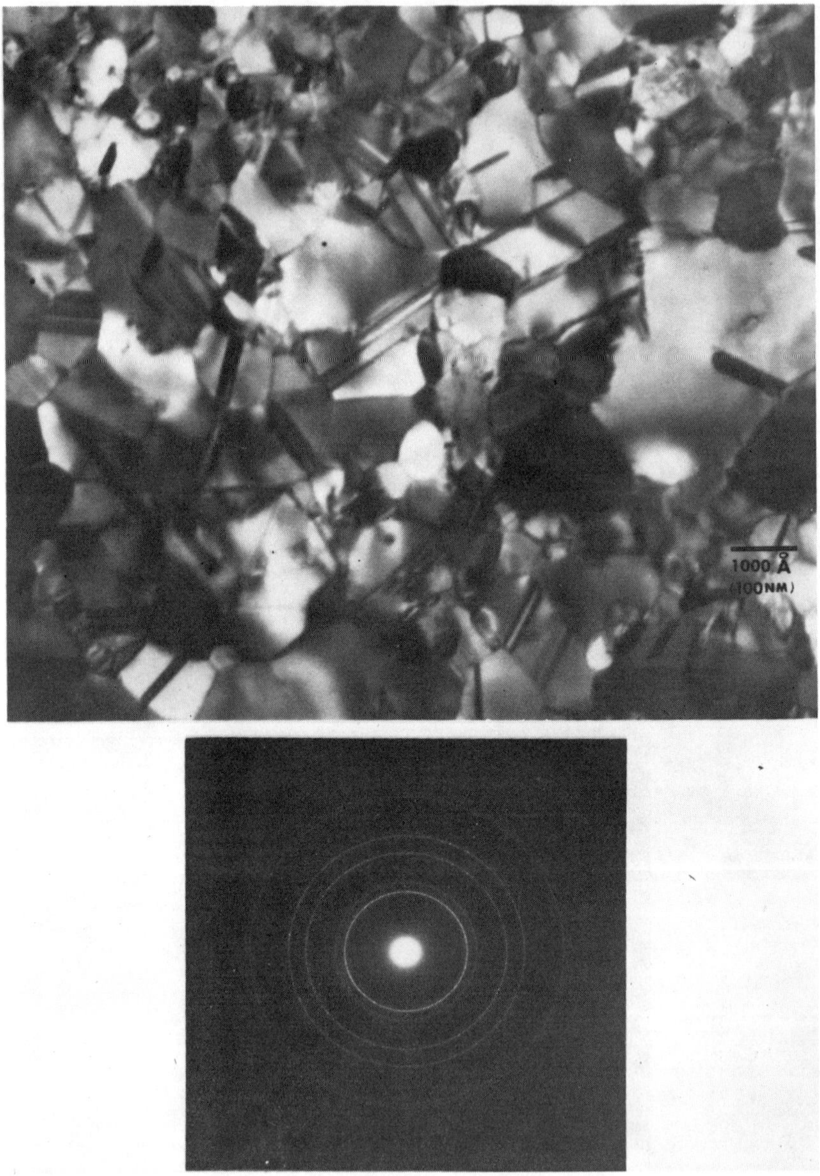

Fig. 3. Transmission electron micrograph and corresponding selected area diffraction pattern from a 40nm gold film sputtered onto a NaCl substrate at 225°C. Average grain size is approximately 115nm. Accelerating voltage is 100 KeV.

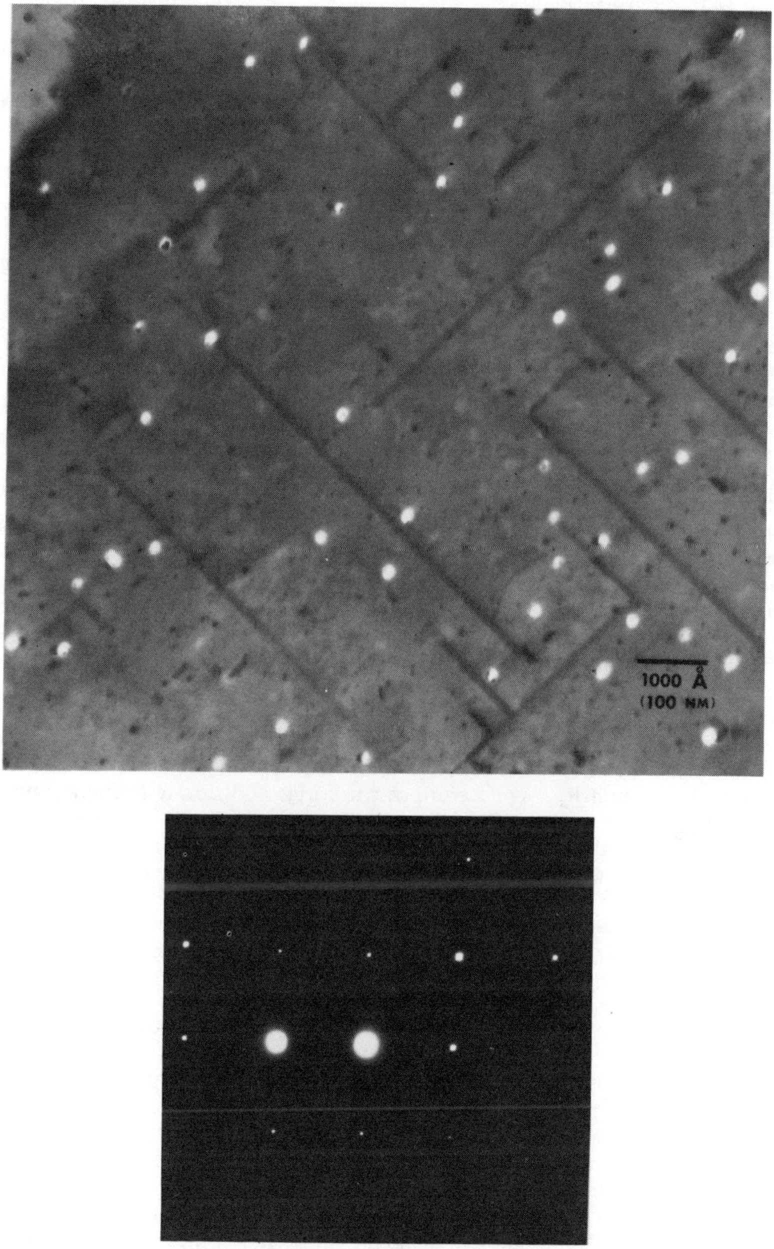

Fig. 4. Transmission electron micrograph and corresponding selected area diffraction pattern from a 40nm gold film sputtered onto a NaCl substrate at 305°C. Film is essentially a single crystal. Accelerating voltage is 75 KeV.

grained film has a measured average grain size of 8nm, with a range of 4 to 30nm. Using the average size and assuming that the grain shape may be approximated by a hexagon in two dimensions and that a grain boundary is several lattice parameters (0.5 to 1.0nm) in width, it is calculated that the boundary region occupies between 12 and 23 percent of the cross sectional area. Thus the volume of high diffusivity paths is a very significant percentage of the total volume of the gold film for this structure. The SAD indicates that the grain orientations are nearly random with only a hint of preferred orientation.

The sample deposited at 225°C, Fig. 3, shows greater than an order of magnitude increase in grain size over the 55°C sample. The average grain size is 115nm, with a rather broad range of 15 to 500nm. Using the same method as above, the boundary region represents between 0.5 and 1.0 percent of the cross sectional area. This is greater than an order of magnitude decrease in high diffusivity path volume compared to the smallest grain size sample.

The sample deposited at 305°C, Fig. 4, had an indeterminable grain size as no grain boundaries were observed in the TEM foil. The SAD also indicates a single grain or single crystallographic orientation with a 100 plane parallel to the surface of the foil. The foils were scanned in the diffraction mode and the SAD was invariant. Thus it can be concluded that this high temperature sputtered sample approximates a single crystal. However, several other types of defects are apparent. These include the dark straight bands, which exhibit a fringe pattern, and dark and bright circular spots, which are randomly distributed throughout the film. The bright spots are pores in the foil. These were not observed at the other deposition temperatures and are not expected to be present in the thicker gold layers used for the diffusion-oxidation experiment. Dark spot defects in quenched metals were described by Whelan[12] and have been observed in gold when quenched from 750°C and below.[13] The shape and size of the defects leads to the conclusion that they are probably internal voids in the foil.[14] These should have minimal or no influence on the rapid transport of copper through the gold layer.

Relevant defects are the long bands with the fringe pattern. Such fringes are commonly observed for stacking faults, twin boundaries and grain boundaries which lie at an angle to the plane of the foil.[15] The diffraction pattern analysis precludes their being grain boundaries. The nature of the fringe addition with increasing film thickness, as well as the nature of the first and last fringe, can be used to distinguish defect types.[16] Using these methods, it is concluded that the defects are twin boundaries.[14] Similar observations in gold foils have been made by Pashley and Stowell.[17]

A twin boundary may serve as a preferred diffusion path through the foil, similar to grain boundaries. However, the disarray of atoms in a coherent twin boundary is substantially less than that in a grain boundary as indicated by their respective surface energies in gold (γ_{GB} = 380 ergs/cm^2; γ_{TB} = 15 ergs/cm^2) [18]. Thus, they are not expected to be nearly as effective as a grain boundary in defect path diffusion.

Auger Electron Spectroscopy (AES) and Scanning Electron Microscopy (SEM) — Diffusion/Oxidation Analysis

Profiles of the Cu can be constructed from the AES spectra taken at various times during sputtering and are a measure of the amount of material which has been transported through the pipes (diffusion) and has accumulated on the surface (oxidation). Profiles for the 150°C aging in clean air for various times are shown in Figs. 5–7 for the three film structures. As usual, the initial rise in the Cu signal with depth is due to the removal of a carbonaceous contamination layer. For the fine-grained specimens, copper is observed after just 1h and increases for times up to about 3 days. Copper at the surface after just 1h is expected based on the previously measured value of the defect path diffusion rate of copper in gold,[1] which predicts only 19 min. to

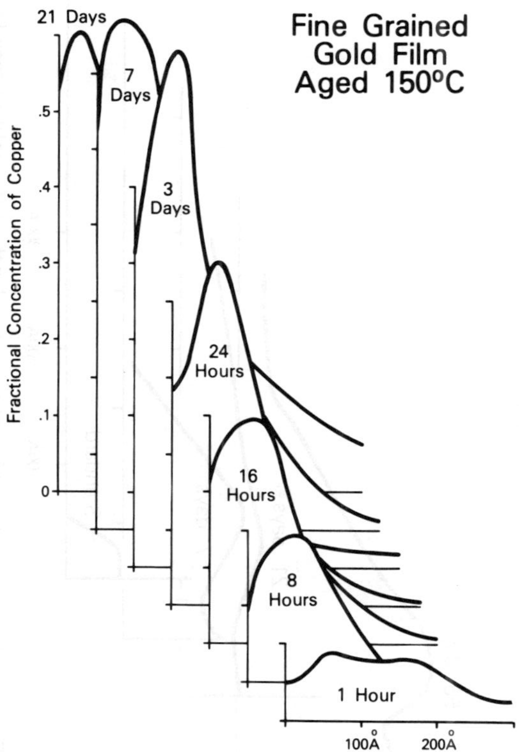

Fig. 5. Fractional concentration of copper versus depth at the gold surface for the fine-grained gold film specimens aged for various times at 150°C.

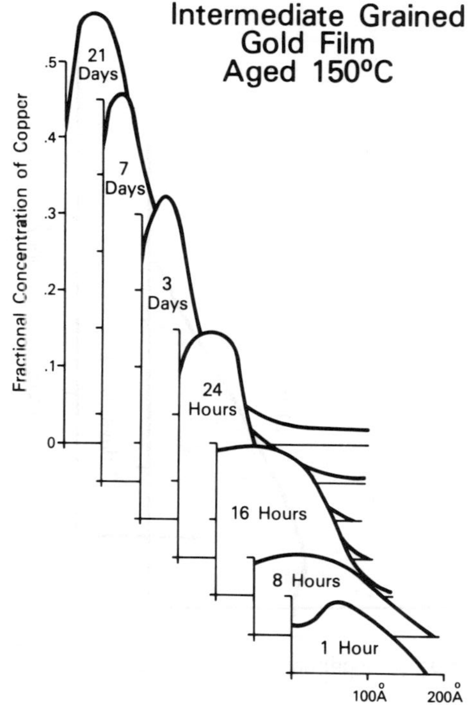

Fig. 6. Fractional concentration of copper versus depth at the gold surface for the intermediate-grained gold film specimens aged for various times at 150°C.

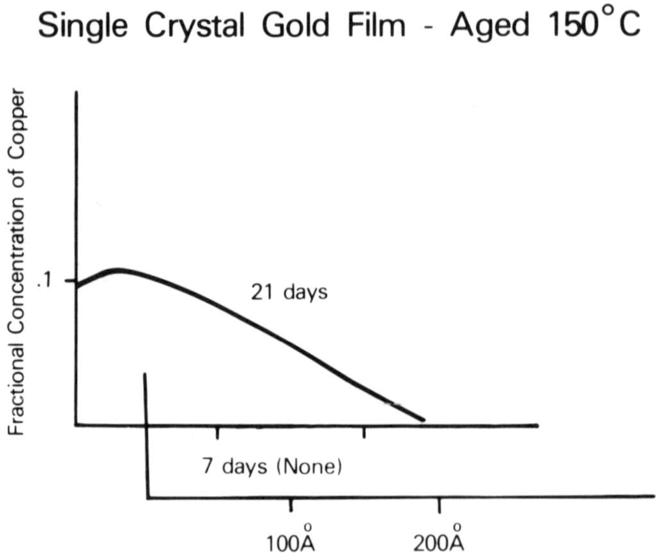

Fig. 7. Fractional concentration of copper versus depth at the gold surface for the single crystal gold film specimens aged for several times at 150°C.

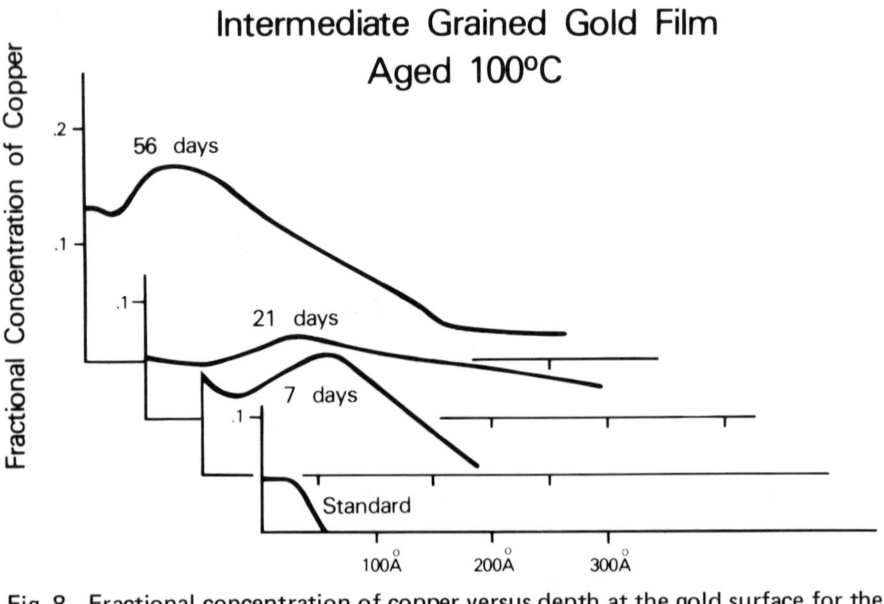

Fig. 8. Fractional concentration of copper versus depth at the gold surface for the intermediate-grained gold film specimens aged for various times at 100°C. Standard represents an "initial condition" after 4 months at room temperature.

diffuse copper through 300nm of gold at 150°C (coefficient of 1.3×10^{-13} cm^2/sec). After 3 days, the amount does not increase significantly for times to 21 days, suggesting either saturation or a change in the controlling mechanism producing a much slower rate. For the intermediate grained specimens, the saturation level is not reached within 21 days. No copper is observed in any of the spectra for the 7 day sample of the single crystal film. For the 21 day single crystal film, a trace of copper is observed.

Typical data for samples aged at 100°C are shown in Fig. 8 for the intermediate grained films. As before, we see a gradual increase with time in the amount of copper present on the surface. The fine grained specimens showed consistent behavior with a more rapid increase. Both films increase at a rate considerably slower than the corresponding 150°C specimens, and the near leveling off in growth observed at 150°C was not seen for the times evaluated at 100°C.

Summarizing the above observations, it is seen that, as expected, higher temperature causes a higher accumulation rate of copper on the surface. The key observation is that the rate is a function of grain size. The fine grained specimens have the highest accumulation rate and the single crystal specimens the lowest.

A visual confirmation of the nature and distribution of the copper on the gold surface was provided by the SEM–EDAX examination of specimens aged in typical laboratory air which, when compared to clean air, can have an accelerating effect on the oxidation rate[1,4,6] believed to be due to the presence of chlorine.[4] Shown in Fig. 9a is the surface of the as-deposited fine grained gold film; the other grain size films appeared identical to this prior to any elevated temperature aging. The only features apparent are cracks and occasional NaCl crystals from the deposition substrate. Following aging in laboratory air at 150°C for one day, the single crystal film (Fig. 9b) shows no change. However, the intermediate grained specimen (Fig. 9c) aged under the same conditions shows a distribution of discrete, nearly equal sized particles on the gold surface. EDAX analysis of the particles showed an enhanced concentration of copper over the background levels of gold and copper, and a trace of chlorine. A precise analysis of the particles was not possible due to their small size relative to the electron beam. Chlorine was not detected on spot analyses at locations off the particles. Auger analysis revealed an increased level of chlorine compared to that found on specimens aged in clean air. However, the chlorine signal typically disappeared in less than two minutes of sputtering, while the copper and oxygen signals persisted for many minutes. These observations lead to the conclusion that the particles are primarily copper oxide, presumably Cu_2O, with possible traces of copper chloride. The primary role of the chlorine appears to be one of accelerating the oxidation process and changing the morphology of the oxide such that the locations where copper reaches the surface are decorated by these particles. The fine grained specimen (Fig. 9d) also aged under the same conditions, shows a much greater density of the same size particles which are rather uniformly distributed over the entire surface. This is consistent with the much greater density of defect paths transporting copper to the surface.

The distances between these oxide caps are greater than the average grain size of the gold films as determined in the TEM analysis. This suggests that in the early stages not all grain boundaries are equally effective in transporting copper. However, as time progresses, the surface becomes more completely covered and the dimensions between oxide caps approach that of the grain size. This is pictured in Fig. 10 as a time sequence for fine grained specimens aged at 150°C. At 8 h the particles are just beginning to appear, while at 48 h fairly general coverage of the surface exists. This coverage approaches that seen at 7 days and is consistent with the Auger results which suggests that this represents a level at which the growth rate is declining very significantly. In this condition total coverage of the surface by oxide has not been achieved. Even though the density of particles increases with time, it is clear that all the particles are of fairly uniform size after the longer anneals. This corresponds

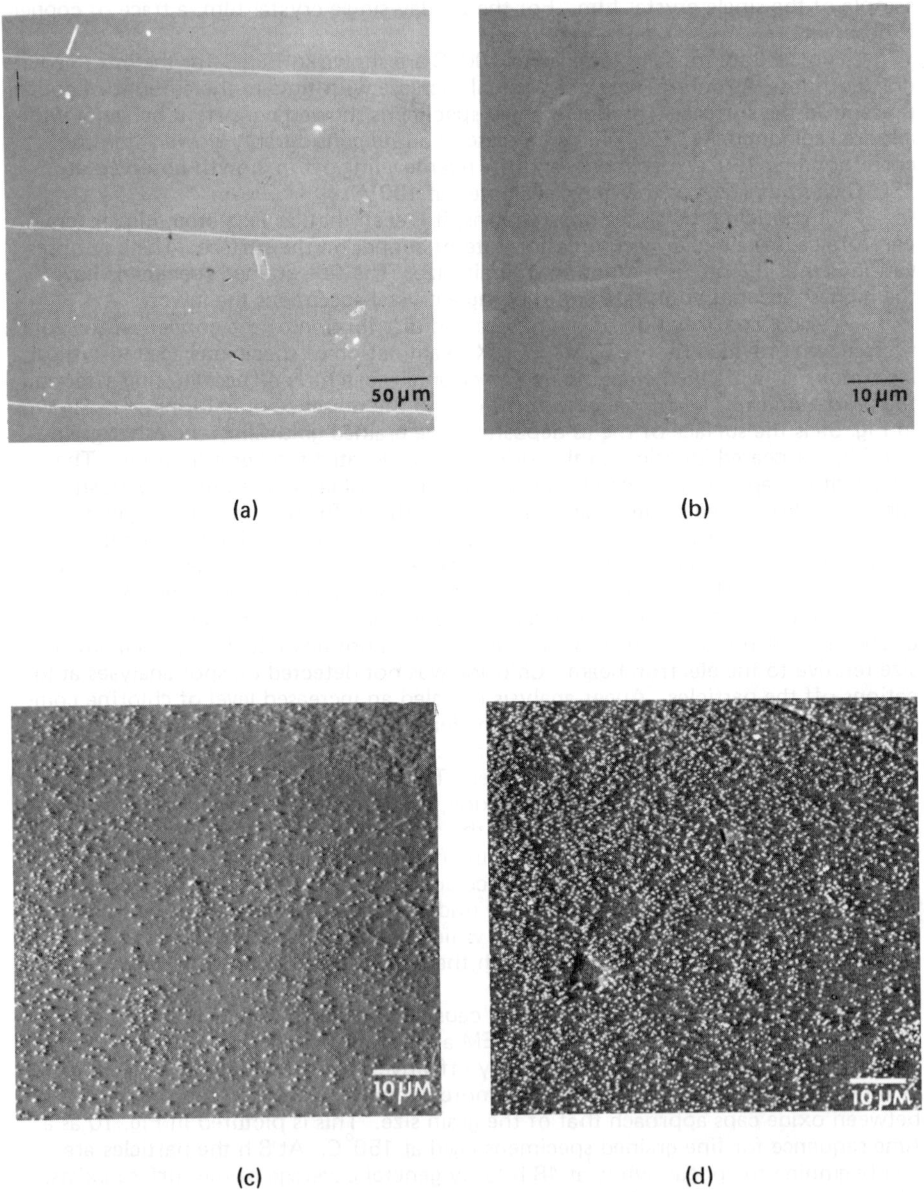

Fig. 9. SEM secondary electron micrographs of as-deposited and aged (150°C–1 day – common room air) gold over copper thin films (differentiated signal). a.) fine-grained gold film; as-deposited; 200X. b.) single crystal gold film; aged; 1000X. c.) intermediate-grained gold film; aged; 1000X. d.) fine-grained gold film; aged; 1000X.

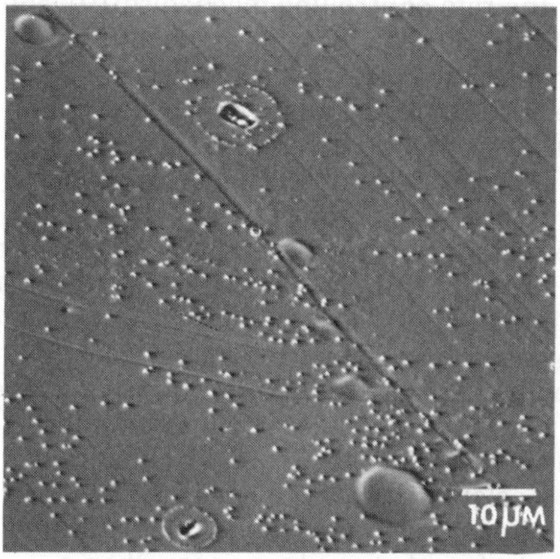

(a)

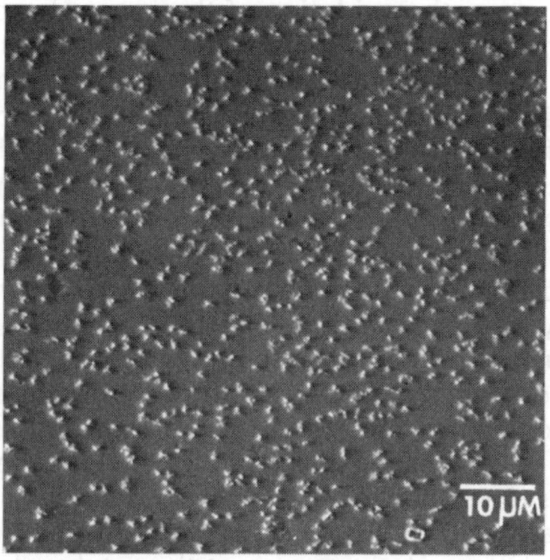

(b)

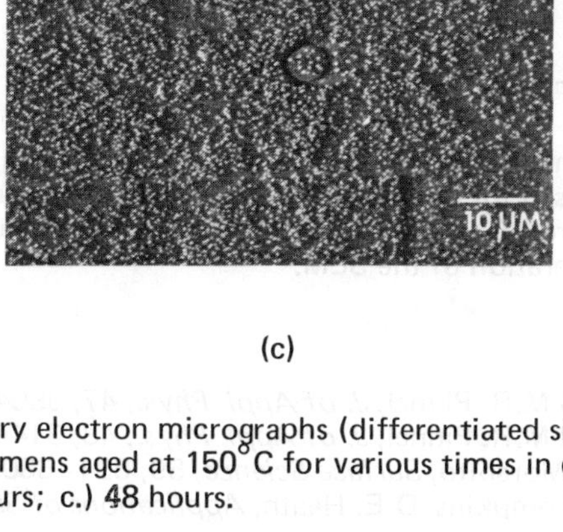

(c)

Fig. 10. SEM secondary electron micrographs (differentiated signal) of fine-grained gold over copper specimens aged at 150°C for various times in common room air. a.) 8 hours; b.) 16 hours; c.) 48 hours.

with the premise that the oxidation step is rate limiting on the spreading of the surface film.

These results are consistent with the model described in Fig. 1 in which copper will rapidly appear at the surface of gold due to diffusion along the grain boundaries and that the rate of accumulation initially is directly related to the area available for diffusion, i.e., a function of the grain size. The rate of accumulation should then slow down as the oxidation of the copper becomes the rate controlling mechanism[8] and limits the lateral spreading. This effect is qualitatively seen in the present results.

The copper has just begun to appear for the single crystal specimens after 21 days. If lattice diffusion alone were operative, one would calculate a diffusion time of 290 days to bring the first copper to the surface of a 300nm gold film at 150°C (coefficient of 6×10^{-18} cm^2/sec.[2,19]). This suggests that other higher transport rate paths, such as noncoherent twin boundaries may be operative.

The Auger and SEM observations are consistent with a very nonuniform surface coverage by the copper oxide film. A further proof of this characteristic was provided by results of Janssen, et al[11] in which the limited lateral spreading of silver over the surface of gold in a vacuum was observed after silver had diffused through the gold along grain boundaries. In this case, the spreading via surface diffusion (a sinking reaction comparable to oxidation) was limited by carbonaceous surface contamination.

A final ancillary point concerns the concept of "saturated" grain boundaries. As noted previously, the grain boundary area is calculated to be between 12 and 23 percent of the total cross sectional area for the fine-grained specimens. If "saturated" grain boundaries represent the total replacement of gold atoms by copper, then a large copper peak should have been observed in the Auger after sputtering well into the gold layer. In most cases the copper signal was below the detectability limit of about 1% and was not observed at all. This suggests that "saturated" gold boundaries, in fact, contain only small amounts of the diffusing element and remain gold-rich.

SUMMARY

The results of these experiments demonstrate that rapid transport of copper through a gold film occurs along the gold grain boundaries and that the initial rate of copper accumulation on the surface of gold is a function of the grain boundary density. The rate of accumulation slows rapidly with time, consistent with the fact that the surface sinking reaction of oxidation becomes rate controlling. The time required for the leveling off in accumulation is a direct function of the temperature and the gold film grain size. The distribution of copper on the surface is highly nonuniform and results indicated that areas of the gold surface are still exposed after the surface accumulation of copper has declined to a very low rate. The results of this and of other cited investigations strongly support the boundary diffusion - rate controlling surface reaction model for the process.

ACKNOWLEDGMENTS

The authors acknowledge C.W. Hoover, Jr. for suggesting the experiment and its basic design, D. Brasen for operation of the TEM, S. Mahajan for interpretation on the TEM microstructures, D. E. Heath for carrying out the diffusion - oxidation anneals, and R.E. Frank for operation of the SEM.

REFERENCES

1. H.G. Tompkins and M.R. Pinnel, *J. of Appl. Phys.*, 47, 3804–3812 (1976).
2. H.G. Tompkins and M.R. Pinnel, *J. of Appl. Phys.*, 48, 3144–3146 (1977).
3. P.M. Hall and J.M. Morabito, *Surface Science*, 59, 624–630 (1976).
4. M.R. Pinnel, H.G. Tompkins, D.E. Heath, *Applications of Surface Science*, 2, 558–577 (1979).

5. M.R. Pinnel, H.G. Tompkins, and D.E. Heath, *J. of Electrochemical Society,* 126, 1274–1281 (1979).
6. M.R. Pinnel, H.G. Tompkins, and D.E. Heath, *J. of Electrochemical Society,* 126, 000–000 (1979).
7. R.W. Balluffi and J.M. Blakely, *Low Temperature Diffusion and Applications to Thin Films,* A. Gangulee, P.S. Ho and K.N. Tu, eds., Elsevier Sequoia, New York (1975) p. 363.
8. H.G. Tompkins and M.R. Pinnel, *J. of Appl. Phys.,* 50, 000–000 (1979).
9. J.C.M. Hwang and R.W. Balluffi, *Thin Film Phenomena: Interfaces and Interactions,* J.M. Poate and J.E.E. Baglin, eds., The Electrochemical Society, Princeton, N.J., (1978).
10. A.P. Janssen, J.A. Venables, J.C.M. Hwang, and R.W. Balluffi, *Phil. Mag.,* 36, 1537–1540 (1977).
11. J.A. Augis, C.C. Lo and M.R. Pinnel, *Proceedings of the 4th International Thin Films Congress,* Loughborough, England (1978).
12. G. Thomas and J. Washburn, eds., *Electron Microscopy and Strength of Crystals,* J. Whelan, Proceedings of the First Berkeley International Materials Conf., Interscience Publishing, New York (1963) p. 19.
13. R.M.J. Cotterill, *Phil. Mag.,* 6, 251 (1961).
14. S. Mahajan, Bell Telephone Laboratories, Murray Hill, N.J., personal communication, October 2, 1978.
15. D. Hull, *Introduction to Dislocations,* Pergamon Press, Oxford, England (1965) p. 42.
16. L.E. Murr, *Electron Optical Applications in Materials Science,* McGraw-Hill Book Co., New York (1970) p. 347.
17. D.W. Pashley and M.J. Stowell, *Phil. Mag.,* 8, 1605–1632 (1963).
18. L.E. Murr, ibid., p. 387.
19. P.M. Hall, J.M. Morabito and N.T. Panousis, *Thin Solid Films,* 41, 341–361 (1977).

Stevens, Vander Voort, McCall, eds. Microstructural Science, Volume 8

QUANTITATIVE ANALYSIS OF 300 AND 400 SERIES STAINLESS STEEL BY ENERGY DISPERSIVE X–RAY FLUORESCENCE

Bradner D. Wheeler* and Nancy Jacobus*

ABSTRACT

Recent developments in analytical techniques and software has allowed the accurate quantitative determinations of both the major and minor elements in stainless steels by energy dispersive x-ray fluorescence. The successful analysis of 300 and 400 series stainless steel is reported utilizing this technique. The analysis of this type of material represents one of the most severe tests of the method due to numerous peak overlaps and interelement effects such as absorption and enhancement.

Sixteen standards of ASTM 300 series and ten 400 series were prepared by polishing on a 220 grit aluminum oxide belt and subsequently washing the surface in absolute methanol. Analyses were performed with an EG&G ORTEC 6110 Tube Excited Fluorescence Analyzer utilizing a dual anode (Rh/W) x-ray tube. Peak deconvolutions and interelement corrections were made with a 16K PDP–11/05 computer utilizing the program FLINT[1].

In conclusion, the energy dispersive x-ray fluorescence technique is capable of providing the analyst with rapid and accurate method analysis of stainless steels. The relative accuracy on the major elements is under 1% of the measured concentration.

INTRODUCTION

Quantitative analysis of 300 and 400 series stainless steels has been traditionally performed by wet chemical techniques, optical emission, atomic absorption, and wavelength x-ray fluorescence. Recent developments in solid state detectors and interelement correction programs[1–6] has allowed energy dispersive x-ray fluorescence to be applied in the rapid and accurate analysis of these alloys. The major and minor constituents can be determined with an accuracy equal to conventional chemical or optional instrumental techniques. The x-ray technique also has the significant additional advantage of being non-destructive and generally less time consuming.

INSTRUMENTATION

An EG&G ORTEC 6110 Tube Excited Fluorescence Analyzer was utilized for this study and operated under the following instrumental parameters:

Element:	Al	Si,P,S,Ti,V,Cr, Mn,Fe,Co,Ni,Cu	Nb, Mo	Sn
Anode:	W	Rh	W	W
Anode Voltage:	5kV	15kV	45kV	45kV
Anode Current:	400μA	20μA	200μA	200μA
Filter:	None	None	Cd	Mo
Energy Scale:	0–10keV	0–20keV	0–40keV	0–40keV
Atmosphere:	Vacuum	Vacuum	Air	Air
Counting Time:	200 sec	100 sec	40 sec	40 sec

*EG&G ORTEC, Oak Ridge, Tennessee 37830 USA.

Representative spectra obtained under the various operating conditions as previously stated, which illustrate the resolution, are graphically represented in Figs. 1 through 4.

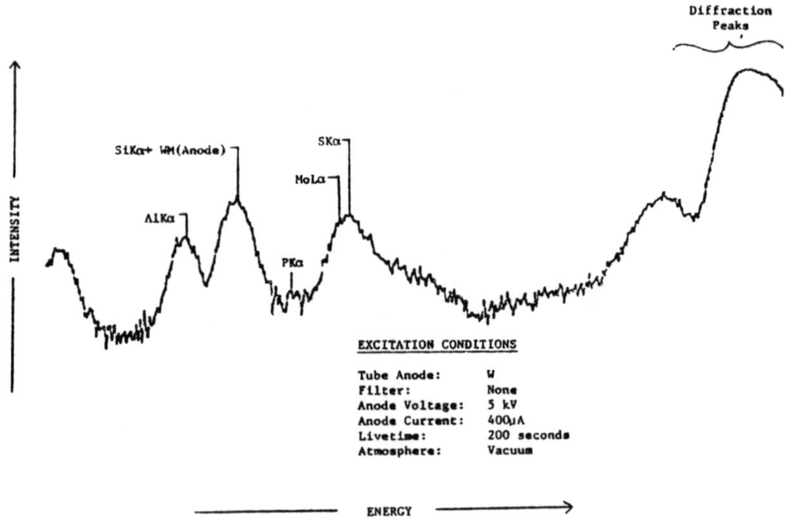

Fig. 1. TEFA spectrum of stainless steel, Sample No. 80C, log scale — expanded view (0 — 10 keV).

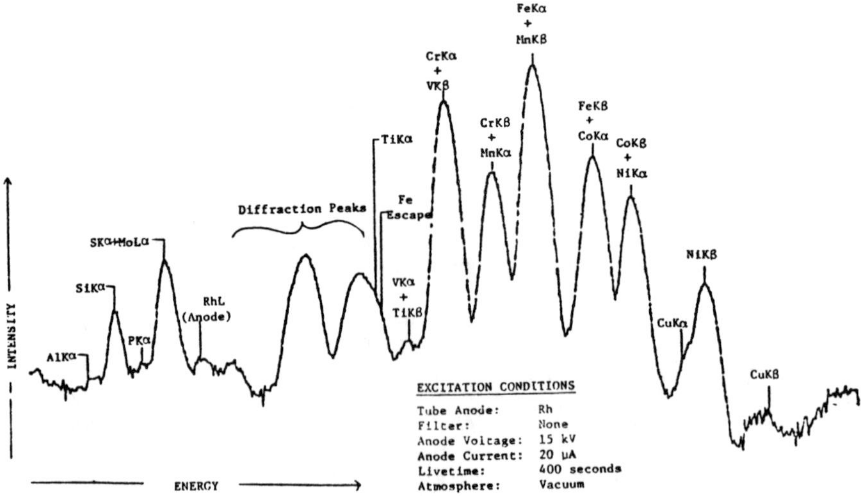

Fig. 2. TEFA spectrum of stainless steel, Sample No. 80C, log scale expanded view (0 — 20 keV).

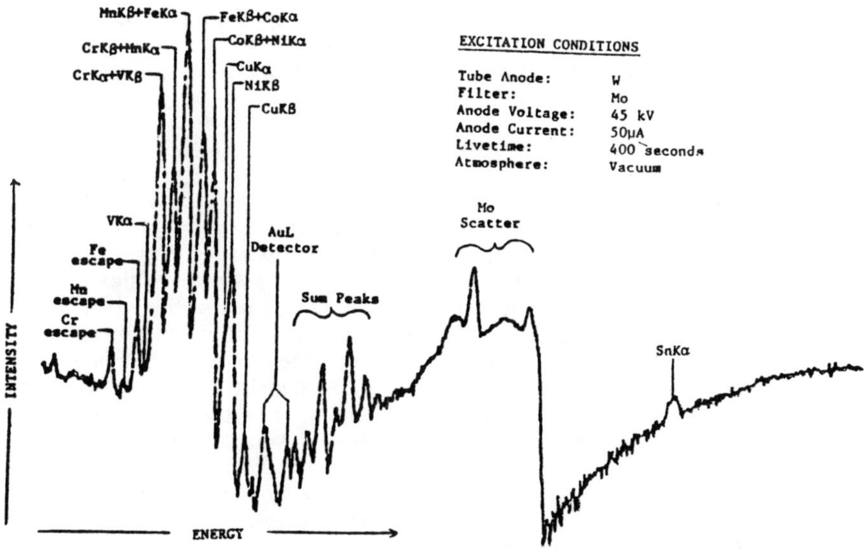

Fig. 3. TEFA spectrum of stainless steel, Sample No. 80C, log scale (0 – 40 keV).

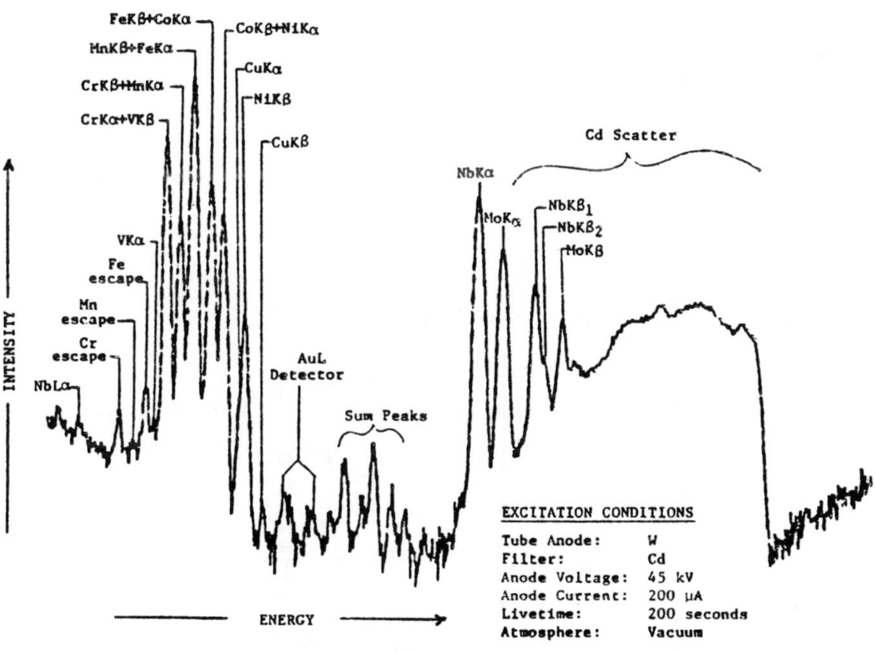

Fig. 4. TEFA spectrum of stainless steel, Sample No. 87C, log scale (0 – 40 keV).

SAMPLE PREPARATION

The standards utilized in this study were obtained from the British Bureau of Standards, National Bureau of Standards, and Carpenter Steel Corporation. Each sample was prepared by first grinding on a 60 mesh belt, then final grinding on an aluminum oxide belt of 220 mesh size. Following the final grinding, the sample was washed in anhydrous ethanol and ether.

EXPERIMENTAL PROCEDURE AND RESULTS

Accurate quantitative chemical analysis by x-ray fluorescence is primarily a comparative technique in which it is desirable that the standards and samples be of a similar composition. Minor variations require that mathematical corrections be applied to the intensities of specific elements in order to correct compositional differences. Systems such as stainless steel may or may not display intensities which are directly proportional to the concentration due to the effect of another element within the sample. This is particularly true in the materials analyzed due to the presence of periodic neighbors in the energy range of 4.5 to 9 keV consisting of titanium through copper. Figure 5 illustrates this problem area in which numerous peaks overlap, such as the TiKβ occurring at the same energy level as VKα. A further complication is present as a result of enhancement and absorption effects. An example of this effect would be the absorption of nickel by iron since the NiKα occurs just on the high energy side of the FeK edge. A similar problem exists at the low Z end of the spectra but is also aggrevated by the presence of L lines of higher energy elements such as niobium and molybdenum.

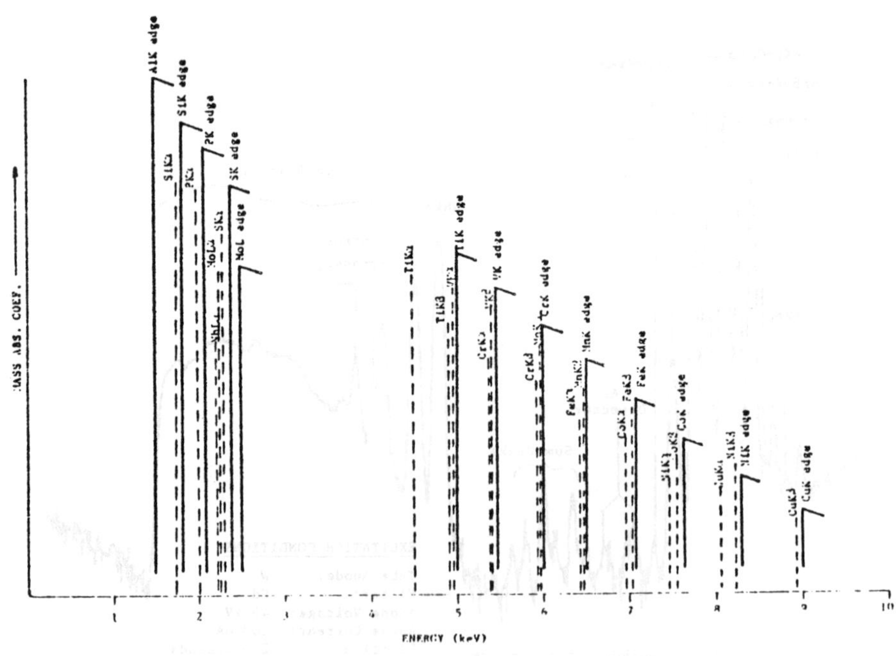

Fig. 5. Mass absorption coefficient versus energy.

Since severe spectra overlaps occur in the region of titanium through copper, the individual peaks involved must be spectrally stripped or deconvoluted prior to any interelement correction procedure for absorption and enhancement. Zemany[6] determined vanadium, chromium, and manganese in low alloy steels by a correction method utilizing the spectrally free peak of VKα and stripping the VKβ interference from the CrKα and then subsequently stripping the corrected CrKβ from the MnKα. In 300 and 400 series stainless steels, the procedure was started with titanium since the TiKβ interferes with the VKα. The spectral deconvolution routine utilized in this analysis is as follows:

$$C_{Ti} = a + bI_{TiK\alpha} \tag{1}$$

$$C_V = a + b\left[I_{VK\alpha+TiK\alpha} - (TiK\beta/TiK\alpha)I_{Ti(1)}\right] \tag{2}$$

where:

C_V = concentration of vanadium

a = x intercept

b = slope of calibration curve

$I_{VK\alpha+TiK\beta}$ = measured intensity of VKα and the interfering TiKβ

(TiK_β/TiK_α) = ratio of TiKβ:TiKα as measured on pure titanium

Ti(1) = interference free TiKα intensity from equation (1)

$$C_{Cr} = a + b\left[I_{CrK\alpha+VK\beta} - (VK\beta/VK\alpha)I_{V(2)}\right] \tag{3}$$

$$C_{Mn} = a + b\left[I_{MnK\alpha+CrK\beta} - (CrK\beta/CrK\alpha)I_{Cr(3)}\right] \tag{4}$$

$$C_{Fe} = a + b\left[I_{FeK\alpha+MnK\beta} - (MnK\beta/MnK\alpha)I_{Mn(4)}\right] \tag{5}$$

$$C_{Co} = a + b\left[I_{CoK\alpha+FeK\beta} - (FeK\beta/FeK\alpha)I_{Fe(5)}\right] \tag{6}$$

$$C_{Ni} = a + b\left[I_{NiK\alpha+CoK\beta} - (CoK\beta/CoK\alpha)I_{Co(6)}\right] \tag{7}$$

$$C_{Cu} = a + b\ I_{CuK\beta} \tag{8}$$

$$C_{Si} = a + b\left[I_{SiK\alpha+WM\alpha} - (IWM\alpha/IWL\alpha)IWL\alpha\right] \tag{9}$$

$$C_S = a + b\left[I_{SK\alpha+NbL\alpha+MoL\alpha} - (NbL\alpha/NbK\alpha)I_{NbK\alpha} - (MoL\alpha/MoK\alpha)IMoK\alpha\right] \tag{10}$$

One of the most severe effects with spectral overlaps occurs in the analysis of sulfur where the Nb$L\alpha$ and the Mo$L\alpha$ are at the same energy level as the S$K\alpha$ as illustrated on Fig. 5. Figure 6 illustrates the improvement in the calibration following the spectral deconvolution routine as described in equation [10].

The previously stated equation only relate to spectral deconvolution and do not correct for absorption and enhancement effects which are also prevalent in these types of alloys. Referring to Fig. 5, it is apparent that several elements are seriously affected—an example being the enhancement of the P$K\alpha$ by the Mo$L\alpha$ and Nb$L\alpha$ due to their proximity to the PK edge. If the analysis is confined to samples of similar composition, the observed intensity will be approximately linear. A generalized linear calibration would be as follows:

$$C_i \approx a + bI_i \tag{11}$$

where: C_i = the concentration of the ith element

a = the x intercept of the calibration curve

b = the slope of the calibration curve

Intensities will often be reduced by absorption or increased by enhancement due to the presence of other elements in the matrix. The concentration of the ith element in this model is given by the following equation:

$$C_i = a + b'I_i \exp^{m_{ij}(I_j - N_j)} \tag{12}$$

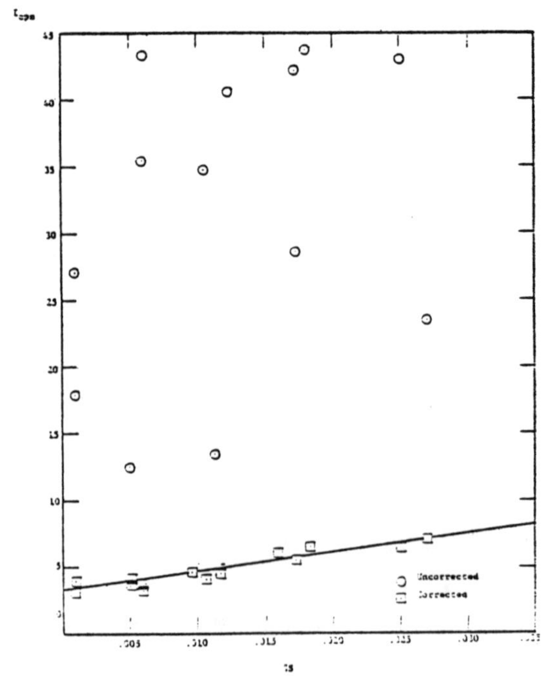

Fig. 6. Sulfur calibration curve.

where: m_{ij} = the interaction coefficient for element j on element i

I_j = the intensity of the jth element

N_j = the average jth element intensity of the standard

The interaction coefficients are determined by a non linear multiple least squares fit of the standard concentration – intensity data which requires a minimum of n + 3 standards where n is the number of interfering elements. Utilization of this technique provides results summarized on Table 1 and listed on Tables 2 and 3.

SUMMARY

Energy dispersive x-ray fluorescence has been shown to be a rapid and accurate method of analysis for stainless steels. Obvious advantages are that the method is non-destructive and that little sample preparation is required. Numerous peak overlaps are present but can easily be deconvoluted and resolved. Absorption and/or enhancement effects can then be solved utilizing the deconvoluted peaks.

REFERENCES

1. J.A. Cooper, B.D. Wheeler, D.M. Bartell, and D.A. Gedcke, *Advances in X-Ray Analysis,* Vol. 19, p. 213, (1976).
2. H.J. Lucas-Tooth and B.J. Price, "A Mathematical Method for the Investigation of Inter-Element Effects in X-ray Fluorescent Analysis", *Metallurgia,* Vol. LXIV, No. 383, (Sept. 1961).
3. G.R. LaChance, "A Simple Method for Converting Measured X-Ray Intensities into Mass Compositions", Geological Survey of Canada, Paper 64–50, (1964).
4. M.F. Hasler and J.W. Kemp, "Suggested Practices for Spectrochemical Computations", ASTM E–2 SM2–3.
5. S.D. Rasberry and K.F.J. Heinrich, "Calibration for Interelement Effects in X-Ray Fluorescence Analysis", *Analytical Chemistry,* Vol. 46, No. 1, (Jan. 1974).
6. P.D. Zemany, "Line Interference Corrections for X-Ray Spectrographic Determination of Vanadium, Chromium, and Manganese in Low-Alloy Steels" *Spectrochimica Acta,* Vol. 16, pp. 736–741, (1960).

Table 1.
Accuracies of 300 and 400 series stainless steel

300 Series

Element	Concentration Range	Average Concentration	Average Error
Al	0.004 - 0.210	0.065	0.006
Si	0.35 - 0.76	0.49	0.02
P	0.005 - 0.035	0.023	0.004
S	0.001 - 0.329	0.050	0.005
V	0.034 - 0.130	0.060	0.011
Cr	17.11 - 27.02	19.56	0.15
Mn	0.41 - 1.78	1.47	0.04
Co	0.06 - 0.50	0.20	0.03
Ni	4.20 - 30.65	12.41	0.15
Cu	0.067 - 0.49	0.23	0.01
Nb	0.005 - 0.66	0.146	0.001
Mo	0.022 - 2.34	0.33	0.015
Sn	0.002 - 0.014	0.008	0.001

400 Series

Element	Concentration Range	Concentration	Average Error
Al	0.004 - 0.015	0.013	0.003
Si	0.13 - 0.80	0.42	0.01
P	0.010 - 0.026	0.017	0.001
S	0.004 - 0.36	0.077	0.006
V	0.025 - 0.09	0.047	0.008
Cr	11.70 - 25.19	15.58	0.08
Mn	0.39 - 1.22	0.60	0.02
Co	0.019 - 0.16	0.051	0.008
Ni	0.24 - 8.22	1.93	0.02
Cu	0.052 - 2.32	0.442	0.014
Nb	0.001 - 0.67	0.238	0.001
Mo	0.027 - 0.80	0.228	0.003
Sn	0.004 - 0.013	0.007	0.002
Pb	0.0001- 0.001	0.0007	0.0003
Ag	0.0002- 0.0013	0.0005	0.0004

Table 2.
Results of 300 series stainless steel analysis

Standard	Al (B) Calc	Al (B) List	Si (B) Calc	Si (B) List	P (B) Calc	P (B) List	S (S) Calc	S (S) List	Ti Calc	Ti List	V (S) Calc	V (S) List
B.S. 80C	0.018	0.010	0.52	0.55	0.027	0.027	0.323	0.329	0.19	-	0.034	0.034
B.S. 81D	-	(0.003)	0.50	0.53	0.025	0.023	0.026	0.027	0.14	-	0.049	(0.05)*
B.S. 82A	0.004	0.002	0.35	0.33	0.023	0.019	0.010	0.005	0.16	-	0.047	0.049
B.S. 83A	0.005	0.002	0.40	0.40	0.020	0.025	0.010	0.012	0.15	-	0.065	0.044
B.S. 84D	0.017	0.011	0.55	0.57	0.027	0.035	0.022	0.025	0.17	(0.005)*	0.050	(0.10)*
B.S. 85A	0.034	(0.034)	0.49	0.50	0.028	0.029	0.016	0.017	0.28	0.143	0.078	0.052
B.S. 86C	0.21	0.21	0.37	0.41	0.019	0.017	0.0129	0.0011	0.26	0.38	0.089	(0.060)*
B.S. 87C	0.077	0.099	0.74	0.76	0.014	0.021	0.007	0.006	0.18	(0.031)*	0.048	0.042
N.B.S. 1155	0.037	-	0.50	0.50	0.027	0.020	0.028	0.018	0.19	-	0.033	0.047
N.B.S. 1185	0.030	-	0.38	0.40	0.025	0.019	0.018	0.016	0.24	-	0.019	-
Cartech 498	0.032	-	0.39	0.35	0.023	0.020	0.013	0.009	0.08	-	0.097	-
" 566	0.16	0.15	0.48	0.42	0.023	0.023	0.001	0.001	0.53	0.54	0.08	0.09
" 969	0.010	-	0.53	0.50	0.024	0.031	-	0.006	0.15	-	0.11	0.13
" 946	0.002	-	0.42	0.42	0.019	0.023	-	0.005	0.13	-	0.060	0.060
" -	0.013	-	0.56	0.58	0.029	0.029	0.31	0.31	0.19	-	0.034	0.044
" 160	0.001	-	0.70	0.67	0.012	0.005	0.010	0.011	0.08	-	0.023	-
Av. Dev.	0.006		0.02		0.004		0.005	-	0.09		0.011	

Standard	Cr (S)(B) Calc	Cr (S)(B) List	Mn (S) Calc	Mn (S) List	Co (S)(B) Calc	Co (S)(B) List	Ni (S)(B) Calc	Ni (S)(B) List	Cu (S)(B) Calc	Cu (S)(B) List	Nb Calc	Nb List
B.S. 80C	17.47	17.24	1.79	1.78	0.14	0.13	8.41	8.16	0.20	0.20	0.007	(0.007)
B.S. 81D	18.55	18.33	1.58	1.58	0.32	(0.12)*	8.55	8.46	0.21	0.23	0.006	-
B.S. 82A	22.87	22.81	1.80	1.73	0.161	0.169	13.00	13.16	0.16	0.18	0.005	0.005
B.S. 83A	24.45	24.46	1.39	1.32	0.203	0.173	19.83	20.14	0.06	0.086	0.014	0.012
B.S. 84D	16.91	17.15	1.68	1.70	0.36	0.37	12.35	12.40	0.23	0.24	0.013	-
B.S. 85A	17.45	17.33	1.66	1.69	0.131	0.141	9.96	9.99	0.173	0.155	0.001	-
B.S. 86C	20.83	20.66	0.85	0.85	0.06	0.06	30.88	30.65	0.58	0.58	0.043	(0.045)
B.S. 87C	17.51	17.44	1.60	1.61	0.26	0.20	10.10	9.82	0.177	0.153	0.66	0.66
N.B.S. 1155	18.70	18.45*	1.56	1.63	0.23	0.101*	12.35	12.18	0.182	0.169	0.005	-
N.B.S. 1185	16.65	17.09	1.37	1.22	0.30	-	13.22	13.18	0.062	0.067	0.004	-
Cartech 498	27.71	27.02	0.40	0.41	0.58	-	4.22	4.20	0.22	0.19	0.021	-
" 566	17.56	17.56	1.51	1.60	0.38	0.37	11.11	11.34	0.34	0.37	0.012	-
" 969	17.01	17.11	1.41	1.44	0.42	0.50	12.83	13.02	0.40	0.39	0.039	-
" 946	24.68	24.52	1.62	1.71	0.047	0.059	12.51	12.69	0.081	0.074	0.001	-
" -	17.89	17.78	1.65	1.64	0.22	0.16	9.05	9.08	0.48	0.49	0.009	-
" 160	17.89	17.95	1.76	1.71	0.05	0.11	10.14	10.05	0.20	0.19	0.734	-
Av. Dev.	0.15		0.04		0.03		0.15		0.01		0.001	

Standard	Mo Calc	Mo List	Sn Calc	Sn List	Type
B.S. 80C	0.31	0.30	0.010	(0.010)	303
B.S. 81D	0.31	0.32	0.012	0.014	304
B.S. 82A	0.15	0.15	0.009	0.010	309
B.S. 83A	0.153	0.153	0.003	-	310
B.S. 84D	2.11	2.14	0.007	0.008	316
B.S. 85A	0.44	0.43	0.008	0.006	321
B.S. 86C	0.21	0.21	0.001	(0.002)	-
B.S. 87C	0.27	0.26	0.008	0.005	347
N.B.S. 1155	2.39	2.38	0.006	-	316
N.B.S. 1185	2.02	2.01	0.004	-	316
Cartech 498	1.50	1.43	0.003	-	329
" 566	0.46	0.47	0.006	-	321
" 969	2.31	2.34	0.006	-	316L
" 946	0.019	0.022	0.002	-	309
" -	0.39	0.41	0.007	0.007	303
" 160	0.20	0.20	0.002	-	347
Av. Dev.	0.015		0.001		

NOTES:

( ) List values not certified
* Value not used in calibration of error.
(S) Stripping equation used for inter-element corrections.
(B) Beer's equation used for inter-element corrections.

B.S. - Brammer Standard Company (Coraopolis, PA.)
N.B.S. - National Bureau of Standards (Washington, D.C.)
Cartech - Carpenter Technology Corporation (Reading, PA.)

Table 3.
Results of 400 series stainless steel analysis

	Al (B)		Si (B)		P (B)		S (S)		V		Cr (S)(B)		Mn (S)	
Sample I.D.	Calc	List	Calc	List	Calc	List	Calc	List	Calc	List	Calc	List	Calc	List
Standards														
BS89D	0.002	0.006	0.33	0.31	0.018	0.015	0.012	0.010	0.02	0.04	11.70	11.70	0.54	0.59
BS90D	0.02	<0.01*	0.56	0.56	0.026	0.026	0.33	0.32	0.046	(0.055)	13.51	13.34	0.49	0.46
BS92A	0.004	(0.004)	0.31	0.325	0.016	0.020	0.011	0.014	0.041	0.032	15.78	15.72	0.70	0.71
BS93C	0.007	0.007	0.37	0.39	0.016	0.015	0.008	0.010	0.063	0.061	16.84	17.10	0.51	0.47
BS94A	0.037	(0.040)	0.80	0.80	0.019	0.018	0.005	0.004	0.066	0.052	25.24	25.19	1.23	1.22
BS93D	0.017	0.008	0.51	0.50	0.020	0.022	0.024	0.016	0.08	(0.09)	16.80	16.80	0.57	0.53
Cartech 410	0.014	0.015	0.24	0.27	0.015	0.015	0.013	0.023	0.030	0.025	11.96	12.04	0.46	0.48
" 416	0.018	-	0.61	0.63	0.018	0.018	0.35	0.36	0.033	0.025	13.18	13.15	0.51	0.52
" 450	0.054	-	0.30	0.29	0.015	0.014	0.003	0.006	0.043	0.043	15.24	15.20	0.40	0.39
" 455	0.050	-	0.15	0.13	0.010	0.010	0.013	0.005	-	-	11.41	11.37	0.066	0.074
Avg. Error	0.003		0.01		0.001		0.006		0.008		0.08		0.02	

	Co (S)		Ni (B)		Cu (S)(B)		Nb		Mo		Sn	
Sample I.D.	Calc	List	Calc	List	Calc	List	Calc	List	Calc	List	Calc	List
Standards												
BS89D	0.02	0.037	0.47	0.49	0.040	0.052	-	-	0.055	0.060	0.003	(0.005)
BS90D	0.049	(0.016)*	0.46	0.48	0.056	0.082	0.002	-	0.12	0.12	0.005	(0.004)
BS92A	0.039	0.036	2.15	2.13	0.103	0.086	0.002	0.001	0.114	0.115	0.008	(0.007)
BS93C	0.094	(0.031)*	0.32	0.29	0.092	0.068	0.009	-	0.49	0.48	0.009	(0.008)
BS94A	0.044	0.045	0.32	0.31	0.097	0.070	0.001	0.001	0.083	0.083	0.009	(0.013)
BS93D	0.028	0.034	0.45	0.44	0.085	0.09	0.020	(0.003)*	0.48	0.48	0.007	0.008
Cartech 410	0.026	0.023	0.32	0.34	0.073	0.079	-	-	0.056	0.053	0.007	0.006
" 416	0.040	0.019	0.25	0.24	0.070	0.084	0.001	-	0.066	0.065	0.006	0.005
" 450	0.157	0.16	6.34	6.36	1.49	1.49	0.67	0.67	0.79	0.80	0.009	0.008
" 455	0.100	-	8.23	8.22	2.32	2.32	0.28	0.28	0.025	0.027	0.006	0.004
Avg. Error	0.008		0.02		0.014		0.001		0.003		0.002	

	Pb		Ag		
Sample I.D.	Calc	List	Calc	List	Type
Standards					
BS89D	0.0004	(0.0006)	-	-	410
BS90D	0.0008	(0.001)	-	-	416
BS92A	0.0007	-	0.0012	-	431
BS93C	0.0007	-	0.0010	-	440C
BS94A	0.0008	-	0.0013	-	446
BS93D	0.0007	(0.0001)	0.0017	-	440C
Cartech 410	0.0008	<0.001	0.0006	0.0002	410
" 416	0.0007	<0.001	0.0001	0.0002	416
" 450	0.0007	0.001	0.0007	0.0013	450
" 455	0.0008	<0.001	0.0006	0.0002	455
Avg. Error	0.0003		0.0004		

NOTES:

- \- Value not listed.
- ( ) List value not certified.
- * Value not used in calibration of error.
- S Stripping equation used for interelement corrections.
- B Beer's equation used for interelement corrections.
- BS Brammer Standard Company (Coraopoles, PA.)

Cartech-Carpenter Technology Corporation (Reading, PA.)

Stevens, Vander Voort, McCall, eds. Microstructural Science, Volume 8

MEASUREMENT OF CATALYST IN PLATED THROUGH HOLES

Ronald L. Wollenberg*

ABSTRACT

Electronics, and consequently the printed circuit board industry, has grown by leaps and bounds in recent years. This fast growth has resulted in many unanswered questions concerning the processes used in the manufacturing of a printed circuit board.

The Laminated and Insulating Business Department of the General Electric Company is making an attempt to answer some of these questions concerned with the chemical processing and drilling of printed circuit boards. This paper deals with one facet of this work and is concerned with the measurement and distribution of the catalyst used in the chemical reduction of a metal in a drilled hole on a printed circuit board.

INTRODUCTION

Modern electronics centers on myriad quantities of electronic components mounted on a printed circuit board. The circuits may exist on both sides of the board or many layers of circuits laminated together as in a multilayer board. A conductive junction must be made between the circuits, and the circuits and the components. This junction is made by drilling holes through the board and various parts of the circuits. The conductive junction is then made by chemically reducing an electroless metal on the surface of the drilled hole wall. The most accepted chemical reaction showing this reduction of copper is:

$$Cu+_{2+}+2e^{-} \longrightarrow Cu^{\circ} \qquad \textbf{(reduction)}$$

$$2HCHO+4\,OH^{-} \longrightarrow 2HCOO^{-} + 2H_2O + 2e^{-} + H_2\uparrow \qquad \textbf{(oxidation)}$$

$$Cu+_{2+}+2HCHO+4OH^{-} \longrightarrow Cu^{\circ}+H_2\uparrow +2H_2O+2HCOO^{-}$$

(the accepted reaction for HCHO reduction of copper)

Electroless copper reduction on a properly catalyzed surface

In order to chemically reduce a metal on an insulator in a reasonable amount of time, a palladium-tin catalyst must be adsorbed in solution by the hole wall surface. After catalyst deposition and during rinsing of the panel, the catalyst undergoes a complex series of redox, hydrolysis, dissolution and precipitation reactions. This produces layers of stannous and stannic hydrous oxides and oxychlorides with palladium. The catalyst must then be activated by etching. This is a critical step and must be done properly in order for optimum activation of the catalyst to be achieved.

*General Electric Company, Laminated and Insulating Materials Business Department, Coshocton, Ohio, 43812 USA.

ESCA data, Fig. 1, shows Sn^{2+} and Sn^{4+} to be present after etching and that the palladium is now present as Pd°. ESCA data also shows a large decrease in the tin peak and a strengthening of the palladium peek with trace amounts of Pd^{2+} after etching. Wet analysis showed similar results as ESCA with a 25% tin reduction and 8% palladium reduction.[1]

Many plaguing questions have risen regarding distribution and levels of catalyst before and after activation that are required to give complete metal coverage with good metal-hole wall adhesion. Therefore, we are making an attempt to measure the levels of catalyst and substrate on the surface of the drilled hole wall and relate this to problems seen with the chemical reduction of metal on the drilled hole surface. Hopefully this data will help us solve customer processing problems routinely as well as develop more compatible laminates.

DISCUSSION

The analyses are being carried out on a fully automated Etec SEM with an energy dispersive spectrometer, a four crystal wavelength dispersive spectrometer, and an Interdata 7–16 computer.

The samples are prepared by cross sectioning the panels above the holes of interest. The sample is then polished on a wet metallurgical wheel until less than half of the hole remains and mounted on a metallurgical mount. The surface of the hole wall remains essentially untouched.

The geometry during analysis is shown in Fig. 2. The sample is tilted 45° from normal beam incidence and turned 45° CCW from the secondary detector. This allows the hole wall surface to be aligned with the collimator to the wavelength spectrometer. The analysis is carried out in the tangent plane to the hole wall surface which is parallel to the 45° tilt plane. This allows for somewhat of a constant take-off angle with some deviations induced by the topography of the hole wall surface.

A laminate is constructed by pressing layers of semicured epoxy coated glass weave through a thermal curing cycle. A finely polished cross section of a laminate is shown in Fig. 3. It is obvious that the topography and composition of the drilled hole surface will vary with drilling conditions and customer processing. The surface of a drilled hole, shown in Fig. 4, can briefly be described as a non-uniform smear of what appears to be degraded epoxy resin and glass particles with varying numbers of exposed glass fibers. This type of surface obviously complicates X-ray analysis.

The composition of E-glass used in the construction of a laminate is shown below. This glass and the brominated epoxy resin will form the substrate or

Composition of E-glass[2]

Chemical Composition	Wt%
B_2O_3	5–10
CaO	16–25
Al_2O_3	12–16
SiO_2	52–56
MgO	0–5
$Na_2$and K_2O	0–2
TiO_2	0–0.8
Fe_2O_2	0.05–0.4
F_2	0.0–1.0

drilled hole surface onto which the catalyst will be adsorbed.

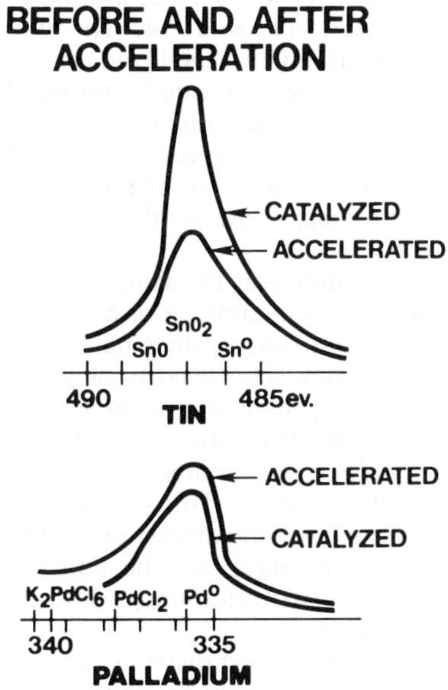

Fig. 1. ESCA data before and after acceleration.

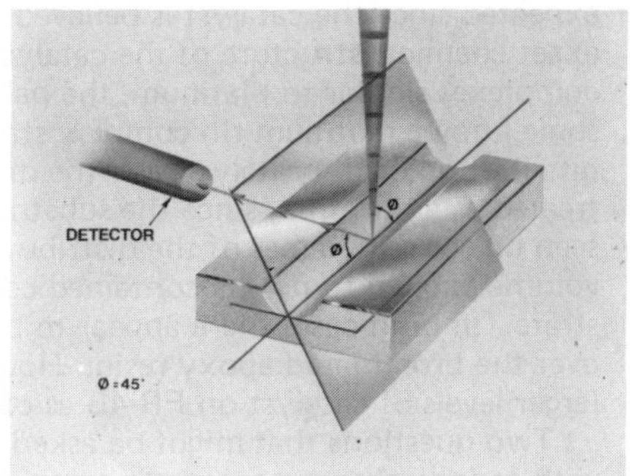

Fig. 2. System geometry.

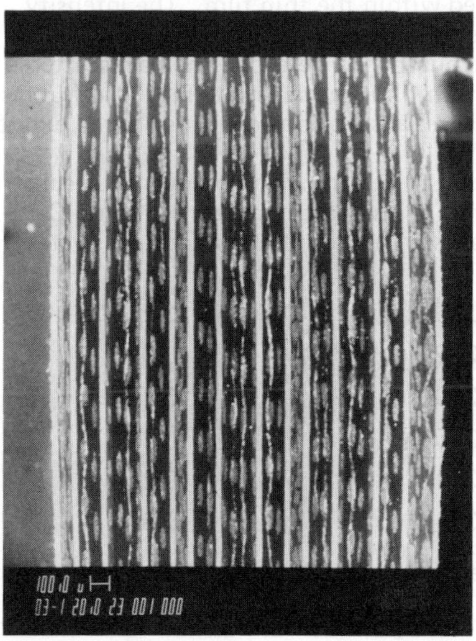

Fig. 3. Cross section of twelve layer multi-layer laminate.

Fig. 4. Drilled hole in multilayer laminate.

Micrographs and X-ray distribution maps of the drilled hole surface at 2,000X of the two samples labeled FR–4A and FR–4B are shown in Figs. 5–14. The two samples represent two different constructions, but were drilled and catalyzed under identical conditions. The distribution maps were done at an accelerating potential of 20 KeV, beam current of 20nA and 250 lines/frame at one line per second.

The bromine map shows the distribution of the brominated epoxy resin which is used as a flame retardant. The calcium map shows the distribution of the electrical or E-grade glass after drilling. The palladium and tin maps show the distribution of the catalyst in each case.

The maps show the palladium and tin to be distributed over the same areas as was expected since the catalyst is believed to be a palladium tin complex or colloid. The exact chemical structure of the catalyst is not known. However, since palladium forms complexes similar to platinum, the palladium tin complex is believed to be similar to some known platinum tin complex structures.[3] These maps also show an uneven distribution of the catalyst over the drilled hole surface. The catalyst should be treated as a thin film, since the substrate that is glass and brominated epoxy resin, is seen on the same areas of the distribution maps as the catalyst. This shows the X-ray volume excited to not be contained completely in the layer of catalyst over the substrate. In both cases there appear to be lower levels of catalyst over the glass than over the brominated epoxy resin. However, the X-ray distribution maps tend to show larger levels of catalyst on FR-4B as compared to FR-4A.

Two questions that might be asked are how does the catalyst distribution vary with topography and substrate composition of the drilled hole surface and consequently the effect of these parameters on electroless copper coverage and bond to the substrate.

Since electroless copper will be reduced at the catalyst sites first followed by autocatalyzation on the freshly reduced copper. Results from distribution maps and semiquantitative analysis of the catalyst before electroless reduction might be correlated to the distribution of copper at various times in the same electroless copper bath. This should give us a time-elapsed sequence of the electroless copper deposition.

In a thin film situation the X-ray intensity increases with thin-film thickness since more of the X-ray excited volume is contained within the thin film. The intensity from the thin film also increases with increasing substrate atomic number, since the backscatter coefficient increases with atomic number. Therefore, the backscattered electrons from the substrate may excite the thin film, the substrate may also fluoresce the thin film if an element in the thin film has an absorption edge close to the excited X-ray energy from the substrate.

Quantitative analysis has been performed on thin films with homogeneous substrates. The following data, taken from another source, shows a comparison of Monte Carlo calculations and experimental data for 500 angstrom Au foil on various substrates;

Substrate	K calc.	K meas.
Unsupported	0.202	0.183
Si	0.220	0.221
Ni	0.290	0.257
Mo	0.311	0.290
W	0.361	0.310

Operating conditions, E_o = 20.0 KV, takeoff angle = 18.0°, Au M-alpha line.[4]

This clearly shows the effect of different homogeneous substrate on the measured intensity of a 500 angstrom thin film of gold.

Fig. 5. Area of FR-4A drilled hole surface.

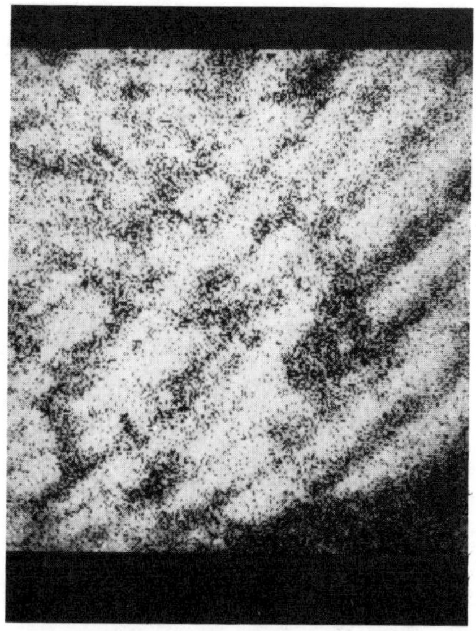

Fig. 6. Calcium distribution map of area shown in Fig. 5.

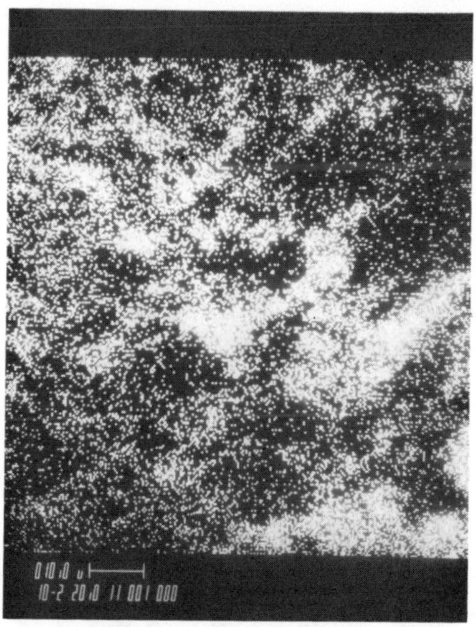

Fig. 7. Bromine distribution map of area shown in Fig. 5.

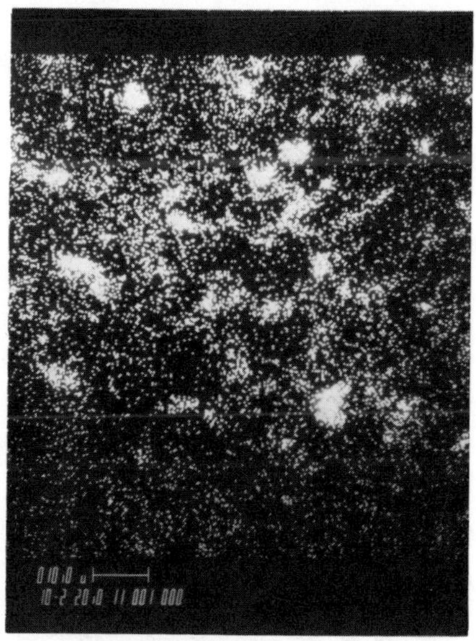

Fig. 8. Tin distribution map of area shown in Fig. 5.

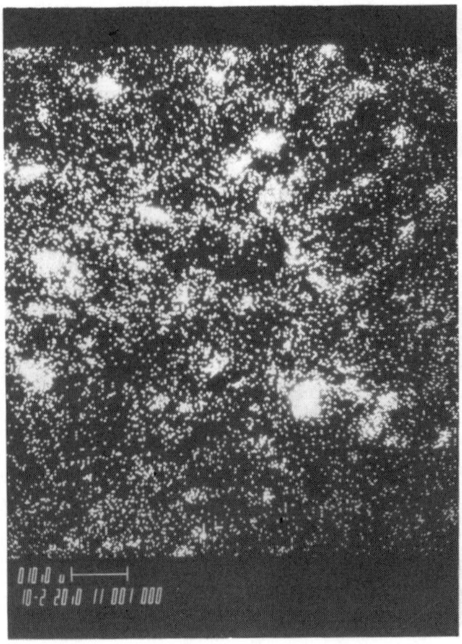

Fig. 9. Palladium distribution map of area shown in Fig. 5.

Fig. 10. Area of FR-4B drilled hole surface.

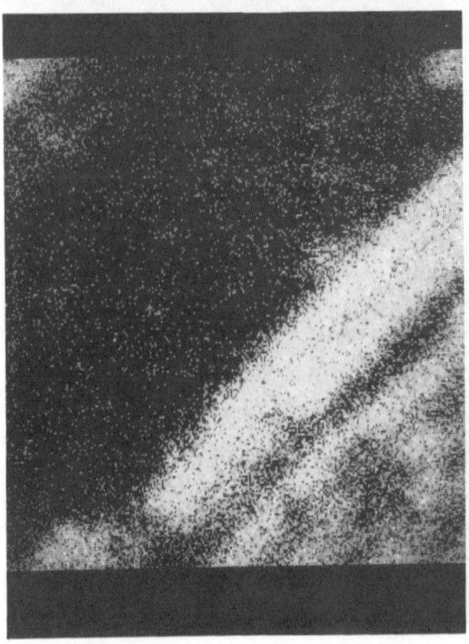

Fig. 11. Calcium distribution map of area shown in Fig. 10.

Fig. 12. Bromine distribution map of area shown in Fig. 10.

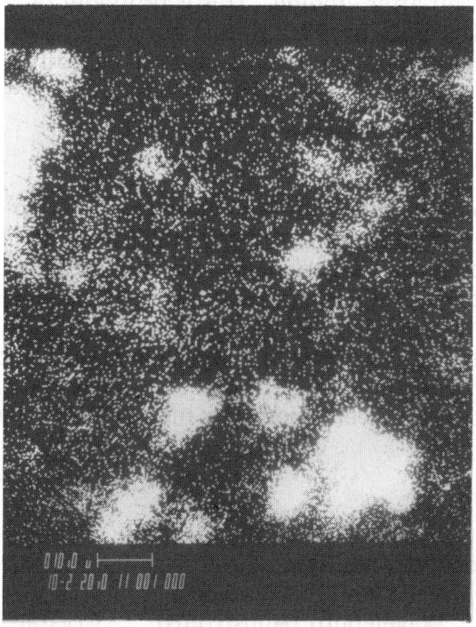

Fig. 13. Tin distribution map of area shown in Fig. 10.

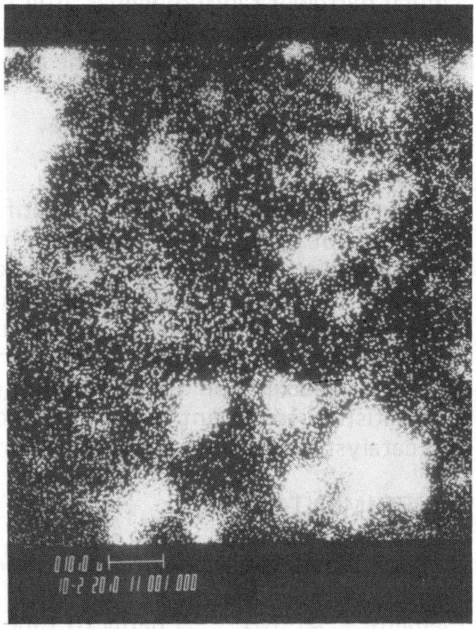

Fig. 14. Palladium distribution map of area shown in Fig. 10.

There are three important factors which make it difficult to obtain a semiquantitative analysis of the palladium tin catalyst on the surface of a drilled hole.

1) The atomic number of the substrate is changing due to the smearing of the glass and resin in drilling.
2) The thin film layer of catalyst appears to be unevenly distributed over the surface.
3) The topography of the drilled hole makes it difficult to perform quantitative X-ray analysis.

Obviously it is difficult, if not impossible, to make thin film standards of a known thickness for this application.

At this point, it is important to note that only a comparative, and not an absolute quantitative analysis is required. What kind of difference in catalyst levels and palladium/tin ratios can be observed for good and bad electroless depositions providing all other parameters are held constant?

In order to obtain a nearly flat area on the drilled hole surface, a small reduced scan area at 2,000X is used. This helps considerably in reducing the relative errors introduced by the interaction of X-rays with a rough topography. However, more analyses are necessary in order to sample a given area.

Obviously, if the second factor is not considered the results may show large relative errors that are caused by an actual uneven distribution of the catalyst over the surface. This problem might be eliminated somewhat by doing a palladium or tin distribution map. This map would help in identifying the sites where there are high and low levels of catalyst. Therefore, maximum and minimum levels of catalyst might be obtained.

The first factor can also be eliminated to some degree by only comparing areas of high resin and low glass and areas of low resin and high glass. This would also enable the use of a quantitative thin film ZAF correction program.

However, regardless of the levels of catalyst the palladium/tin ratio appears to have a much narrower range of values after catalyst and after etching. This is probably resulting from the fact that palladium and tin differ only by four atomic numbers. Both elements essentially experience the same X-ray and back scattered effects since both elements occupy the same sites.

DISCUSSION SUMMARY

If similar geometrical substrate configurations are characterized for catalyst absorption, some of the experimental errors involved with quantitative x-ray analysis on an irregular surface can be minimized. In order to create similar geometrical substrate configurations of glass and smeared resin, standard drilling conditions should be established which will reproducably create these geometries. This type of approach would hopefully enable an irregular surface present in a drilled hole to be characterized for catalyst adsorption under different plating conditions. After numerous analysis and statistical treatment of the data, a correlation between electroless metal deposition and catalyst absorption might be obtained.

EXPERIMENTAL

The following wavelengths are considered for the analysis;

Element	At. No.	Energy of Line (KeV)	Wavelength (Angstroms)*	Crystal
Al_1	13	1.486	1.2859	RAP
Si_1	14	1.739	3.2824	PET
Ca_1	20	3.690	3.35948	LiF
Br_3	35	I.526	1.2525	RAP
Pd_2	46	2.838	2.838	PET
Sn_2	50	3.443	3.59994	LiF

*LiF units 1 = K-alpha, 2 = L-alpha, 3 = L-Beta[5]

The Al intensity can only be measured in the presence of no brominated epoxy resin since the Al K-alpha and bromine L-alpha are separated by only 6 eV. Therefore, in order to measure the intensity from bromine, the L-Beta line for bromine is used. These elements represent the major inorganic components present in the substrate and catalyst.

The following standards are used, but perhaps better standards are being looked into;

Element	Standard	Purity
Al	Al	≈100%
Si	SiO_2	≈100%
Ca	$Ca_5FP_3O_{12}$	≈100%
Br	KBr	≈100%
Pd	Pd	≈100%
Sn	Sn	≈100%

A beam current of 20–30nA at 20 KeV is used. Higher beam currents tend to melt the epoxy resin. A count time of 10 sec. is used to measure the intensity of the background and the peak. Therefore, it takes approximately four minutes to do one analysis. For this reason it is important to have a stable electron column.

The semi-quantitative results listed below were obtained on the samples used in the previous X-ray distribution maps. Three analyses of each sample were performed. This number of analyses is obviously not enough for a good statistical distribution.

Material	Process Step	Pd Wr %	Sn Wt %	Pd/Sn Ratio
FR-4A	catalyst	.66 ± .16	1.21 ± .25	.54 ± .04
FR-4B	catalyst	10.05 ± 3.01	22.91 ± 8.96	.46 ± .06
FR-4A	Accelerator	1.37 ± .15	.99 ± .28	1.48 ± 33
FR-4B	Accelerator	6.66 ± 3.60	5.87 ± 2.93	1.12 ± .15

Accelerator = catalyst etch or catalyst activation step.

These results obviously show that FR-4B has much higher and uneven distribution of catalyst after catalyst and accelerator. The large standard deviation in some instances may be due to factor 1 and 2 mentioned in the section on discussion. Only factor 3 was considered during these analyses. The palladium/tin ratio does increase after accelerator. This agrees with ESCA data compiled from earlier sources. This ratio also appears to follow a trend which may be a function of electroless deposition.

FUTURE WORK

In order to better help us characterize the catalyst as well as other problems in this industry other surface instrumentation is being investigated. This technique obviously has some advantages and disadvantages. One instrument that is seriously being considered is an ISS/SIMS with ESCA and SAM. This is truly a surface analysis technique. Hopefully, this will help us better understand the distribution and chemical bonding in the catalyst.

REFERENCES

1. Krulik, Gerald A.; *The Catalytic Process in Electroless Plating,* presented at American Electroplaters Society Convention, p. 7, (June 30, 1976).
2. U.S. Government; *Military Specifications Glass Cloth Resin Preimpregnated (B-stage),* Department of Defense, MIL-G-55636B Amendment 2, p. 3, (May 25, 1978).
3. Krulik, *Op. Cit.,* p. 3.
4. J.W. Colby, J.R. Coleman, J.I. Goldstein, E. Lifshin, D.E. Newbury, H. Yakowitz, *Practical Scanning Electron Microscopy,* J.I. Goldstein (ed.), H. Yakowitz (ed.) 3rd ed., Plenum Press, pp. 310–317, (1977).
5. G.G. Johnson, Jr., E.W. White, *X-ray Emission and Absorption Edge Wavelengths and Interchange Setting for LiF Geared Crystal Spectrometer,* The Pennsylvania State University, Earth and Mineral Sciences Experiment Station, Special Publication No. 1–70, (July 1970).

Stevens, Vander Voort, McCall, eds. Microstructural Science, Volume 8

METALLOGRAPHIC ASPECTS OF THE SCANNING AUGER MICROPROBE

W.C. Johnson*

ABSTRACT

By combining the surface sensitivity of the Auger electron spectroscopy (AES) technique with a focussed electron beam, the scanning Auger microprobe (SAM) provides a surface sensitive microprobe capability. The advantages and limitations of utilizing the SAM on metallographic problems as a complement to the EPMA and SEM/EDX techniques are explored, and the comparison between X-ray and AES chemical spatial resolution discussed. The SAM technique is then applied to several metallographic problems, including interface definition in a bimetal strip, sulfide inclusion characterization in a ferrous alloy, and the characterization of the intermediate compound structure in an Al-Zn coated steel. Finally, it is concluded that the SAM has potential as a complementary tool for solving routine metallographic problems.

INTRODUCTION

Classical metallographic analysis involves sample selection, preparation for analysis, optical microscopy, scanning electron microscopy (SEM), electron probe micro-analysis (EPMA) and other techniques. The routine application of such techniques often provides important clues for solving practical metallurgical problems. The practical limitations of utilizing characteristic x-rays for chemical analysis, such as limited light element sensitivity, limited chemical spatial resolution due to subsurface backscattered electrons (usually limiting the spatial resolution of the electron microprobe to approximately 1 μm) as well as the limited capability to detect true surface composition variations (sub monolayer to 500 Å), have provided a stubborn barrier to solving certain practical problems.

The purpose of this study was to investigate the feasibility of applying the high resolution scanning Auger microprobe (SAM) technique to typical metallography problems. This was done to provide unique information which would complement and complete the normal metallographic characterization of the material utilizing optical metallography, SEM analysis, including EDX, and EPMA.

After the basic characteristics of the SAM technique are briefly reviewed and several aspects of sample preparation for SAM analyses considered, two examples are presented which explore the potential of the SAM technique for characterizing practical metallographic samples.

EXPERIMENTAL PROCEDURE

The Scanning Auger Microprobe

The Auger electron spectroscopy (AES) technique has been adequately reviewed elsewhere.[1–5] It is important to note that the technique utilizes the characteristic

*Perkin-Elmer Corp., Physical Electronics Division, Eden Prairie, Minnesota, USA.

Auger electron emission from atoms under electron bombardment. The Auger electron emission process is complementary to the emission of characteristic x-rays, and the characteristic kinetic energy of Auger electrons identify the atoms giving rise to the emissions in a manner analogous to the EDX technique.

However, in contrast with EDX or EPMA, AES data reflects the chemical composition of the outermost surface regions.[5] Since the mean free path of electrons with kinetic energies in the range usually utilized for AES analysis (50 eV – 2000 eV) is only on the order of a few atom distances in most solids, Auger electrons emitted from deeper than the first few atom layers undergo inelastic scattering (energy loss) before escaping from the sample. Electrons which have lost some of the characteristic Auger kinetic energy do not contribute meaningful AES data, but contribute to the background continuum of secondary electrons. This fundamental difference between the transparency of most solids to x-rays and the limited mean free path for electrons in the same solids, provides the unique surface sensitivity of the AES technique. Because of this surface sensitivity, AES has been routinely applied to problems where the chemical composition of the top few atom layers (5–20 Å) is of concern, and early AES instruments were designed to optimize this unique aspect of the technique. [6,7]

However, in practical applications to metallurgical problems, it is apparent that the spatial distribution of a surface chemical variation can be as crucial as elemental detection. By designing an AES instrument with a focussed electron beam which can be rastered across a specimen surface, this instrument, a scanning Auger microprobe (SAM), can provide the surface sensitivity of the Auger technique in the chemical mode, along with the micrograph imaging capability of the SEM.[4] Commercially available SAM instruments now routinely provide electron probe diameters less than 500 Å, with highly sophisticated data collection and manipulation capabilities.

In addition to the ability to image specimen surfaces and obtain the spatial variation of surface chemistry, the SAM technique potentially provides a chemical microprobe capability with better spatial resolution than that provided by the x-ray techniques. A comparison of the composition profile obtained in cross section at an infinitely sharp interface between two different materials, obtained by alternately analyzing characteristic x-rays and Auger electrons, Fig. 1, illustrates this potential advantage. Because of subsurface backscattering of the primary electrons, the x-rays analyzed in EPMA are emitted from a relatively large volume beneath the surface of the solid.[5,8,9] For probe dia. $< 1\mu m$, the spatial extent of this subsurface volume of excitation depends primarily on the beam potential, rather than the primary electron beam diameter. Hence, the compositional profile obtained by analyzing characteristic x-ray emission will show a "chemical broadening", for small electron beam sizes. Fundamentally, this generally limits the chemical spatial resolution of EPMA and SEM/EDX techniques to approximately 1 μm. [5,10]

Although Auger electrons are emitted from a similar subsurface volume of excitation, only those electrons emitted from the top few atomic layers of the surface can escape without undergoing inelastic scattering (energy loss), and contribute meaningful chemical information. In principle, this implies that the SAM technique should provide microprobe capability with chemical spatial resolution dependent on the primary electron beam diameter. In practice some chemical broadening should be expected due to the immediate electron scattering occurring even within the top few atomic layers. However, the ultimate chemical spatial resolution should be substantially less than 1 μm. [5,8,10,11] With proper consideration of the limitations of the technique, the SAM should provide a useful complement to EPMA for certain metallographic problems.

Normal metallographic preparation techniques were applied to prepare several specimens for SAM analysis. However, it was found advantageous to perform a final lap on the polishing medium *parallel* to the interface represented in cross section. This final step was found to minimize the effects of elemental smearing across the interface.[12]

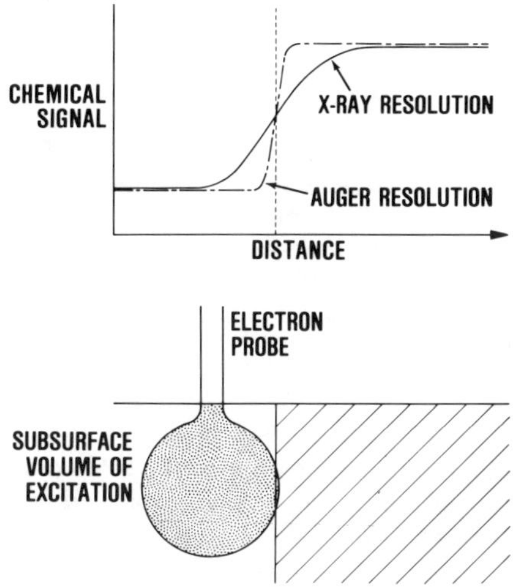

Fig. 1. Schematic of Auger and x-ray chemical resolutions at a sharp interface.

The specimens then were mounted on a standard sample holder in a SAM system.* Most of the SAM analyses were performed with an electron probe dia. of 2000 –3000Å and a beam voltage of 10 KeV, which provides optimum cross sections for AES analysis. Inert ion sputter etching was employed to remove surface contamination deposited during the sample preparation step, as well as to remove any traces of elemental transfer across the interface.

The SAM instrument utilized in this study was equipped with an instrument computer which permits preprogrammed digital operation during analysis. For a typical experiment, the electron beam was focussed to minimum dia., a secondary electron image obtained from the sample area to be analyzed, and a digital line scan programmed to traverse the interface in cross section. In this mode, the electron beam is stepped from point to point along a line across the sample. The density of points across the line, the dwell time per point, as well as the total number of cycles per line are parameters chosen to obtimize a particular measurement. During residence at each point, the difference between the signal strength at the maximum excursion of an Auger peak obtained in the N(E) pulse counting mode and the background corresponding to the base of the peak is determined. The data is normalized by dividing the difference spectrum by the background. This 'background normalization' reduces the effects of surface topography on the relative Auger peak intensities. After an experiment is completed, the data is displayed on a graphics terminal, and various manipulation routines such as smoothing, line superposition, and expansion utilized to assist in data analysis. Data for up to 20 elements can be collected automatically during the sequencing, although it is most common to carefully monitor a maximum of 3 to 4 elements to optimize analysis time.

If required, AES point analyses can be obtained from localized areas on either metallographically prepared or in-situ fractured surfaces. This data is usually presented in the first derivative form (dN(E)/dE) to facilitate peak identification. Once again, computer manipulation of the data such as smoothing, element substraction, or expansion are convenient for ease of analysis.

*Model 590 Scanning Auger Microprobe manufactured by Physical Electronics Division, Perkin-Elmer Corporation.

EXAMPLES

Test of AES Chemical Spatial Resolution

A simple AES line scan analysis across a metallographically prepared interface in a bimetal strip (Fe-Ni/Cu) was used to test the chemical spatial resolution of profiles obtained with the SAM technique. Measuring the 10%–90% fall-off of SAM line scans across the interface, Fig. 2, indicates an effective beam size of 0.25 μm, which corresponds to the actual electron beam spot size. Hence, chemical broadening due to sursurface backscattering of the primary beam is not significant in this sample. It is likely that a similar line scan obtained with EPMA would show a breadth of greater than 1 μm, regardless of the electron probe diameter. This data also indicates that the bimetal strip interfaces are extremely sharp, with little if any interdiffusion across the interface.

Sulfide Inclusions

A practical metallographic application of the SAM technique is to characterize inclusion in metal alloys. As an example, an as-polished ferrous alloy sample was examined in the SAM instrument. A secondary electron micrograph, Fig. 3a, clearly indicates inclusions on the sectioned surface. Auger chemical maps, obtained by rastering the electron beam over the same area as in the micrograph, while modulating the intensity of the CRT with the output of the Auger detector and setting the detection energy to correspond to a particular Auger peak, produced a series of images representing the spatial distribution of the different elements on the surface, Figs. 3b–d. This data identifies the inclusion as a duplex structure, consisting of a small Mn sulfide particle encased in a larger Fe sulfide shell.

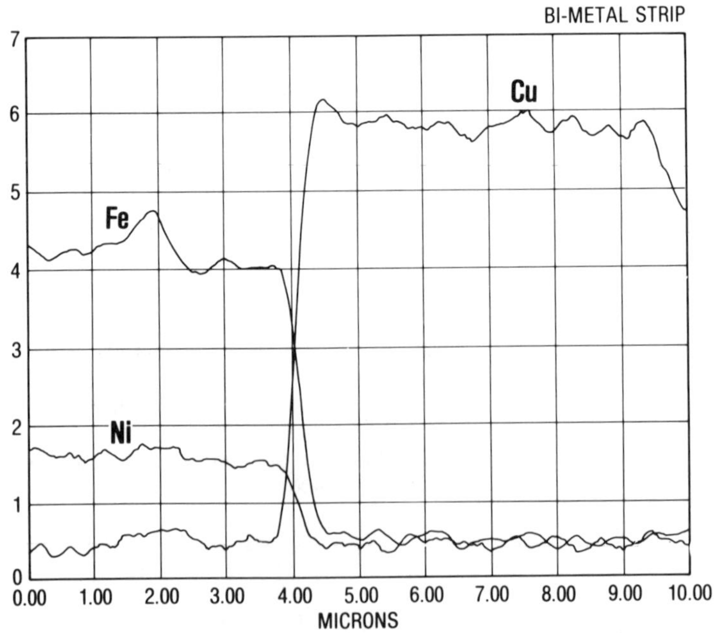

Fig. 2. Fe, Ni, and Cu SAM line scans across sharp interface in bimetal strip cross sectioned for AES line scan analysis.

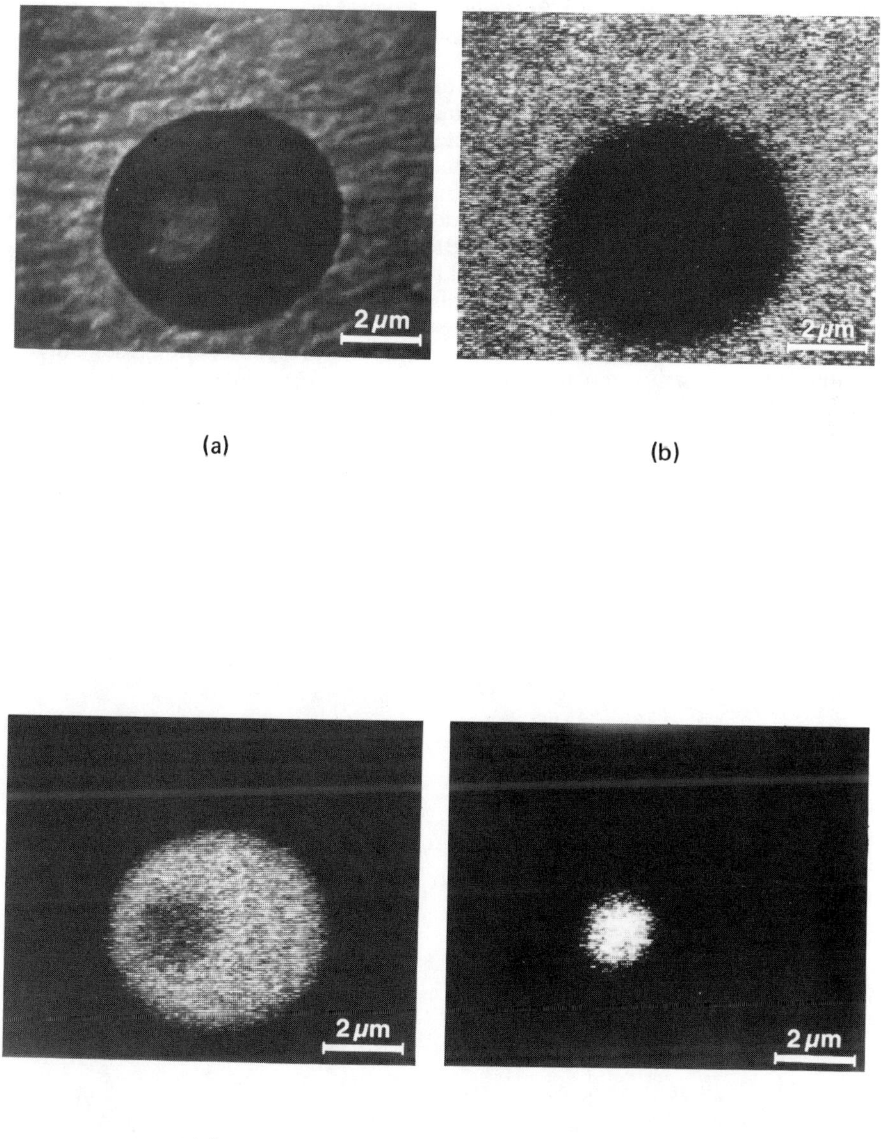

Fig. 3. SAM analysis of sulfide inclusions from metallographically prepared ferrous alloy surface. (a) Secondary electron micrograph, (b) Fe AES chemical map, (c) S AES chemical map, (d) Mn AES chemical map.

The line scan technique can also be applied to this inclusion to obtain an estimate of the relative amounts of different elements in the structure. The results, Fig. 4, confirm the Auger map results and also indicate that the large precipitate has a finite manganese concentration. Examination of the fall-off of the line scans once again indicates interface definition consistent with the electron beam spot size of 2000 Å. For example, the Fe (47 eV) line scan across the edge of the Fe-S precipitate, Fig. 5 shows negligible broadening. Auger point analysis was performed on each of the regions in the structure to obtain a semiquantitative estimate of the relative amounts of the different elements.

A second type of SAM analysis was performed on the same sample in order to fully characterize the inclusion structure. Small fracture bars, 3mm x 3mm x 10mm, were machined from a ferrous alloy and fractured in-situ in the SAM under ultra high vacuum (10^{-10} Torr). Inclusions were observed on the fracture surfaces, Fig. 6, and AES point analysis was performed on both the surface of the inclusion, where the particle separated cleanly from the metal structure (point A), as well as from a cleavage fracture face through the inclusion (point B). The results indicate that residual copper impurities (estimated at 100 ppm bulk) have segregated strongly to the interphase boundary of the inclusion and the metal matrix, Fig. 7. Further, the AES peak to peak height ratios obtained from the cleavage fracture surface should correspond to the bulk composition for the inclusion, since the cleavage fracture presumably occurs along a random path through the inclusion. This permits internal calibration of the AES data without making questionable assumptions concerning the alteration of the surface stoichiometry due to the sputter etching process used on the polished surfaces.

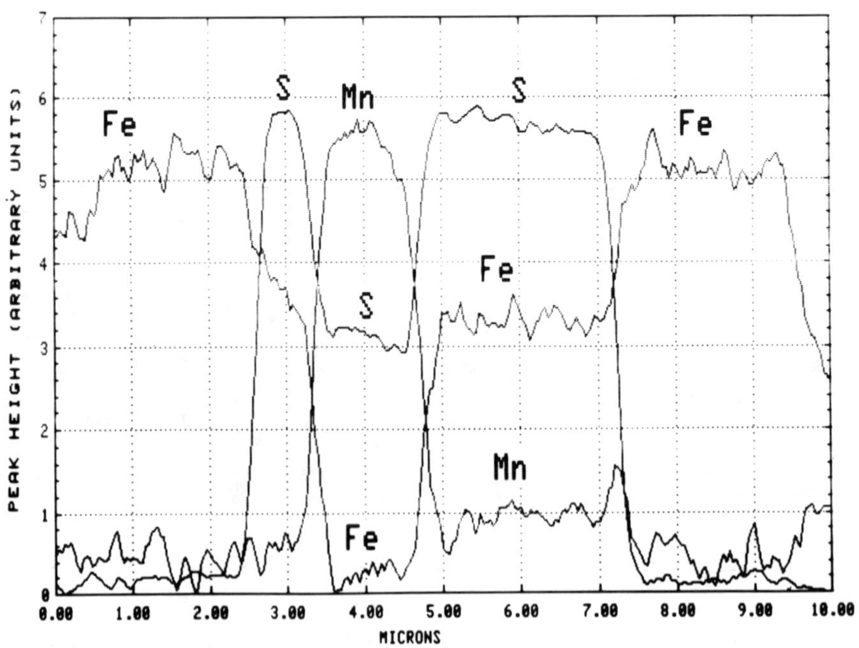

Fig. 4. Fe, S, and Mn SAM horizontal line scans across duplex sulfide inclusion from metallographically prepared ferrous alloy surface.

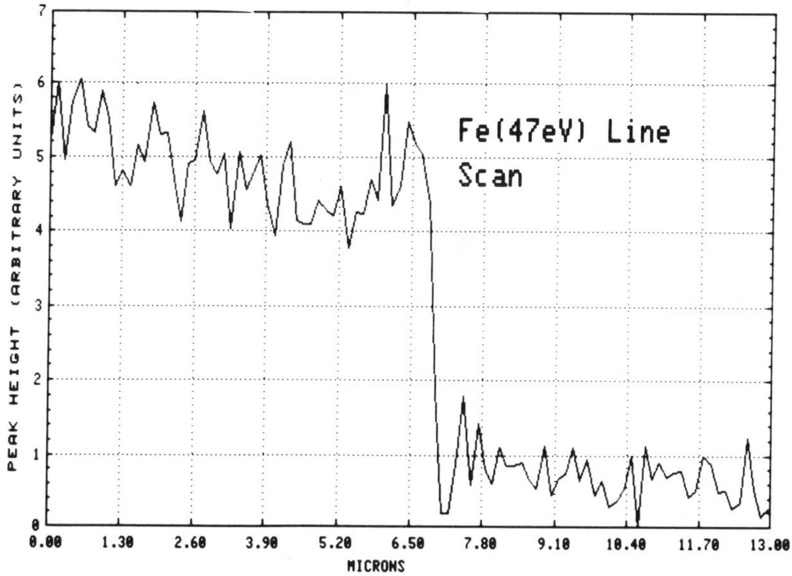

Fig. 5. Fe SAM line scan across edge of sulfide inclusion from metallographically prepared ferrous alloy surface.

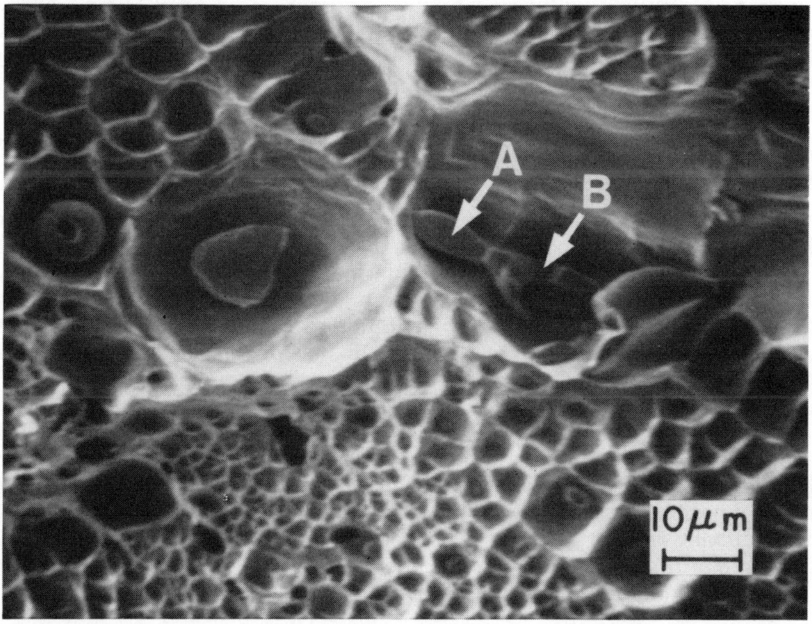

Fig. 6. SAM electron micrograph of in-situ fracture of ferrous alloy (Point A — interphase boundary; Point B — cleavage fracture).

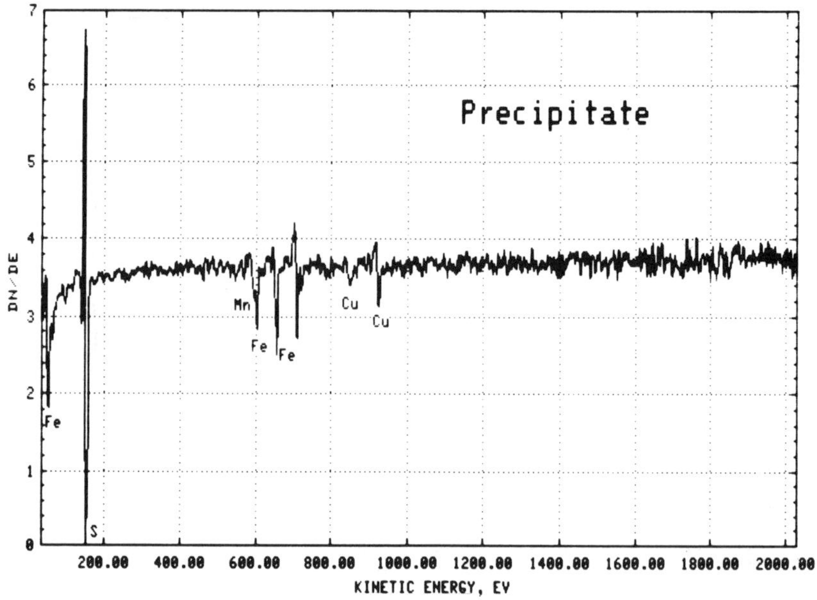

Fig. 7. AES point analysis from interphase boundary from sulfide inclusion.

Combining the results from both the cross section and the fracture analysis provides a fairly complete chemical characterization of the inclusions. The cross section analysis is useful for evaluating the structure of the inclusions, as well as providing a semi-quantitative indication of their composition. The fracture analysis provides detailed information concerning the interfaces in the system, as well as permitting a more quantitative calibration standard for determination of compositions.

Intermetallic Compounds in Coated Steel

As a final application, SAM was used to characterize the interface regions in a commercial Al-Zn alloy coated steel. As above, a sample was sectioned and metallographically prepared for SAM analysis. A low magnification micrograph from the SAM, Fig. 8, shows a 20 μm Al-Zn coating, a thin intermediate compound at the interface, and the steel substrate. A higher magnification micrograph, Fig. 9, shows the intermetallic region, and SAM line scan analysis, Fig. 10, indicates that the compound contains Fe and Al, with little if any Zn. Furthermore, the narrow breadth of both interfaces in the line scan data indicates that there is little interdiffusion between the three distinct layers in the structure, and little if any chemical broadening.

By utilizing pure elemental standards,[13] the data from AES point analysis on the intermetallic compound was found to correspond to a stoichiometry of: 40 at.% Fe—56at.% Al—4at.% Zn. Since altering the energy of the ion beam used for inert ion sputtering did not change the ratio of the Auger peak heights, it was assumed that preferential ion etching was not a problem in this alloy.[14] A fracture-type analysis could also be performed on this material to obtain an internal standard for the quantification of the compound, as in the sulfide study.

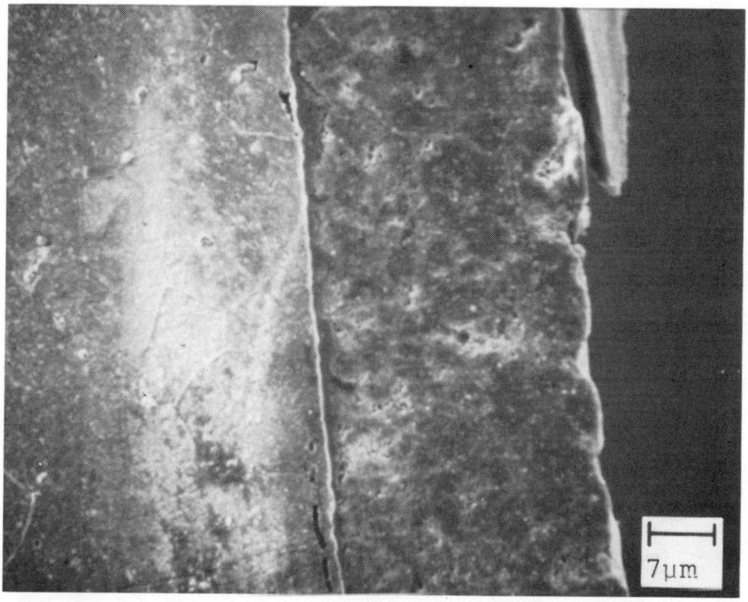

Fig. 8. SAM electron micrograph of Al-Zn coated steel cross sectioned for AES line scan analysis.

Fig. 9. SAM electron micrograph of intermediate compound in Al-Zn coated steel.

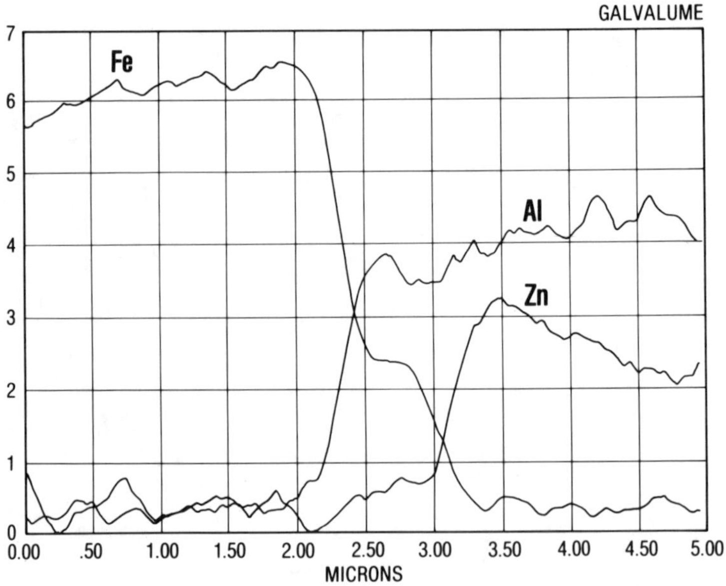

Fig. 10. Fe, Al, and Zn SAM line scans across intermediate compound in Al-Zn coated steel.

CONCLUSIONS

The results of this study demonstrate the utility of the high resolution SAM in providing complementary information from standard metallographic techniques such as EPMA, SEM with EDX, and optical microscopy. The advantages include light element sensitivity, high chemical spatial resolution, and the ability to chemically analyze surface or interface compositions. Finally, the results indicate that the technique can be applied to practical problems, and suggest further experiments to extend the applicability of the SAM to routine metallographic problems.

ACKNOWLEDGEMENTS

The author gratefully acknowledges the assistance and advice of S.P. Clough, B.A. Bruns, D.G. Jean, and J.F. Moulder.

REFERENCES

1. A. Joshi, L.E. Davis, and P.W. Palmberg: *Methods and Phenomena – Methods of Surface Analysis,* edited by A.W. Czanderna, Elsevier, pp. 159, New York (1975).
2. C.C. Chang: *Surface Sciences,* Vol. 25, pp. 53, (1971).
3. W.C. Johnson, D.F. Stein, and A. Joshi: *Can. Jour. of Spec.,* Vol. 17, pp. 88, (1972).
4. N.C. MacDonald, G.E. Riach, and R.L. Gerlach: *Res. and Dev.,* Vol. 27, No. 8, pp. 42, (1976).
5. A. Joshi: "Experimental Techniques to Study Segregation", in *Interfacial Segregation,* edited by W.C. Johnson and J.M. Blakely, ASM Press, pp. 55, Metals Park, Ohio (1979).
6. P.E. Palmberg, G.K. Bohn, and J.C. Tracy: *Appl. Phys. Letters,* Vol. 15, pp. 254, (1969).

7. D.R. Stein, R.E. Weber, and P.W. Palmberg: *J. of Metals,* Vol. 23, No. 2, pp. 39, (1971).
8. J. Kirschner: *Appl. Phys.,* Vol. 14, pp. 351, (1977).
9. N.C. MacDonald: *Proc. IITRI/SEM–1971,* pp. 89, (1971).
10. H.S. Dudek and R. Borath: *Scanning,* Vol. 2, pp. 39, (1979).
11. A.P. Janssen and J.A. Venables: *Surf. Sci.,* Vol. 77, pp. 351, (1978).
12. J.F. Moulder, D.G. Jean, and W.C. Johnson: *Thin Solid Films,* Vol. 64, pp. 427, (1979).
13. L.E. Davis, N.C. MacDonald, P.W. Palmberg, G.E. Riach, and R.E. Weber: *Handbook of Auger Electron Spectroscopy,* 2nd edition, Physical Electronics Ind. Inc., Eden Prairie, MN, (1976).
14. G.K. Wehner: *Methods and Phenomena – Methods of Surface Analysis,* op.cit., pp. 6.

METALLOGRAPHIC CHARACTERIZATION OF SURFACES: II

Chairpersons: J.I. Goldstein and M.R. Pinnel

Stevens, Vander Voort, McCall, eds. Microstructural Science, Volume 8

AN INTRODUCTION TO SCANNING AUGER MICROSCOPY AND ITS APPLICATIONS

Richard J. Blattner*

ABSTRACT

During the past ten years Auger electron spectrometry has matured from an experimental curiosity to a routine analytical tool for characterizing surfaces, interfaces and thin films. It is a very versatile technique when applied to problems of metallurgical interest and has been used in studies of grain boundaries, corrosion, passivation, bonding, diffusion, etc. Early instrumentation employed relatively large primary electron beams (~100 μm) which gave good sensitivity, but provided little spatial information. The first generation of scanning Auger microprobes (SAM's) were generally limited in spatial resolution (by spot size) to 3–10 μm. The demands for improved spatial resolution and the development of high brightness electron sources led to a second generation of scanning instruments with resolutions in the 1000–2000 A range. Very recently, machines have become available which constitute the true marriage of the SAM with scanning electron microscopy (SEM). Auger resolutions of a few hundred Angstroms and SEM resolutions down to ~ 40 A have been achieved in these combined instruments.

This talk will introduce the basic concepts of Auger spectroscopy as related to microbeam instrumentation. Representative applications will be presented which illustrate the capabilities and limitations of the various generations of instrumentation through the present.

INTRODUCTION

In recent years, considerable progress has been made in the development of analytical techniques and instrumentation for materials characterization, particularly in areas related to surfaces and thin films. Consequently, Auger electron spectrometry (AES), electron spectroscopy for chemical analysis (ESCA), Rutherford backscattering spectrometry (RBS) and secondary ion mass spectrometry (SIMS) have joined optical and electron microscopy and the electron microprobe as powerful and commonplace tools in the arsenal of the materials scientist. Each of the surface analytical techniques derives much of its utility from an intrinsic in-depth localization of the analysis by virtue of the physical processes involved. With the exception of the stigmatic imaging ion microanalyzer SIMS instruments, lateral resolutions are limited essentially by the spot size of the probing beam (electrons in AES, x-rays in ESCA and ions in both RBS and SIMS). The demand for greater lateral localization in surface analysis, e.g., metallurgical grain boundary and corrosion product studies or LSI and VLSI in the microelectronics industry have provided the impetus for continued improvement in this regard. Because of the advanced state-of-the-art in electron beam production and focusing technologies, AES is by far the leader in lateral surface microanalysis.

*Materials Research Laboratory, University of Illinois at Urbana-Champaign, Urbana, Illinois, 61801 USA. Present address: Charles Evans & Associates, 1670 S. Amphlett Boulevard, Suite 120, San Mateo, California 94402 USA.

This work supported in part by the National Science Foundation Grants DMR-77-23999 and CHE-76-03694.

In this paper, the general principles of Auger spectroscopy will be reviewed, the unique capabilities and problems associated with microbeam Auger instruments will be discussed in the context of selected examples performed by the author at the University of Illinois using a Physical Electronics model 545 scanning Auger microprobe (SAM), and finally some perspectives on the current status and future trends for microbeam Auger instrumentation will be given.

BASICS OF ANALYTICAL AUGER SPECTROSCOPY

When an energetic (1–10 keV) electron beam impinges upon a solid, various inner shell ionization events occur leaving the target atoms in an excited state. Deexcitation will result in either x-ray emission (used for analysis in the electron microprobe and energy dispersive x-ray systems) or Auger electron emission via a radiationless transition (the Auger effect). In both cases, the energy of the transition product (x-ray or electron) carries with it an elemental "identity". However, if the Auger electron emanates from a depth greater than the inelastic mean free path of the solid (escape depth), there is a high probability that the electron will experience an energy loss collision and thereby lose its analytical utility. Energy and intensity analysis of the electron emission from the sample provides qualitative and quantitative elemental (and in some cases chemical) information about the composition within the volume analyzed (escape depth x area irradiated). Escape depths vary with electron energy, target material, etc., but typically range from 3–30 A in most cases. It is this shallow escape depth that makes Auger spectroscopy a surface sensitive analytical technique.

Figure 1 shows typical Auger spectra for the metals Cr, Mn and Fe. Each element has a characteristic spectrum or "fingerprint" (commonly taken in derivative form as shown) useful for qualitative identification. Peak-to-peak heights in derivative spectra can provide quantitative information, although the integral of the peak in the undifferentiated N(E) spectrum is more truly representative of concentration. Auger spectroscopy is well suited for light element analysis ($Z > 2$) as evidenced by the presence of residual carbon and oxygen in these spectra even after the samples had been sputter cleaned using an argon ion gun configured into the analytical system.

An ion gun is a vital system component because it allows the shallow analytical volume to be advanced into the specimen in a controlled way either to clean the sample surface as indicated above or to permit depth profiling, i.e., to perform in-depth distributional analysis of thin films or interfaces. After argon sputtering, an Ar Auger signal will appear in the surface spectrum due to implantation of the ion beam and the sensitivity of the technique to entrained noble gases.

A typical depth profiling study is illustrated in Fig. 2. Two samples of electropolished Ni were sputter coated with about 1000 A of Pd. In one case the deposition was performed directly with no substrate cleaning, whereas the other nickel sample was sputter cleaned in the deposition apparatus prior to coating. Both depth profiles exhibit a layered structure of Pd over Ni, although there is an apparent loss in depth resolution for the sputter cleaned sample presumably due to induced surface roughness (texturing) from the cleaning process. The chemical effect of substrate cleaning can be seen in the oxygen distributions. Considerable interfacial oxygen appears in the sample without cleaning, but does not appear in the profile of the sample which had been sputter-cleaned prior to Pd deposition.[1]

Because Auger spectroscopy is a surface sensitive analytical technique, ultrahigh and ultraclean vacuum conditions in the experimental chamber are imperative. This fact is exemplified by an in-situ fracture study of Zr_3Al alloys. Samples can be fractured in an Auger spectroscopy vacuum system (ion and Ti sublimation pumped in this case) at an ambient pressure $\leqslant 5 \times 10^{-10}$ Torr. Figure 3 shows the measured carbon and oxygen surface concentrations expressed as fractions of a monolayer as a function of time after fracture assuming ordinary first order adsorption kinetics. There normally is a minimum delay of 60–300 sec between the time of fracture and the onset

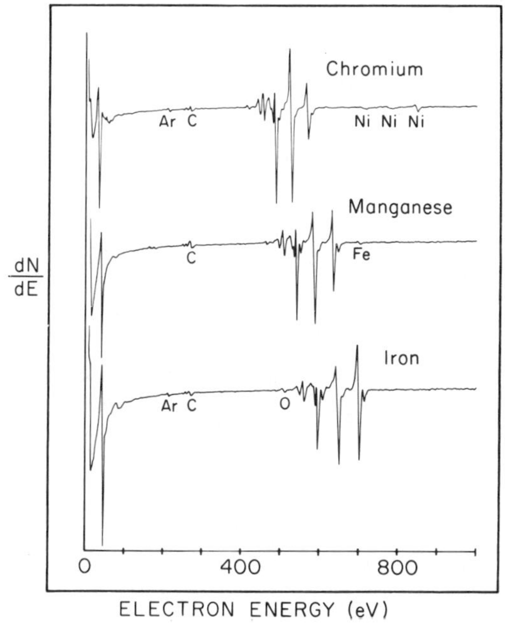

Fig. 1. Typical derivative Auger spectra of Cr, Mn, and Fe taken under essentially constant experimental conditions.

of data collection. As the figure indicates, adsorption of active gases begins immediately with coverage growing rapidly during the initial stages. Carbon and oxygen will thus be observed on the fracture surface after a brief exposure to the ambient vacuum albeit at relatively low levels. The question remains: were carbon and oxygen present on the fracture surface prior to fracture, or are they simply a residual vacuum artifact? We can provide a reasonable answer by extrapolating the data in Fig. 3 to the time of fracture, i.e., t = o. At that point we observe the carbon and oxygen concentrations to vanish within experimental limits, and therefore, we can conclude that carbon and oxygen were not native to the grain boundaries, and more importantly, are probably not involved in the intergranular embrittlement mechanism for these alloys. A small amount of Fe was also observed on the fracture surface and subsequently identified as the embrittling agent via grain boundary segregation.[2]

FUNDAMENTALS OF AUGER MICROSCOPY

An electron gun is obviously an essential component in an Auger spectroscopy system. If the instrument includes provisions for focusing and rastering the electron beam, then with the appropriate readout and control electronics, it is capable of performing electron and Auger microscopy. In fact, some of the early analytical Auger microscopy was performed in a scanning electron microscope configured with an electron energy analyzer.[3,4] Commercial instruments of this type have been available for about six years and are becoming increasingly popular as lateral resolutions are improved.

A scanning Auger microprobe is analogous to an SEM/EDX system or an electron microprobe except that Auger signals are monitored rather than x-rays. In addition to absorbed current and/or secondary electron imaging, Auger microprobes provide laterally localized surface point analyses, elemental line scans and Auger images.

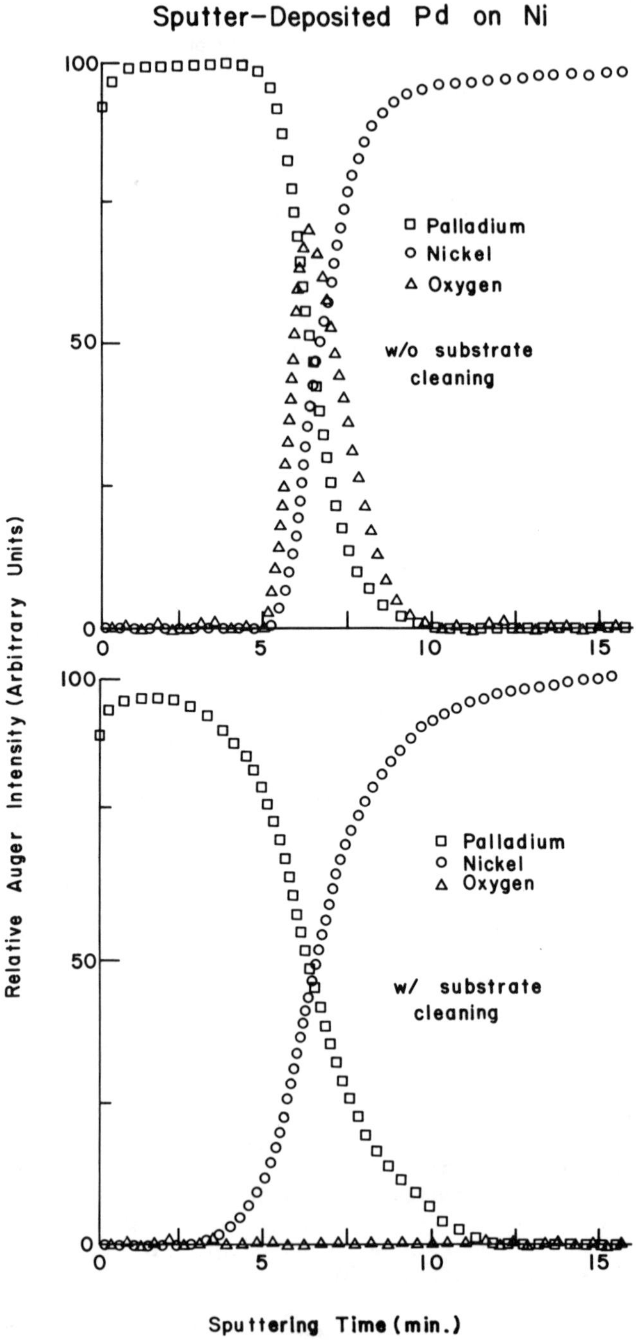

Fig. 2. Typical 2 keV Ar+ ion sputter depth profiles of sputter deposited Pd on nickel without substrate sputter cleaning (upper) and with substrate sputter cleaning (lower) prior to deposition.

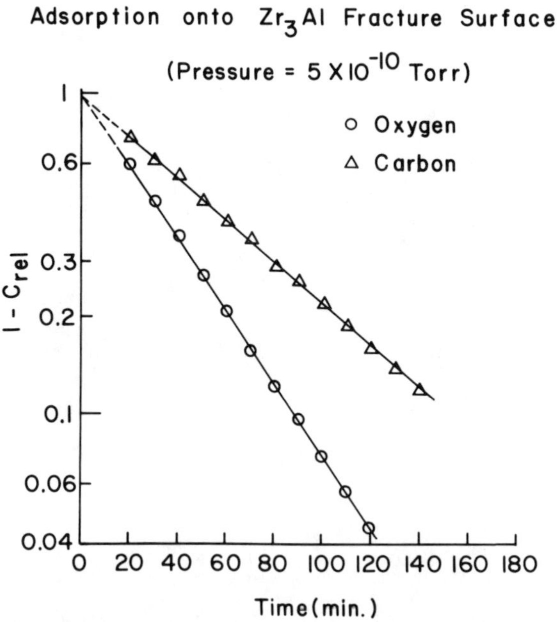

Fig. 3. Normalized carbon and oxygen Auger signals (proportional to coverage) measured as a function of exposure time to a residual vacuum of $\leqslant 5 \times 10^{-10}$ Torr for a Zr_3Al fracture surface.

Combined with ion beam sputtering, they are capable of complete three dimensional elemental characterization of a solid sample. Figure 4 shows typical absorbed current and Auger images taken from a Nb — 5% V alloy specimen which had been charged above the solubility limit with oxygen at high temperature and quenched in H_2O. Large precipitates can be seen in the grain boundaries in the absorbed current micrograph. The oxygen Auger image reveals the presence of high levels of oxygen in the precipitates (presumably oxides) while the niobium Auger image indicates a reduced niobium level compared to the bulk. It is interesting to note that just as the different grain orientations produce varying contrast levels in the absorbed current image, so to the Auger signals are seemingly influenced by the different grain orientations even though the bulk compositions are the same.

There are numerous physical and goemetrical factors which influence Auger signal levels which explains why quantitation is difficult to achieve in the imaging mode. However, reasonable quantitation can be obtained in phase segregated materials using line scans or point spectra. The equation relating Auger current to concentration can be written as: [5]

$$I_s(E_i,XYZ) = I(E_o)N_i\sigma_i(E_o,E_X)\, r_{s,i}(E_o,E_X)\, v_{s,i}(W_j,X)\gamma_i(XYZ). \qquad (1)$$

$$\lambda_s(E_i)\, F(E_i,E_a)\, T(E_i,E_a)\, D(E_a)\, G_b R,$$

where: $I_s(E_i,XYZ)$ = measured Auger signal at energy E_i for the transition XYZ from sample s,

$I(E_o)$ = incident beam current at energy E_o,

N_i = volume density of atoms of the i-th species,

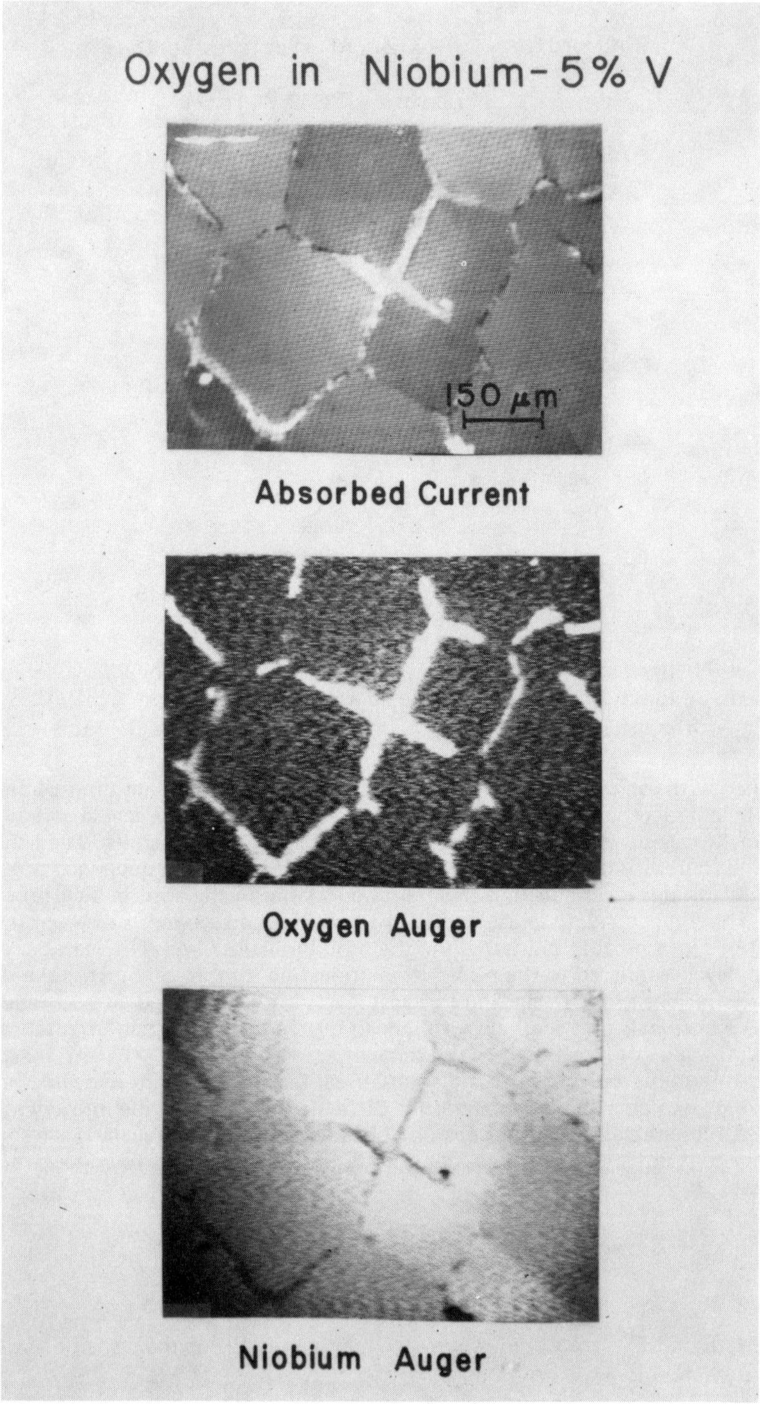

Fig. 4. Typical absorbed current and Auger images taken using a Physical Electronics model 545 SAM from an oxygen doped Nb – 5% V sample.

$\sigma_i(E_o,E_X)$ = ionization cross section of level X with binding energy E_X at the beam energy E_o,

$r_{s,i}(E_o,E_X)$ = correction for additional ionization of electrons in level X of the i-th species due to backscattered electrons,

$v_{s,i}(W_j,X)$ = correction for additional ionization in level X due to Auger and Coster-Kronig transitions that can occur after ionization of electrons in other levels W_j,

$\gamma_i(XYZ)$ = probability that an atom ionized in the X level will decay through an XYZ Auger transition,

$\lambda_s(E_i)$ = total inelastic mean free path for electrons of kinetic energy E_i,

$F(E_i,E_a)$ = radiative energy loss factor important if electrons are decelerated from an initial energy E_i to E_a before analysis

$T(E_i,E_a)$ = analyzer transmission function

$D(E_a)$ = detector efficiency at energy E_a,

G_b = geometrical factor that accounts for the area of the sample irradiated, area viewed by the analyzer and the acceptance solid angle,

and R = surface roughness factor.

In the typical analytical situation most of these factors are unknown but can be eliminated at least in part through the use of internal and/or external standards. It is customary to rewrite equation (1) as:

$$I_{s,i} = K_s N_i \tag{2}$$

where K_s is a collective "constant" for a given element in a given matrix under a given set of experimental conditions. This constant can be evaluated using pure elements or a reference collection of "standard" compounds, and hence the relationship between $I_{s,i}$ and N_i can be established.

ILLUSTRATIVE APPLICATIONS

Three specific examples have been chosen to illustrate the use of scanning Auger microscopy, viz.: 1) the understanding of a surface contamination phenomenon, 2) elucidation of a three dimensional metallurgical interaction in a thin film system, and 3) a quantitative microanalysis of oxide precipitates in an oxygen-doped metal.

Surface Contamination Produced During High Dose Ion Implantation [6]

A silicon wafer was ion implanted with 100 keV $^{28}Si^+$ to a dose of 7.6×10^{15} atoms-cm^{-2} in a conventional 10^{-6} Torr oil diffusion pumped ion implanter. During implantation the sample was held at -110°C by liquid nitrogen cooling. Analysis of the silicon surface by AES after implantation indicated complete coverage by a carbonaceous layer, the thickness of which could be estimated to be ~ 100 Å from the depth profile shown in Fig. 5. Virtually no silicon could be detected at the outer surface of the sample, while the original silicon surface is identified in the profile by the interfacial native oxide. An attempt to remove the layer by a dilute HF etch produced etch pits which could be seen in a SAM absorbed current micrograph (Fig. 6a). The bright stripe near the top of the photograph corresponds to the edge of the sample. Figs. 6b and c show carbon and silicon Auger images, respectively. Apparently the etch pits correspond to "weak" points in the otherwise tenacious carbon coating which permit etching of the underlying substrate revealing the silicon.

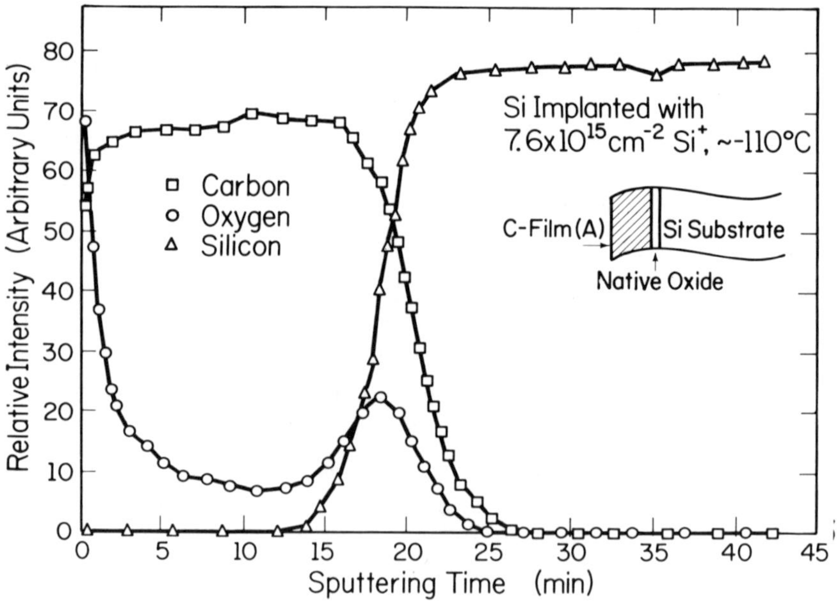

Fig. 5. Auger depth profile of carbon contaminated silicon after high dose ion implantation of 100 keV $^{28}Si^+$ at a substrate temperature of -110°C.

(Note the silicon signal from the sample edge which had not been exposed to the implantation beam.) In all other regions of the surface, the silicon is masked by the carbon film. Figure 6d shows a carbon Auger image of this sample after depth profiling through the carbonaceous layer. The etch pits are no longer visible due to the absence of carbon and the lack of another contrast mechanism.

The complete results of this study indicated that the carbon contamination occurred as a result of adsorption of hydrocarbons from the implanter residual vacuum followed by radiation-induced polymerization. This mechanism is quite similar to that observed in a conventional oil diffusion pumped SEM. By replacing the normal diffusion pump oil with a perfluorinated polyether fluid (recommended by the SEM literature), the carbon contamination problem in high dose ion implantation was found to disappear.

When applied to generalized materials characterization problems the various features / capabilities of an SAM must be viewed symbiotically. As in the previous example, surface spectroscopy and imaging as well as ancillary techniques such as ion beam sputtering all tend to contribute valuable information. We are really dealing with a complete multifaceted analytical system which can also include the ability to perform in-situ fracture and in-situ heat treatment. The latter capability is featured in the following example.

Thermal Annealing Study of the Au/Ti–W Metallization on Silicon [7]

Gold is an excellent outer electrical contact for a silicon based electronic device but is a disastrous impurity in the silicon itself. Furthermore, gold is a fast diffuser in silicon, even at modest temperatures (250–300°C). Therefore, it is essential that a diffusion barrier be incorporated between the Au layer and the underlying silicon. One popular barrier material is sputter deposited $Ti_{0.3}W_{0.7}$ alloy. There exists some controversy, however, as to the intrinsic diffusion barrier properties of the pure alloy film and the

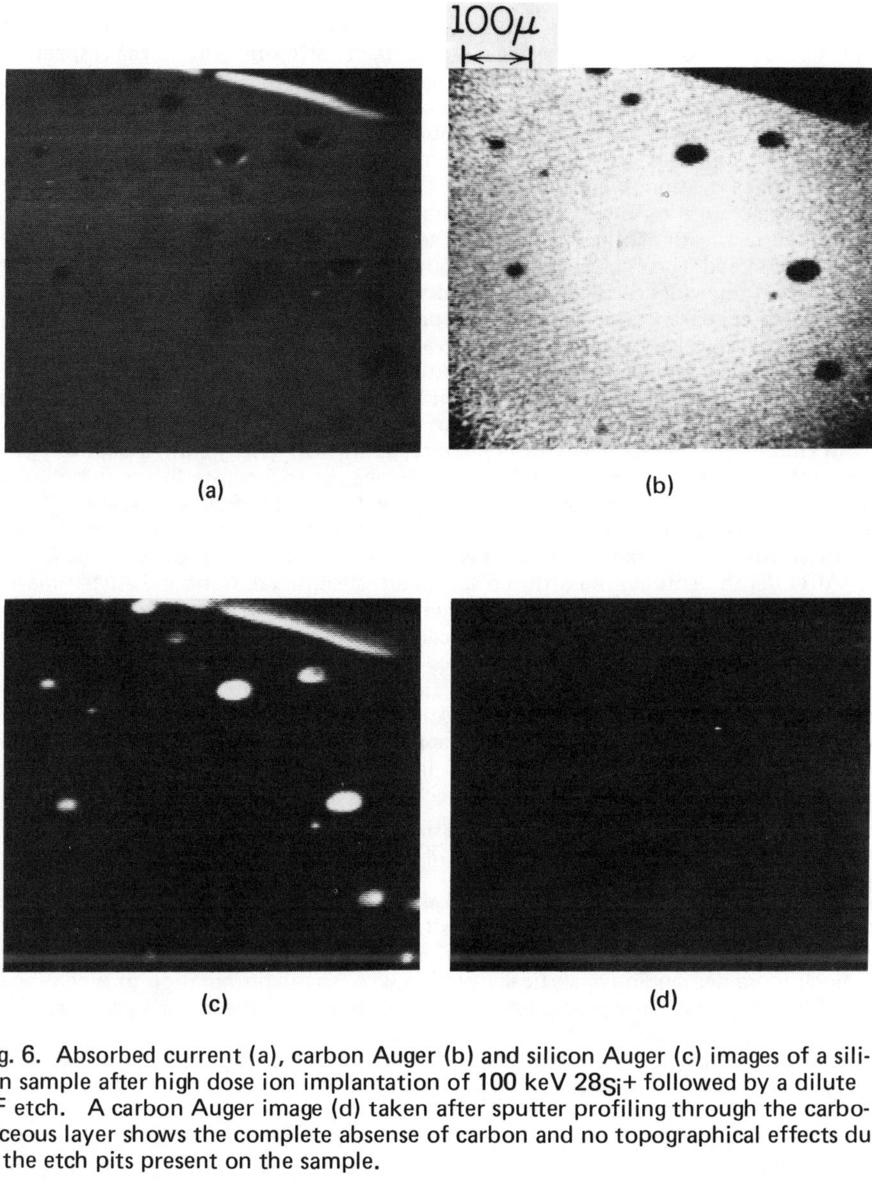

Fig. 6. Absorbed current (a), carbon Auger (b) and silicon Auger (c) images of a silicon sample after high dose ion implantation of 100 keV $^{28}Si^{+}$ followed by a dilute HF etch. A carbon Auger image (d) taken after sputter profiling through the carbonaceous layer shows the complete absense of carbon and no topographical effects due to the etch pits present on the sample.

role of accidentally incorporated impurities such as nitrogen or oxygen. Two samples were prepared with 1000 A layers of Au and $Ti_{0.3}W_{0.7}$ sequentially sputter deposited onto <111> silicon at two different partial pressures of residual H_2O in the sputtering vacuum chamber, viz. 10^{-7} Torr (type 1 samples) and 10^{-8} Torr (type 3 samples) as measured by a residual gas analyzer. The silicon substrates were sputter cleaned prior to deposition to remove the native oxide which itself would act as a diffusion barrier if Au diffusion through the $Ti_{0.3}W_{0.7}$ layer subsequently occurred. Sputter cleaning does produce surface and near surface point defects which can normally be annealed away but were not for the purposes of this study.

Auger depth profiles for the two as-deposited sample types revealed the presence of discrete layered structures (as we would expect) differing only in the respective oxygen levels in the $Ti_{0.3}W_{0.7}$ layers. The measured values for oxygen concentration were ~5% and 2%, with the higher value obtained for the sample deposited at the higher H_2O partial pressure. These semiquantitative results are in good agreement with nuclear microanalysis data taken from essentially the same samples in which the oxygen concentrations were found to be 7.0 and 2.5%, respectively. These samples were annealed in the Auger vacuum chamber at 300°C for 30 min at a pressure of 2 x 10^{-9} Torr. Depth profiles taken after the anneal are shown in Fig. 7. Although some silicon and Ti can be seen to have outdiffused to the surface in the type 1 sample, there is no evidence for Au indiffusion. The depth profile basically retains its initial layered appearance. However, the type 3 sample with lower oxygen in the $Ti_{0.3}W_{0.7}$ layer shows almost all of the gold transported through the "diffusion barrier". A long diffusion tail can be seen on the Au distribution extending well into the silicon substrate. Evidently there exists a threshold oxygen concentration between 2.5 and 7 at% below which $Ti_{0.3}W_{0.7}$ is a nonbarrier to Au diffusion. Based on subsequent TEM studies of the grain size of the polycrystalline $Ti_{0.3}W_{0.7}$ film, a bulk oxygen concentration between 5 and 7 at% would be needed to saturate the grain boundaries assuming monolayer coverage. Thus it was concluded that grain boundary diffusion is probably the basic Au transport mechanism, and that diffusion can be precluded by saturation of the grain boundaries with a reactive impurity such as oxygen.

After depth profiling the diffused specimen, absorbed current and Auger images were taken from the bottom of the sputtered crater (silicon substrate) as shown in Fig. 8. Equilateral triangular structures can be seen randomly distributed throughout the exposed substrate area in the absorbed current micrograph. Auger imaging indicated that these triangular regions are gold rich. Point analyses from several different triangles yielded a composition of Ca. $Au_{0.67}Si_{0.33}$ suggestive of the eutectic composition $Au_{0.69}Si_{0.31}$. The triangular appearance of these "second phase" clusters is indicative of < 111 > point defect nucleation and growth. Since Au is known to be gettered by damage, the formation mechanism appears to consist of: 1) grain boundary diffusion of Au through the $Ti_{0.3}W_{0.7}$ layer, 2) gettering of the Au to the residual point defects produced by the initial sputter cleaning of the single crystal silicon wafer, and 3) nucleation and growth of a Au-Si eutectic phase at the damage sites.

In the previous example it was shown that reasonably quantitative Auger results could be obtained for major constituents (Au-Si) and for a minor impurity (O) in cases where lateral resolution was not a serious issue. The following example illustrates a much more demanding analytical situation, viz., the determination of precipitate stoichiometry where the precipitate size is comparable to the probe diameter.

Quantitative Analysis of Oxide Precipitates in Niobium Using Scanning Auger Microscopy [9]

A Nb-O master alloy was prepared by doping a high purity (99.99+% pure) Nb strip with a known quantity of oxygen in a high vacuum Sieverts' apparatus at 1560°C. Subsequently, the strip was vacuum annealed at the doping temperature to ensure homogeneity. Specimens for Auger analysis were encapsulated in quartz and annealed for 26.3 hrs at 1000°C and water quenched to room temperature. The terminal oxygen solubility at 1000°C is 2.03 ± 0.05 at% (bulk niobium grains). The total oxygen concentration was determined by vacuum fusion analysis to be 5.63 ± 0.08 at%.[10] If thermodynamic equilibrium was achieved during the final anneal, the oxide phase formed should be NbO. The object of this study was to determine if thermodynamic equilibrium had, in fact, been achieved.

Figure 9 shows a typical set of images acquired in the course of this study. Figure 9a is an absorbed current micrograph taken on the SAM using a 3–5 μm diameter electron beam. Figure 9c, in contrast, was taken from the same sample area using a

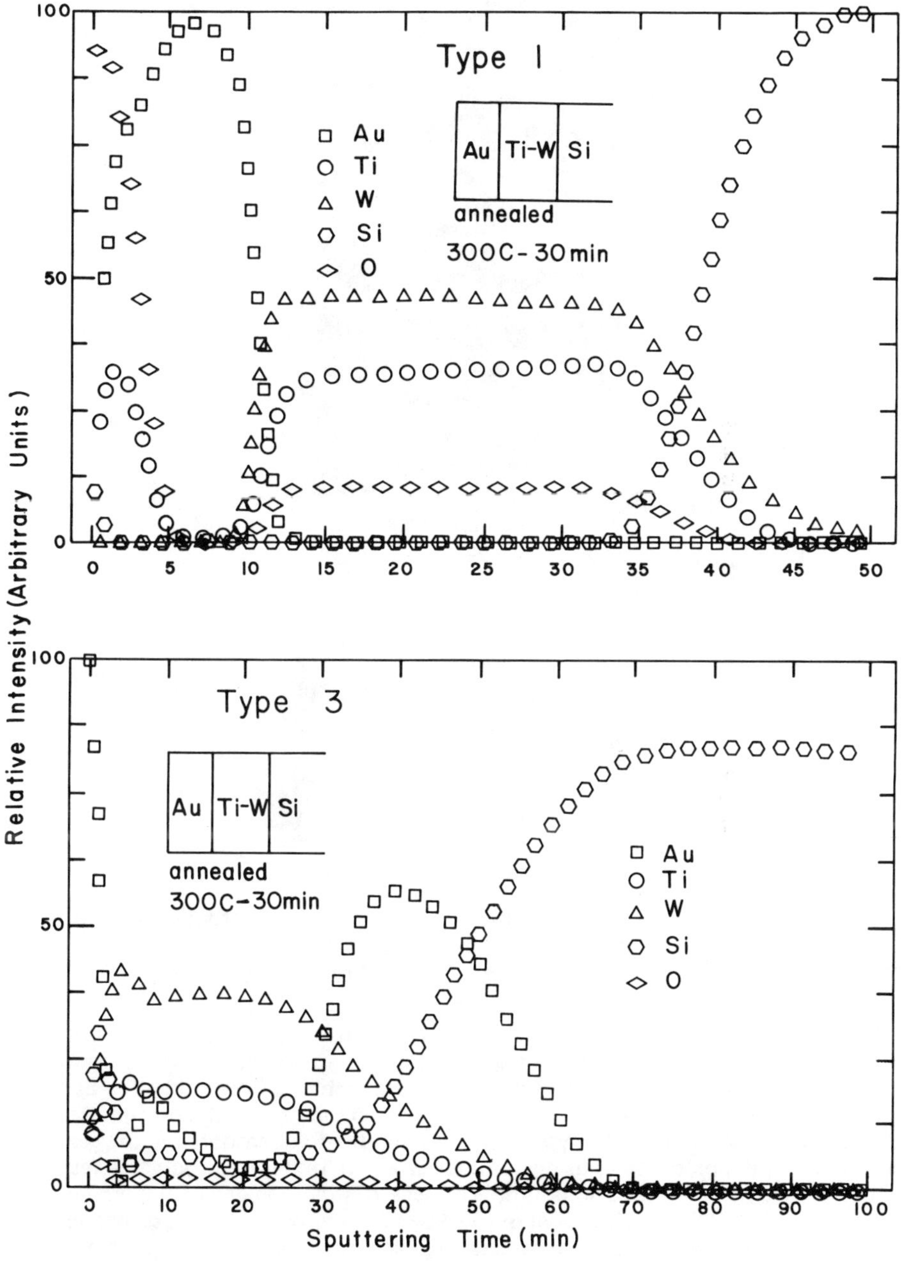

Fig. 7. Auger depth profiles of Type 1 (upper) and Type 3 (lower) Au/Ti-W metallization layers on silicon after in-situ annealing at 300°C for 30 min.

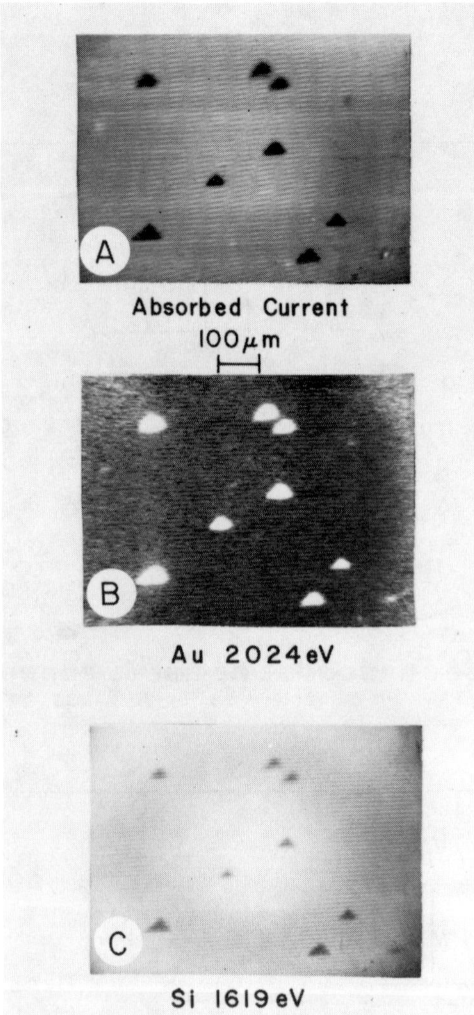

Fig. 8. Absorbed current and Auger images taken from the bottom of the sputter crater after depth profiling the annealed Type 3 Au/Ti-W specimen. Similar imaging for the Type 1 sample revealed no discernable features.

200 Å diameter beam in a JEOL–U3 scanning electron microscope. The spot size effect on lateral resolution is readily apparent in these two images. Furthermore, the size of the precipitates can be seen to be comparable to or less than the SAM spot size. The oxygen Auger image (Fig. 9b) does indicate that the precipitates contain a substantial amount of oxygen. Initial attempts to quantitate the oxygen concentration in the precipitates using point spectra gave rather irreproducible results with no discernable systematic trends. This lack of precision can be understood by viewing the Auger line scans shown in Fig. 9d, superimposed on an absorbed current image. The line scans were taken horizontally across the viewed area along a line marked by the oxygen Auger z-axis (brightness) modulation scan across the center of the photograph. The y-axis modulation Auger line scans for niobium (upper) and oxygen (lower) were taken across this same central line, with their displacement in

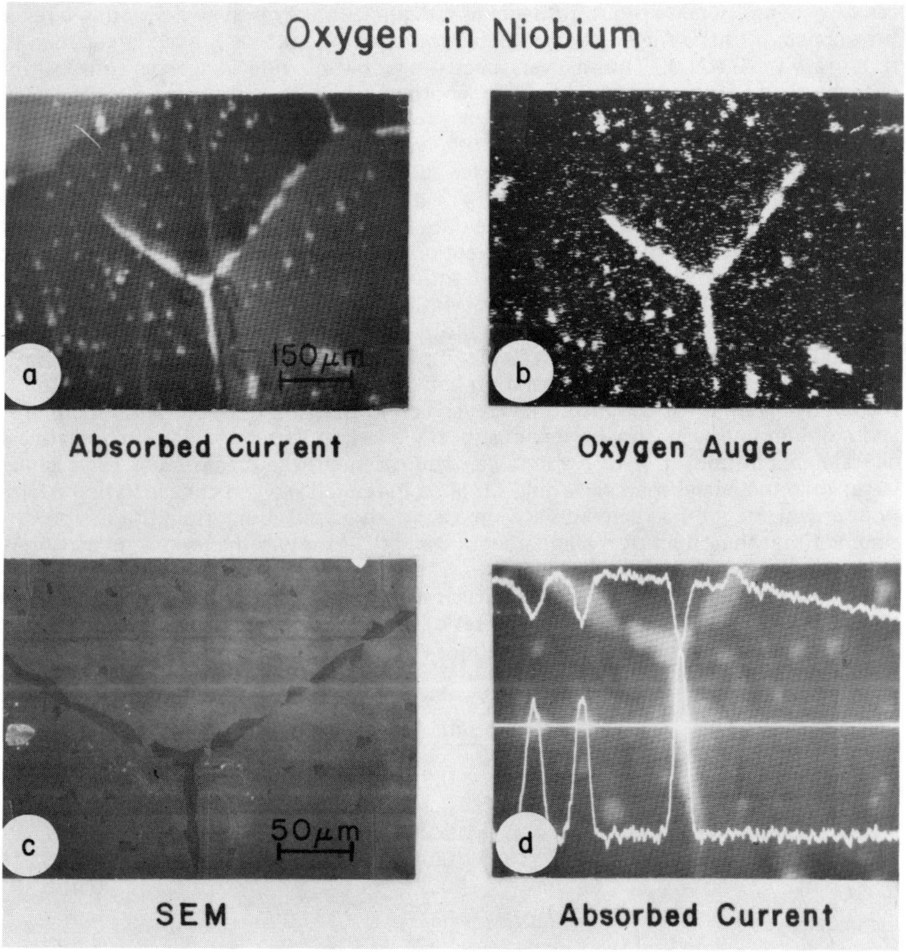

Fig. 9. Representative SAM and SEM images taken from a typical oxygen doped niobium sample indicating the presence of oxide precipitates.

the vertical dependent upon intrinsic signal strengths and various amplifier settings. Clearly, as the beam traverses the two smaller bulk precipitates to the left of the central grain boundary, the oxygen peak intensity does not reach the same level as it does as the beam traverses the large grain boundary precipitate in the middle of the image. The converse situation is observed for the decrease in the niobium signal at the corresponding points. Thus, there appears to be a correspondence between the measured signal intensity and the size of the precipitate. In an effort to obtain a spectrum representative of the composition of an isolated precipitate, spectra were taken using the TV image mode of beam scanning over an area encompassing a single precipitate (or segment thereof) to avoid the random amounts of grain overlap due to the extended beam diameter. This can be accomplished without influencing the quantitative results by choosing a response time constant of the Auger electron detection system (~0.1 sec) much larger than the raster rate (~10 KHz). Thus a given spectrum would provide an average composition over the field of view (measurable from a hard-copy image at the same magnification). An areal deconvolution must be performed to account for the relative contributions to the spectrum of precipitate and grain. This was done using an oxygen Auger image taken at the same magnification as the TV raster and the known oxygen concentration of the grains.

This quantitative approach led to the data for seven different precipitates shown in Fig. 10. Precipitate areas are listed as fractions of the total rastered area, and oxygen concentrations are expressed as atomic fractions. There are two serious problems with these data. First, the relative standard deviation is very poor (nearly 20%), and perhaps more importantly, the mean value seems to be metallurgically unrealistic. A careful inspection of these data indicate a systematic trend, graphically illustrated by the plot in Fig. 10. Here we see a direct correlation between relative precipitate area (one minus the relative grain area) and measured oxygen concentration reminiscent of the oxygen and niobium line scans shown previously. If we extrapolate these data to a relative precipitate area of unity (relative grain area of zero), i.e., if the precipitate was as large or larger than the beam, then we would obtain a measured oxygen concentration near 50% (y intercept) in good agreement with the thermodynamic equilibrium prediction. It was assumed that the origin of the apparent systematic error was the use of oxygen Auger images (taken with the 3–5 μm beam) for measuring precipitate areas. More exact measurements could be made using SEM images of the same areas used to acquire the AES data. These measurements produced a set of corrected area data and in turn led to a new set of measured oxygen concentrations shown below.

Precipitate #	Precipitate Area (Corrected)	Atomic Fraction Oxygen
1	0.32	0.455
2	0.14	0.438
3	0.26	0.457
4	0.08	0.443
5	0.30	0.486
6	0.35	0.464
7	0.26	0.489
	$\bar{x}$ =	0.642
	$\frac{\sigma_{\bar{x}}}{\bar{x}}(\%)$ =	4.3

Dramatic improvement in precision (<5% relative standard deviation) and accuracy (~ 10% assuming a stoichiometry of NbO) is evidenced by these results. Furthermore, no spatially dependent systematic variations can be detected. The mean value below

Precipitate Number	Area(uncorrected)	O Concentration
1	0.54	0.301
2	0.32	0.217
3	0.48	0.278
4	0.21	0.186
5	0.53	0.313
6	0.57	0.316
7	0.50	0.288

$\bar{X} = 0.271$

$\frac{\sigma_{\bar{X}}}{\bar{X}}(\%) = 18.5$

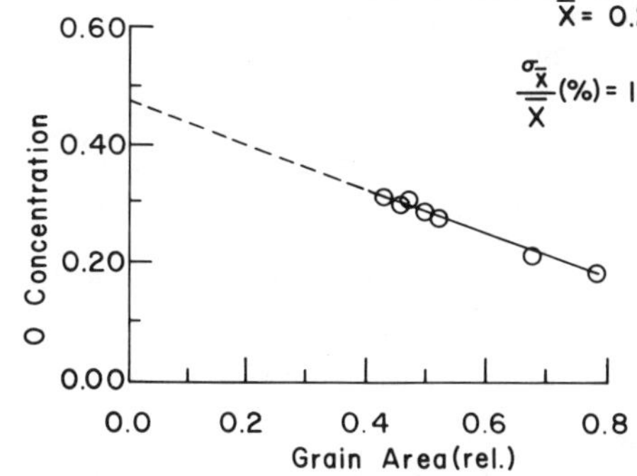

Fig. 10. Quantitative results for relative precipitate areas (taken from oxygen Auger images) and oxygen concentrations (atomic fractions) for seven different oxide precipitates in niobium, and a linear plot of these parameters extrapolated to a relative precipitate area of 100%.

0.50 may be due to preferential sputtering effects (the samples were continuously sputtered during data acquisition to provide a "clean" surface), and/or electron beam damage such as surface reduction via oxygen desorption. Subsequent electron diffraction studies on TEM specimens did verify the oxide phase to be NbO although the exact stoichiometry could not be ascertained by that technique.

The previous quantitative data is more or less typical of what can be done using Auger spectroscopy for quantitative analysis. The spatial deconvolution problems are obviously not present when the features of interest are sufficiently large compared to the probing beam diameter. As was demonstrated, however, the situation is still manageable when the two dimensions are comparable. This latter observation does not seem to be limited to the 3–5 μm regime, but rather appears to be a general result applicable whenever the comparability condition is satisfied.

CURRENT STATUS AND FUTURE TRENDS

The examples given above were taken from a relatively low lateral resolution instrument by contemporary standards (3–5 μm). Since 1975, tremendous strides have been made by commercial manufacturers in producing instruments with significantly smaller probe diameters (and hence better lateral resolutions). Such development is not without its problems. Auger signal is directly proportional to beam current as indicated in Equation 1, and therefore, as the beam is focused to improve resolution, Auger signal strengths decrease due to the reduced beam current requiring longer analysis time and/or poorer detection limits. The use of pulse counting detection and signal averaging partly make up for the lost signal, but high brightness electron sources such as LaB_6 or ultimately, field emission (coupled with pulse counting detection) seem to be the best solution to the low signal problem. However, the high current densities produced by these sources can lead to severe sample damage effects (dependent upon dose —— electrons/unit area) such as desorption, dissociation, or diffusion. There are also geometrical problems associated with sample topography such as shadowing, angular dependent Auger yields, etc. which become quite important on these smaller scales (⩽ 1000 Å). Lateral surface composition may vary dramatically compared to the underlying material making depth profiling a much more complex task.

Although there are a number of unique problems associated with microbeam Auger spectroscopy, the detailed information it can afford far outweighs the potential shortcomings. This writer has recently been engaged in an evaluation of several instruments presently on the market for possible purchase by the University of Illinois. (These machines all carry lofty six digit price tags so that such an evaluation cannot be conducted casually!) Most of the instruments are closely matched in overall performance, differing mainly in specific capabilities and details. This fact greatly complicates a comparative evaluation, particularly because of the inherent instrumental duality of SEM/SAM. Thus, ratings tend to reflect the persuasion of the person doing the rating (electron microscopist or surface analyst). Consequently, the intended use and potential users of the instrument must be considered along with total performance.

However, this is a very complex issue, and each potential buyer must perform his own evaluation and make a judgement in the context of his organization's needs, goals and funds available.

The following microbeam Auger data was collected during the course of this writer's evaluation and is intended only to provide a "flavor" of what is available and not as a critical review. Consequently, results will be limited to the two Physical Electronic instruments, the 590 "Super SAM" and the 595 "Multiprobe".

One of the test specimens used in this comparison was an evaporated thin film structure consisting of a silicon wafer substrate, a 1000 Å thick Pd layer, and a 1 μm overcoat of amorphous silicon. The sample was cleaved in liquid N_2 and examined in cross section. Electron beam evaporation of the silicon layer was accomplished using a graphite boat to hold the silicon charge which led to the incorporation of homogeneous carbon contamination (1–3% atomic) throughout the silicon film. This contaminant proved useful in assessing detection limits in a microanalysis mode. These various sample details could be easily detected by conventional depth profiling in a "big beam" Auger instrument[12]. Although we initially thought this particular sample would yield a relatively "easy" analysis, such did not prove to be the case. In fact, it provided a severe instrumental test and led to numerous insights into the problems associated with the microbeam Auger technique itself.

Figure 11 shows conventional SEM images taken using a JEOL JSM–35C electron microscope (lefthand photographs). The 1000 Å Pd layer appears as a bright stripe between the substrate and the evaporated Si overlayer. The images on the right were taken using a Physical Electronics 590 "Super SAM" in the SEM (upper) and Pd Auger (lower) imaging modes for the same sample area. One problem encountered with SEM imaging of this sample results from the large secondary electron emission from the edge. To see the Pd stripe it was necessary to tilt the specimen to the proper orientation with respect to the incident electron beam and the detector. Considerable brightness can still be seen in the JSM–35C images. Figure 12 shows a point Auger spectrum taken with the 590 beam centered on the Pd layer, and a Pd line scan across the sample. In the line scan, the sample edge is marked by the sharp decrease in background signal. Based on the silicon layer thickness (1 μm) the Pd layer thickness (1000 Å) can be estimated to be between 1500 and 2000 Å . This result is consistent with the probe

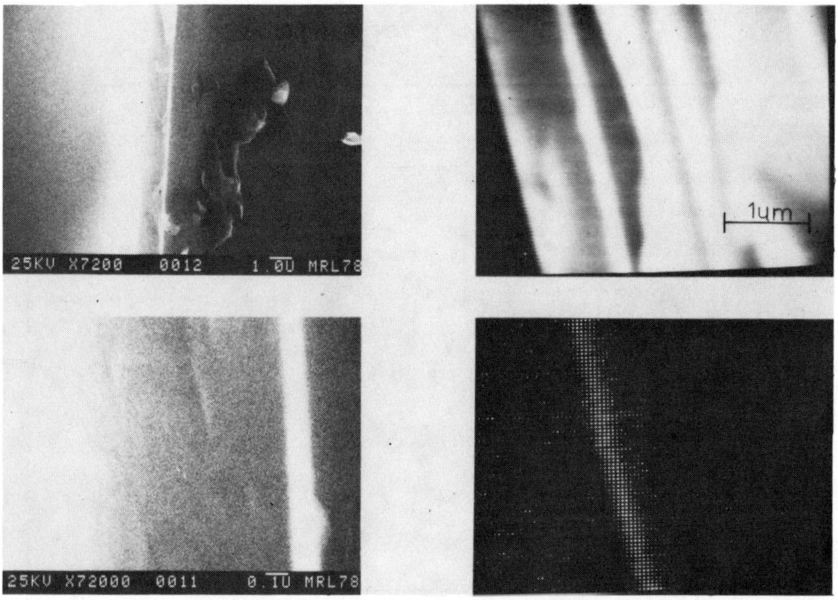

Fig. 11. Micrographs taken using a JEOL JSM–35C scanning electron microscope (left) and the Physical Electronics model 590 SAM (Right) from the Si/Pd/Si evaporated thin film test sample viewed in cross section. The upper 590 image is a secondary electron micrograph whereas the lower 590 photograph is a Pd Auger image.

diameter specified for the 590 instrument (1500 – 2000 Å) and explains the presence of a large silicon signal in the point spectrum (recall the NbO study discussed earlier).

Figure 13 shows point spectra from the evaporated Si layer and the silicon substrate after sputter cleaning, both taken with the 590 instrument. The 2% carbon impurity in the evaporated silicon film is clearly detectable with no evidence for carbon contamination in the wafer (from the residual vacuum for example). These spectra were collected at a beam current of about 0.5 namp (~ 2000 Å spot) with a spectrum integration time of 5 min each. No Pd could be detected in the silicon point analysis because, in this case, the layer thickness is much greater than the beam diameter. A small amount of oxygen (510 eV) can be seen in both spectra either due to incomplete sample cleaning, residual vacuum contamination, or both.

Figure 14 shows SEM images for the same sample as above taken using the PHI model 595 "Multiprobe." These micrographs were acquired at a beam current of $\leqslant 1 \times 10^{-10}$ amp into a spot size < 500 Å . Considerably more detail can be seen in these photographs compared to the corresponding 590 image due to the significant improvement in probe diameter. Some question exists, however, as to the viability of the available probe current for performing an Auger analysis. Figure 15 shows 595 Auger data using the smallest spot size (and hence current) analytical conditions in modes similar to those used with the 590. The critical test is a point analysis in the Pd layer. There is about a factor of 3 loss in signal-to-noise ratio in the 595 instrument, but a factor of 2–3 reduction in silicon signal relative to Pd. Although the Pd distribution in the line scan does not have a flat top (suggesting that the beam is not substantially smaller than 1000 Å) the half width gives a value of about 1000 Å indicating that the beam is probably $\leqslant$ 500 Å. The presence of silicon in the spec-

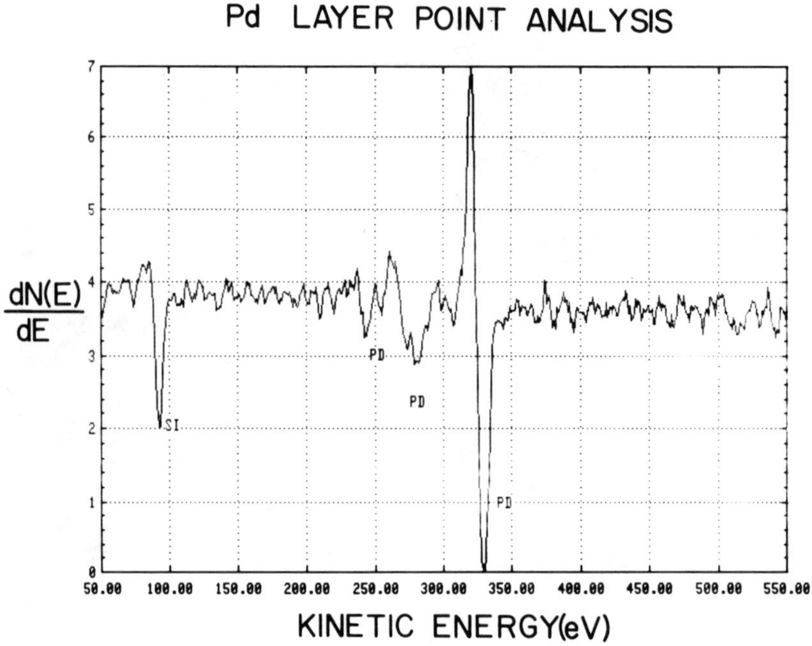

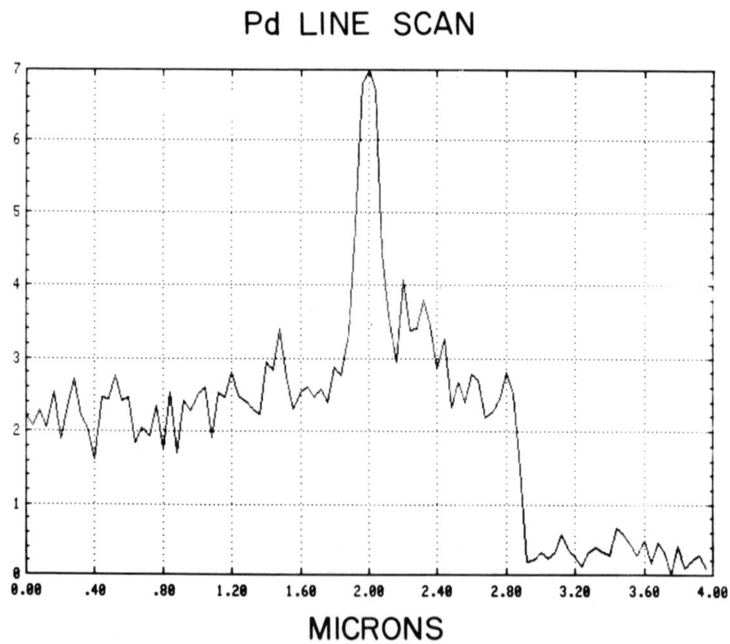

Fig. 12. Point Auger spectrum (upper) taken with the 590 SAM beam centered on the 1000 Å Pd layer, and a Pd line scan (lower) across an area containing the Pd and evaporated silicon layers.

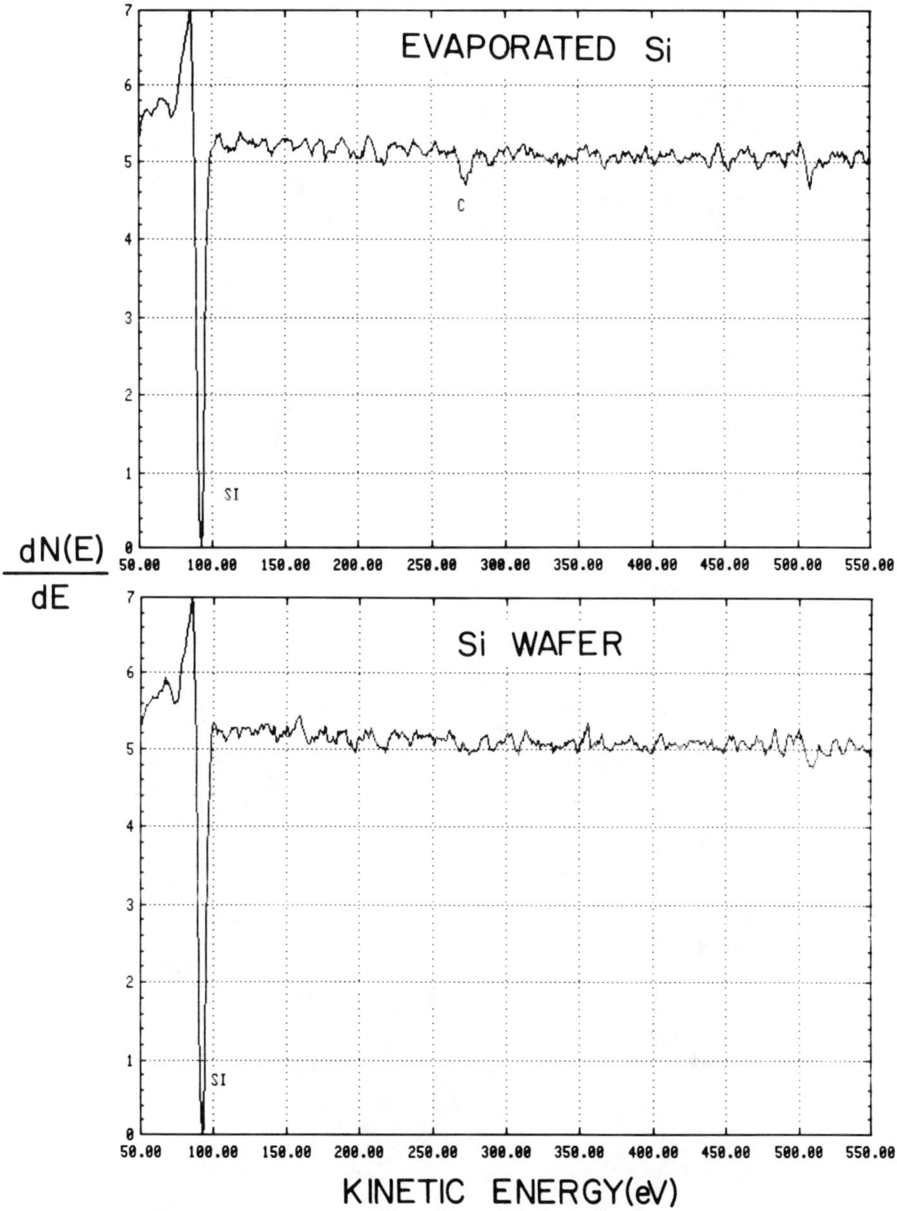

Fig. 13. Point Auger spectra taken with the 590 SAM beam on the evaporated silicon layer (upper) and silicon wafer substrate (lower). (Note the presence of carbon contamination in the evaporated silicon which does not appear in the substrate.)

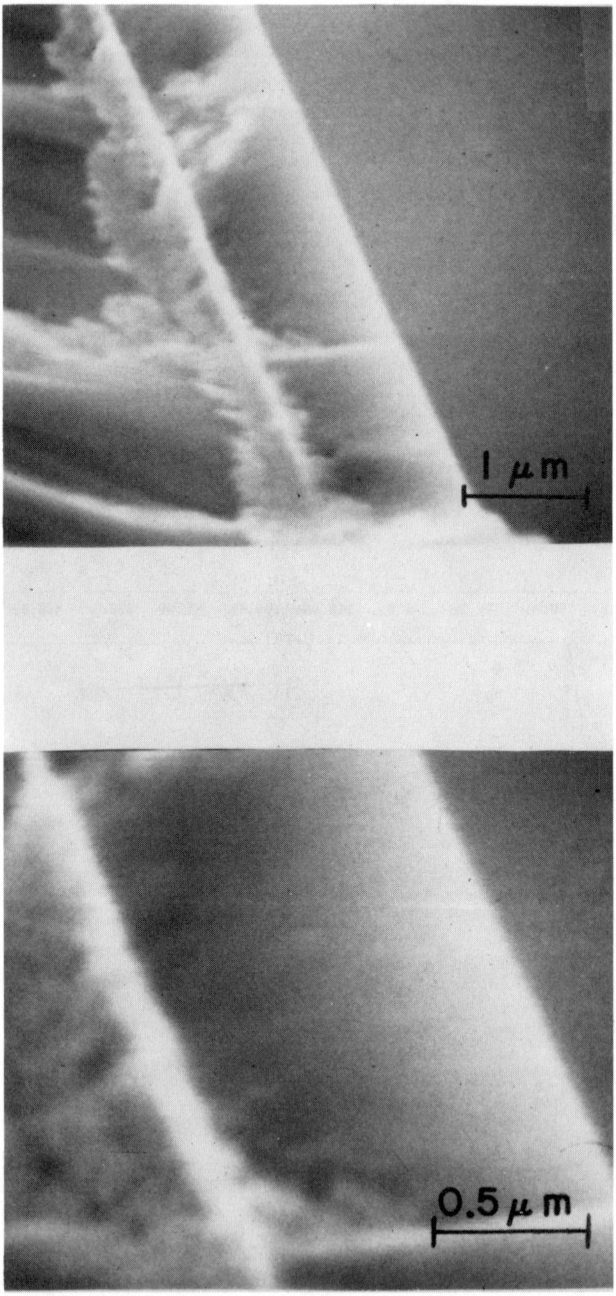

Fig. 14. Secondary electron micrographs taken from the Si/Pd/Si test sample using the Physical Electronics 595 "Multiprobe" instrument.

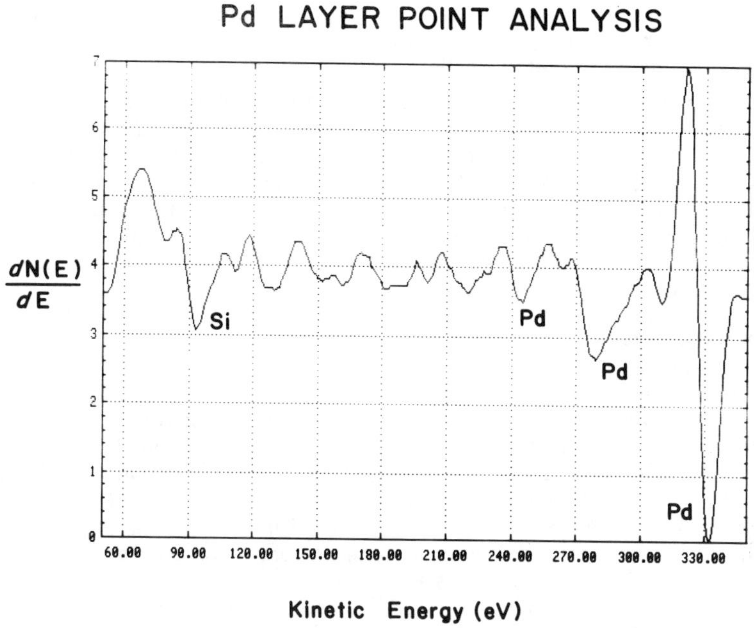

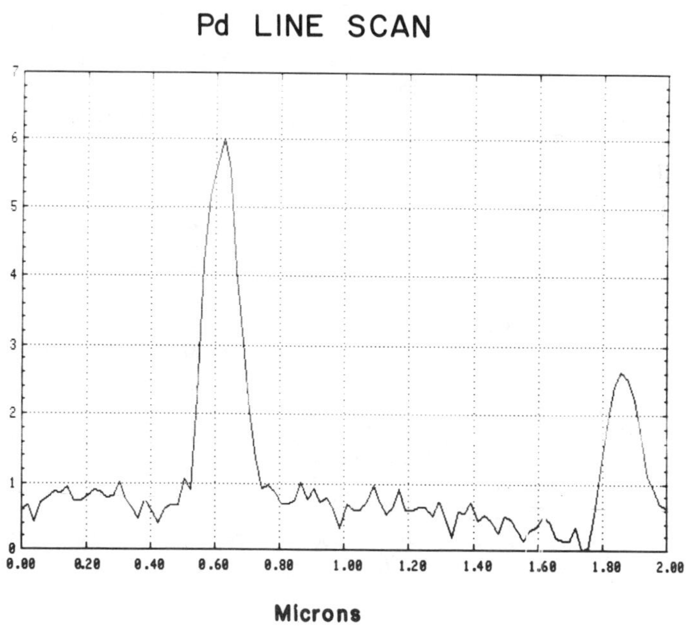

Fig. 15. Point Auger spectrum (upper) taken with the 595 Multiprobe beam situated on the 1000 Å Pd layer, and a Pd line scan (lower) across an area containing the Pd and evaporated silicon layers.

trum may arise from a variety of sources (beam overlap, stage instabilities, contamination during sample preparation, backscattered electrons,[11] etc.) and points out some of the difficulties associated with the technique itself. The 2% carbon impurity could not be detected using the smallest probe diameter in the 595.

At the present time it is not clear that meaningful data can be routinely obtained at the highest possible lateral resolution levels. Considerable "standardization" of the technique will have to be done before much can be said in this regard. Microbeam Auger will spend some time in the future simply as a research topic until many of the problems are fully evaluated. Along the way, insights and solutions to many materials problems will be achieved. When the full potential of the technique is realized it will open whole new worlds to the materials scientist and constitute a major step forward for the entire scientific community.

ACKNOWLEDGEMENTS

The author thanks Dr. D.W. Stevens, Technical Program Chairman, for the opportunity of presenting this work as an Invited Paper at the 12th Annual Meeting of the IMS. Many thanks are also due to several of my colleagues at the University of Illinois, in particular, Dr. Peter Williams, Professor Howard Birnbaum, and Mr. John Woodhouse, and to all the collaborators who have participated in the work presented as indicated in the reference list. This author is grateful to Mr. John Molder and Mr. Claire Hovland, both of Physical Electronics, for their efforts in obtaining the test data from the 590 and 595 SAM instruments, respectively. Finally, the author is indebted to the manufacturer representatives, Mr. J. Callaghan (Physical Electronics), Mr. G. Cogswell (JEOL), and Mr. G. Kearns (VG Scientific) for their cooperation and many helpful discussions during the complex instrument procurement process.

REFERENCES

1. B. Jones and H. Birnbaum, Department of Metallurgy and the Materials Research Laboratory, University of Illinois, Urbana, Illinois (private communication).
2. E. Schulson, Metallurgy Division, Chaulk River National Laboratory, Canada (Private communication).
3. N. MacDonald, *Appl. Phys. Lett.* 16, 76 (1970).
4. *Idem., Appl. Phys. Lett.* 19, 315 (1971).
5. C.J. Powell, in *Proceedings of the 7th International Vacuum Congress and the 3rd International Conference on Solid Surfaces,* Vienna, p. 1 (1977).
6. M.Y. Tsai, B.G. Streetman, R.J. Blattner, and C.A. Evans, Jr., *J. Electrochem. Soc.* 126, 98 (1979).
7. J.E. Baker, R.J. Blattner, S. Nadel, C.A. Evans, Jr. and R.S. Nowicki, "Thermal Annealing Study of Au/Ti–W Metallization on Silicon", (submitted to Thin Solid Films).
8. T.W. Sigman, L. Csepregi, and J.W. Mayer, *J. Electrochem. Soc.* 123, 1116 (1976).
9. R.J. Blattner, G. Steckel, and C.J. Altstetter, "Quantitative Analysis of Oxide Precipitates in Niobium Using Scanning Auger Microscopy," (to be published).
10. Analysis by LECO Corp., St. Joseph, MI.
11. J. Kirschner, *Appl. Phys.* 14, 351 (1977).
12. Z.L. Liau, S.S. Lau, M.–A. Nicolet, J.W. Mayer, R.J. Blattner, P. Williams, and C. A. Evans, Jr., *Nucl. Instru. Meth.* 149, 623 (1978).

Stevens, Vander Voort, McCall, eds. Microstructural Science, Volume 8

ELECTRON SPECTROSCOPIC ANALYSIS OF LUBRICANT–TREATED STEEL SURFACES AFTER SOLVENT CLEANING

Marianne K. Bernett and Harold Ravner*

ABSTRACT

Effective cleaning of miniature instrument bearings prior to assembly in guidance and other critical systems is necessary for the satisfactory performance of such systems; even traces of surface contaminants can contribute to their ultimate failure. Current solvent-cleaning procedures employed by Navy facilities vary with respect to materials and solvents. A recent study critically examined the specific effects of some of these variations on specially cleaned reference surfaces. In the present study, chemically characterized surfaces of 440C bearing steels were coated with very thin films of a formulated synthetic hydrocarbon. These coated surfaces were subjected to standard cleaning procedures at two Navy facilities, and then reanalyzed. The results are discussed in terms of cleaning efficiency, recontamination, and types of random contamination.

INTRODUCTION

The performance and reliability of stainless steel precision miniature bearings for guidance and other critical applications are significantly affected by the chemical composition of the bearing surfaces. The most important changes are those caused by chemical reaction (salt formation due to organic or inorganic acid attack on the metal surface, etc.) and by the incomplete removal of overlaying contaminants (lubricants, additives, airborne pollutants etc.). Bearing surface changes which result in decreased wettability by liquid lubricants are especially detrimental since they generally presage early bearing failure. Effective cleaning of these bearings is therefore necessary before they can be used safely.

Navy precision miniature bearings are cleaned and assembled in Naval Air Rework Facilities (NARF). Since each NARF has considerable leeway in its process, a study was instituted to determine the effect of each cleaning procedure on bearing surfaces. In the first phase of this study[1] reference cleaned and analyzed 52100 and 440C steel bearing surfaces were subjected to standard NARF cleaning procedures and the surfaces reanalyzed. Although the results among the NARFs differed somewhat, the treated surfaces generally were found to contain traces of esters and hydrocarbons, presumably from contaminated solvents, and atmospheric-derived contaminants such as sulfur and halides. In many instances wettability of these surfaces was decreased.

The present report summarizes the results when thin films of a formulated synthetic hydrocarbon lubricant were laid down on reference 440C bearing steel surfaces which were then subjected to both the standard sequential cleaning cycles and to individual solvents at two NARFs.

---

*Naval Research Laboratory, Washington, D.C. 20375 USA.

EXPERIMENTAL

Preparation of the 440C steel cylinder surfaces as representatives of miniature precision bearings has been described in detail earlier[1]; the process includes special cleaning and glow-discharging procedures to assure reproducible reference surfaces free from carbonaceous residues. Each test surface in the present study was covered, immediately upon removal from the glow-discharge chamber, with a thin layer of a synthetic hydrocarbon (oil derived from the polymerization of α-olefins), and containing small amounts of antioxidant and antiwear additives. Such oils are employed to lubricate certain precision bearings in guidance and other critical systems.

The oil layers were carefully wiped down to a thickness of interference fringes (2 to 8 X 10^{-7} m) and the specimens thus prepared were either characterized for reference purposes or shipped to two NARFs (NARF a and NARF b) in special containers[1] designed to maintain the specimens relatively contamination-free. Specimens subjected to complete solvent cleaning processes (i.e., exposure to a sequence of solvents) provided information on their efficacy in oil-removal, while those exposed to individual solvent treatments provided additional information on possible contaminant types and levels from each solvent.

Each steel specimen was analyzed for elemental and chemical consistency with a Physical Electronics combined XPS/AES spectrometer using a double pass cylindrical mirror analyzer. XPS spectra were recorded using A1 $K_{\alpha 1,2}$ radiation (1486.6 eV). Binding energies (BE) were referred to the measured XPS $Au_{4f7/2}$ line at the published standard of 83.8 eV[2]; reproducibility of the Au BE was ±0.2 eV. The 3 KeV electron beam for AES was maintained at 15 μA. The spectroscopic approach used on any one selected area for each specimen was: a) a full XPS spectrum to observe elemental species; b) high resolution XPS spectra for selected peaks, e.g., Fe, O, C, S, N and other identified elements of specific interest, to determine their abundance and chemical state; c) a full AES spectrum.

Wettability was determined by contact angle (θ) measurements made with a Rame-Hart contact angle goniometer using the slowly advancing sessile drop method [3] with three diagnostic liquids: triply distilled water, and carefully purified and percolated methylene iodide (CH_2I_2) and hexadecane ($C_{16}H_{34}$) whose surface tensions were 72.0, 50.8, and 27.6 mNm^{-1}, respectively. Each θ reported was the average of at least two measurements on separate sessile drops, taken in room air at 20–25°C.

RESULTS AND DISCUSSION

Since the results in the specimen sets from the two facilities were essentially similar within the limits of similar processes we will report on only one set; any significant variations will be discussed.

Table 1 summarizes typical elements detected by AES and XPS on the surface of a processed steel specimen coated with a thin layer of the hydrocarbon oil. For reference purposes Table 1 also shows the analyses on a freshly glow-discharged steel surface and on an oil-coated surface prior to the solvent cleaning process. The peak intensities of the various elemental signals are shown, normalized with respect to the signals of Fe (Fe = 10), but uncorrected for the AES or XPS sensitivities to the various elements. Although uninformative as to the actual quantities or percentages of each element present in the surface, these ratios provide useful comparisons among the different specimens examined.

The main elements observable on all specimens were Fe, O (the surface iron oxide), and C whose line intensity and positions are the key indicators of the amount and type of C-containing overlayer. Other elements detected on various specimens in trace amounts were S, N, Na, and Si. Signals of the minor component elements

Table 1
Relative elemental intensities on 440C steel coated with hydrocarbon oil (NARF a)

(Normalized to Fe = 10)

Specimen	AES				XPS				
	S	C	O	Other	S	N	C	O	Other
Glow Disch.	--	1.9	48.0	-	-	-	3.0	23.1	-
Oil-Coated	?	105.0	65.0	-	-	-	540.0	10.0	-
Cleaned	-	3.0	46.0	Na?	0.1	1.0	5.3	24.3	-
Stoddard	?	7.5	42.5	Na; Si	0.6	0.7	13.6	26.0	Si
Freon US*	0.3	4.8	52.4	Na; Si	0.4	0.8	9.2	26.7	-
Trichloroethane	0.8	8.6	64.3	Na	0.6	1.2	26.0	51.8	-
Freon VD*	2.5	15.0	55.0	Na	0.7	0.7	20.0	30.0	-

*US = ultrasonic; VD = vapor degrease

of the steel alloys, such as Cr, were not considered. There were minor variations of elemental intensities in the specimen surfaces among the two NARFs.

Small C signals indicate a surface relatively free of C contamination as shown by the glow-discharged reference specimen (Table 1, Fig. 1). The oil-coated specimen, on the other hand, exhibited a C signal so large (Table 1) as to mask the underlying Fe and O and thus almost completely suppress their spectral signals (Fig. 2). Use of the resulting Fe signals as a basis for normalizing the data thus results in an exaggeration of the C overlayer component with respect to the other substrate elements; however, overall trends would still be apparent. Bracketed between these two extreme reference conditions, the C signals indicated that most of the original oil layer had been removed by the complete cleaning process (Table 1) and also that the process at NARF b was less efficient than at NARF a as shown by the slightly more suppressed Fe and O signals (Fig. 1). Individual solvent treatments turned up such diverse elements as Si, clearly a contaminant peculiar to the specific solvent, but removed subsequently by solvents later in the process. Na was detected in trace amounts on every specimen from both NARFs and is therefore suspected to be an integral component of the original hydrocarbon oil acquired as a by-product during its synthesis or formulation. The absence of Na signals in the reference oil-coated specimen could be attributed to the all-masking large C signal, while the minimal Na trace in the cleaned specimen was an indication of the efficiency of the total process.

The C_{1s} line positions in the XPS spectra of all specimens were predominantly those of aliphatic C–C bonds, at a BE of 285.0 eV[4] (Figs. 1 and 2). Additional small but measurable signals at a BE of ≈289.0 eV, characteristic of carboxy $C{=O \atop -O}$ bonds[5], were detected from several solvents (Figs. 1a and 2).

The dominant O_{1s} signal on all specimens occurred at a BE of ~530.5 eV (Figs. 1 and 2), attributable to iron oxide[6]. A broadening of line width (Figs. 1a and 2) indicated additional oxygen linkages characteristic of organic compounds whose oxygen is bonded to carbon, such as simple carbonyls or esters[7]. These shifts toward higherO binding energy in conjunction with the C signals at ≈289.0 eV were clear evidence that organic contaminants, such as esters were acquired from the solvents.

No significant differences were noted among the various specimens for the BE of $Fe2p_{3/2}$ at ~711.5 eV (Figs. 1 and 2) which is characteristic of an iron oxide[6,8,9]. Thus, there was no evidence of chemical reactions involving iron or of changes from iron oxide to salt.

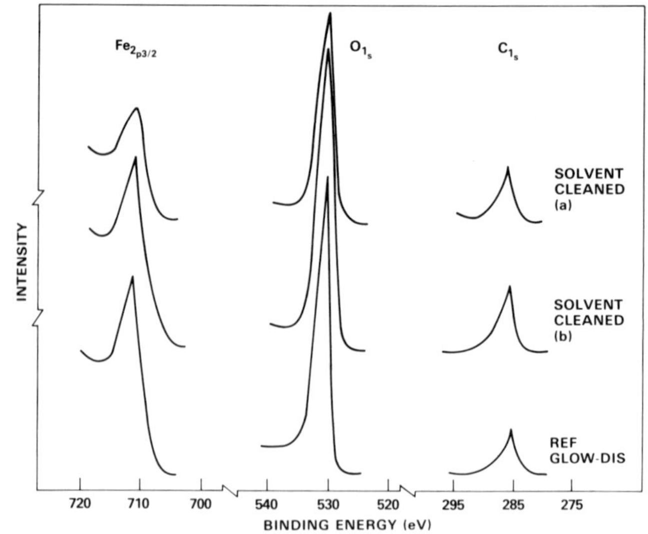

Fig. 1. Relative intensities of Fe, O, and C XPS spectra on 440C steel specimens. Ref: glow-discharged reference. Solvent-cleaned (b): after complete process at NARF b. Solvent-cleaned (a): after complete process at NARF a; note suppressed Fe and O intensities and trace of $C{=O \atop -O}$ peak of C signal.

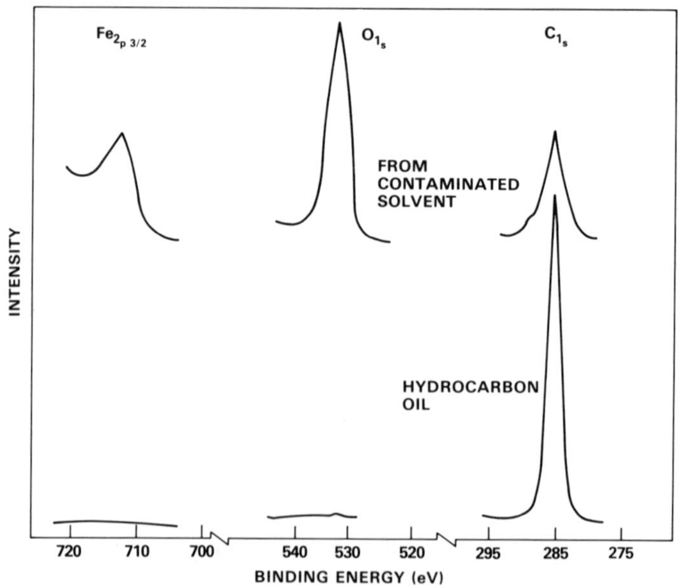

Fig. 2. Relative intensities of Fe, O, and C XPS spectra on 440C steel lubricant-treated specimens. (b) Hydrocarbon oil-covered reference; note large C signal and almost completely suppressed Fe and O intensities. (a) After Stoddard solvent; note suppressed Fe and O intensities, broadened O signal, and small $C{=O \atop -O}$ peak of C signal.

Freshly cleaned metal or metal oxide surfaces are wetted spontaneously ($\Theta=0^\circ$) by $C_{16}H_{34}$ and H_2O, and exhibit a $\theta < 20^\circ$ with CH_2I_2[10,11]; contact angles larger than that are therefore an indication of adsorbed material. Since high surface-energy surfaces readily attract adsorbents from the prevailing environment, such spontaneous wetting is rarely observed under ambient conditions with liquids whose surface tensions exceed ~25 mNm^{-1}. Table 2 lists θ values of H_2O, CH_2I_2 and $C_{16}H_{34}$ on representative sample sets processed at each facility. The contact angles on the oil-coated specimens were indicative of a surface constituted mainly of methyl and methylene groups[12]. The lowest θ's were observed on the completely cleaned specimens; these values were typical of random hydrocarbon coverage[13] signifying a substantial removal of the original oil film. Ranging in between were the values obtained after individual solvent treatments, and here the cleaning efficiency varied not only among the solvents, but also among the two facilities. The increased θ of CH_2I_2 after some solvent treatments (Stoddard, Freon US, Freon VD) suggested not only incomplete oil removal but also additional contamination acquired from these solvent.

Table 2
Wettability of 440C steel coated with hydrocarbon oil (NARFs a and b)

Specimen	Contact Angle (°)					
	H_2O		CH_2I_2		$C_{16}H_{34}$	
	a	b	a	b	a	b
Oil-Coated	100		44		spr	
Cleaned	70	68	44	44	<5	spr
Stoddard	90	–	55	–	<5	–
Freon US*	85	70	52	52	<5	<5
Trichloroethane	91	71	45	45	<5	<5
Freon VD*	92	88	51	50	spr	spr

* US = ultrasonic; VD = vapor degrease

CONCLUSIONS

Despite the fact that the two techniques, AES—XPS and wettability, sampled only small portions of any one surface, the data show good correlation. Neither the complete cleaning process at either facility, nor any one single solvent completely removed the original oil film. Individual solvents varied in their cleaning ability as well as in their contribution to recontamination to the specimen. The most prevalent recontaminants were carboxy-containing compounds, presumably instrument lubricants or their degradation products dissolved or dispersed within the solvent from previous batches of reworked bearings. Other random contaminants, such as Si, probably have similar origins. Some contaminants may also be attributed to inhibitors added by the manufacturer to selected solvents.

Generally, the greater the surface residues on the specimens the less wettable they were. For precision bearings operating under stringent conditions, non-wettable surfaces, especially from uncontrolled contamination, can lead to lubricant starvation and ultimate failure[14].

The iron oxide surface did not react chemically with the overlayers as evidenced by the unchanged XPS binding energies of iron and oxygen, both characteristic spectra of oxides.

The results to date of this study indicate that standard preparation of critical surface-sensitive devices may be inadequate to assure satisfactory surface composition. While it may not always be necessary to analyze such surfaces to the extent detailed here, such analyses should be considered for truly critical applications, as in aerospace devices.

REFERENCES

1. M.K. Bernett, B.J. Kinzig, J.S. Murday and H. Ravner, "Surface Analysis of Bearing Steels after Solvent Treatments" *ASLE Trans.* (in press)
2. K. Siegbahn, "Electron Spectroscopy – An Outlook", *J. Electron Spectrosc. Relat. Phenom. 5,* 3–97 (1974).
3. M.K. Bernett and W.A. Zisman, "Surface Properties of Perfluoro Acids as Affected by Terminal Branching and Chlorine Substitutions", *J. Phys. Chem. 71,* 2075–82 (1967).
4. D.T. Clark, "The Application of ESCA to Studies of Structure and Bonding in Polymers", Advances in Polymer Friction and Wear, *Polymer Sci. and Technol.* 5A, 241–314, Ed. L.H. Lee, Plenum Press, NY, (1974).
5. D.T. Clark and W.J. Feast, "Application of Electron Spectroscopy for Chemical Applications (ESCA) to Studies of Structure and Bonding in Polymeric Systems", *J. Macromol. Sci., Res. Macromol. Chem. C12,* 191–286 (1975).
6. N.S. McIntyre and D.G. Zetaruk, "X-ray Photoelectron Spectroscopic Studies of Iron Oxides", *Anal. Chem. 49,* 1521–29 (1977).
7. D.T. Clark and H.R. Thomas, "Applications of ESCA to Polymer Chemistry. X. Core and Valence Energy Levels of a Series of Polyacrylates", J. Polyer Sci., *Polymer Chem. Ed. 14,* 1671–1700 (1976).
8. K. Asami, and K. Hashimoto, "The X-ray Photoelectron Spectra of Several Oxides of Iron and Chromium", *Corros. Sci. 17,* 559–570 (1977).
9. C.R. Brundle, T.J. Chuang, and K. Wandelt, "Core and Valence Level Photoemission Studies of Iron Oxide Surfaces and the Oxidation of Iron", *Surf. Sci. 68,* 459–468 (1977).
10. M.K. Bernett and W.A. Zisman, "Confirmation of Spontaneous Spreading by Water on Pure Gold", *J. Phys. Chem. 74,* 2309–2312 (1970).
11. M.K. Bernett and W.A. Zisman, "Effect of Adsorbed Water on the Critical Surface Tension of Wetting on Metal Surfaces", *J. Coll. Interface Sci. 28,* 243–249 (1968).
12. H.W. Fox and W.A. Zisman, "The Spreading of Liquids on Low-Energy Surfaces. III Hydrocarbon Surfaces", *J. Colloid Sci. 7,* 428–42 (1952).
13. K. Bewig and W.A. Zisman, "Surface Potentials and Induced Polarization in Nonpolar Liquids Adsorbed on Metals", *J. Phys. Chem. 68,* 1804–1813 (1964).
14. Shri A. Kumar, "Wettability of Gyro Motor Bearing Races", *Proceedings of International Ball Bearing Symposium,* C.S. Draper Laboratories (1973).

Stevens, Vander Voort, McCall, eds. Microstructural Science, Volume 8

SOME RELATIONSHIPS BETWEEN THE OXIDATION MECHANISM AND ELECTRICAL RELIABILITY OF POLYCRYSTALLINE SILICON FILMS

E.A. Irene* and E. Tierney*

ABSTRACT

The use of polycrystalline silicon for conducting lines between devices on silicon device chips is widespread in the Integrated Circuits industry. One of the primary advantages associated with the use of polycrystalline silicon for electronics applications is that it can be controllably oxidized to form an insulating, passivating and protective SiO_2 film on the surface. However, several reliability problems with the oxide have arisen. The oxide grown on polycrystalline silicon has been found to have higher conductivity, increased number of premature dielectric breakdowns and initial shorts as compared with the oxide grown on single crystal silicon. A TEM study of the morphology of polycrystalline silicon films and the oxide grown therefrom has revealed that the polycrystalline silicon films become rough during oxidation, the oxide has thickness undulations and there is extensive intergranular oxidation. It is shown in the present study how these observations lead to an understanding of the major reliability problems associated with polycrystalline silicon.

INTRODUCTION

Polycrystalline silicon, poly-Si, is presently widely used in the electronics industry to fabricate highly conducting line interconnections between devices on integrated circuit chips. Films of poly-Si are usually doped with B, P, or As to achieve high conductivity and are grown to less than 10^3 nm thick by chemical vapor deposition, CVD, using gaseous SiH_4 as the source of Si. Blanket Si films are delineated using masking and etching techniques. CVD Si films prepared at a temperature of 650°C or lower are usually amorphous but further necessary high temperature processing such as doping, annealing and oxidation cause the films to crystallize to poly-Si.

The primary advantages associated with the use of poly-Si are twofold. First, the material, being Si, is compatible with other high temperature processing steps that are necessary for complete silicon device fabrication. Most metal films would delaminate or react during typical device processing. Secondly, the oxidation of poly-Si yields a self-passivating, protective and insulating SiO_2 film. This oxide called poly-ox, enables the preparation of multi-level devices using poly-Si lines with poly-ox insulation. Multi-level devices increase the device density on a chip.

The present study is primarily concerned with the second advantage, i.e., the oxidation of polycrystalline silicon. Notwithstanding the decided advantages associated with poly-Si, it was reported[1,2,3] that the oxide grown via the oxidation of poly-Si, displays significantly higher conductivity than the oxide grown from single crystal silicon. The higher conductivity of poly-ox leads to premature dielectric breakdown of the poly-ox films. In addition to this problem, a high incidence of initial electrical shorts

*IBM Thomas J. Watson Research Center, Yorktown Heights, New York 10598 USA.

was found with these shorts extending from the poly-Si conducting lines through an underlying oxide to either another poly-Si line or to a single crystal silicon substrate. This problem has occurred when metal contact holes were opened through the poly-ox to the poly-Si by masking and chemical etching procedures. The metal contact was evaporated Al. Apparently, the chemical etchant ($HF+H_2O$) penetrated through the poly-ox, poly-Si, and underlying oxide leaving a hole into which the Al migrated and caused an electrical short.

Based on measurements of both dark and photocurrents, DiMaria and Kerr[2] have shown that the high currents in poly-ox are due to localized field enhancement effects. Since the poly-Si is composed of grains, these authors attribute the field enhancement to asperities on the poly-Si surface as a result of grains protruding from the surface. A field enhancement factor of 1.5–2 would explain their results. Anderson and Kerr [3] reported SEM observations of a rough poly-Si surface with asperities having a height to width ratio of 1.8. According to Lewis' calculations for an isolated protuberance of this dimension, a field enhancement of nearly 10 can be expected. Anderson and Kerr[3] require a factor of more than 3 to explain their data. However, since they show that the poly-Si surface asperities have an areal density of more than 5 x 10^9 cm^{-2}, Lewis'[4] calculations show that there is a reduction in field enhancement from 10 for the isolated case to less than 2 for the case of Anderson and Kerr. Therefore, localized field enhancement effects due only to poly-Si surface asperities do not explain all the poly-ox conductivity data. Furthermore, TEM surface replica studies to be shown later do not show the same high degree of surface roughness as reported by Anderson and Kerr[3].

A detailed study of the oxidation of poly-Si[5] has led to an identification of the characteristic features of this type of oxidation. The results of this previous study[5], to be reviewed in the present paper, yield explanations for the origin of poly-Si surface asperities and identifies another way in which the actual poly-ox electric field is locally higher than anticipated. In addition, the previously mentioned initial shorting problem with poly-Si can be understood from the oxidation results. Therefore, the present study deals firstly with describing the morphological aspects of the oxidation of poly-Si, then use is made of these results to understand the electrical properties of poly-ox. Hence, the structure – property relationships will be demonstrated.

EXPERIMENTAL PROCEDURE

All the polycrystalline silicon films were deposited onto oxidized single crystal silicon slices. The silicon slices were commercially available 2Ω-cm p-type silicon chem-mechanically polished on one side, 3.2 cm in diameter and .02 cm thick with (100) orientation. Prior to any film preparation, the silicon slices were thoroughly cleaned according to a previously published procedure[6]. Oxidations of single crystal and polycrystalline silica were carried out in fused silica tubes inside resistance heated tube furnaces with pure dry O_2 flowing. The O_2 is fed into the oxidation furnaces from a liquid oxygen source. The polycrystalline silicon films were chemically vapor deposited from a mixture of SiH_4 and N_2 onto the oxidized silicon substrates. For phosphorus doped polycrystalline silicon samples, $POCl_3$ was deposited upon the polycrystalline silicon films and driven in for 10 min. at 870°C to homogenize and electrically activate the P atoms.

TEM examination consisted of direct transmission through thinned foils and surface replication to produce a negative representation of the surface. The thinning was done by first dicing the samples into sizes to fit the electron microscope sample holder, secondly coating the entire sample with wax, thirdly, scratching a hole in the wax on the silicon substrate side of the sample. The sample is then exposed to an etchant which readily attacks silicon. We found that mixtures of HF and HNO_3 were very useful, since Si is attacked very vigorously by this solution but the oxide film somewhat more slowly. Therefore, etching from the substrate side slows down naturally when the silicon substrate is penetrated. Mixtures of ethylenediamene, – pyrocatechol and H_2O[7] were also found to be useful since this solution also attacks silicon vigorously but hardly etches oxide at all. For this etchant mixture it is necessary to maintain the solution temperature above 80°C and therefore the wax mask is destroyed. This etchant is particularly useful for making TEM observations of SiO_2 films. In addition to these two chemical etchants, we also used ion milling to thin samples. The results and conclusions in this study are the result of concordant TEM observations using all three of the preparation procedures so as to avoid etching related artifacts.

The surface replicas involve several steps. First is the negative replication of the surface using a collodion solution then the evaporation of a thin electron transmitting layer of platinum – carbon alloy. Finally, the thick collodion is removed leaving a thin electron transmitting negative replica of the surface under investigation.

EXPERIMENTAL RESULTS

The progressive stages of oxidation of a poly-Si sample are shown in Fig. 1. Prior to oxidation (1a) the sample displays large equiaxed silicon grains of nearly random orientation. Upon oxidation several morphological features develop. Firstly, oxide forms in the grain boundaries, i.e. intergranularly, as well as on the free polycrystalline silicon surface. Fig. 1(b) and (c) show that the grains (which are the darker outlined objects) are shrinking laterally as oxidation proceeds. In fact there is as much oxide formed on the edge of the grains as on the free surface. Secondly, the oxide formed from the oxidation of poly-Si does not have a uniform thickness. The poly-ox is thinner at the previous grain boundaries and thicker near the mid-grain regions. Furthermore, it is seen that as the oxidation proceeds the thickness undulations become more severe. Thirdly, on some areas of partially oxidized samples there was actual cracking of the oxide at previous grain boundaries.

Figures 2 and 3 show that as deposited poly-Si samples are rather smooth. The sample shown in Fig. 2 was deposited at 800°C, then a $POCl_3$ layer was deposited and the P driven in at 870°C for 10 minutes. After doping, the samples were given a short 10 min. oxidation at 1100°C. The oxide was then stripped and the surface was as shown in Fig. 2a. It can be seen that the grains have smooth surfaces except that there are ridges which outline the grain boundaries. The ridges are in fact grooves on the poly-Si surface, since the replicas are negatives of the real surface. These grooves arise as a result of integranular oxidation and subsequent removal of the oxide from the grain boundaries. Upon further oxidation the surface of the oxide is rough as shown in Fig. 2b. When the oxide is removed, Fig. 2c shows that the poly-Si surface displays even larger and sharper surface features as compared with Fig. 2b. Fig. 3 shows a sample which was prepared at a lower temperature of 650°C, $POCl_3$ doped identically to the sample in Fig. 2 but then mildly oxidized at 800°C and the oxide was stripped prior to Fig. 3a. Again this series of micrographs shows that the poly-Si surface stripped of oxide is rougher than the poly-ox surface.

In order to determine if the oxidation temperature is important in altering the poly-ox surface morphology, the same poly-Si sample as shown in Fig. 1 and 2 was oxidized to about the same extent at both 800°C in Fig. 4a and 1150°C in Fig. 4b. These micrographs show that the thickness undulations in the poly-ox are more severe for the lower oxidation temperatures.

MODEL FOR POLY–SI OXIDATION

The combined experimental results above provide the basis for a descriptive model for the oxidation of poly-Si. This model will now be presented and following the model the various reliability aspects of poly-ox will be discussed.

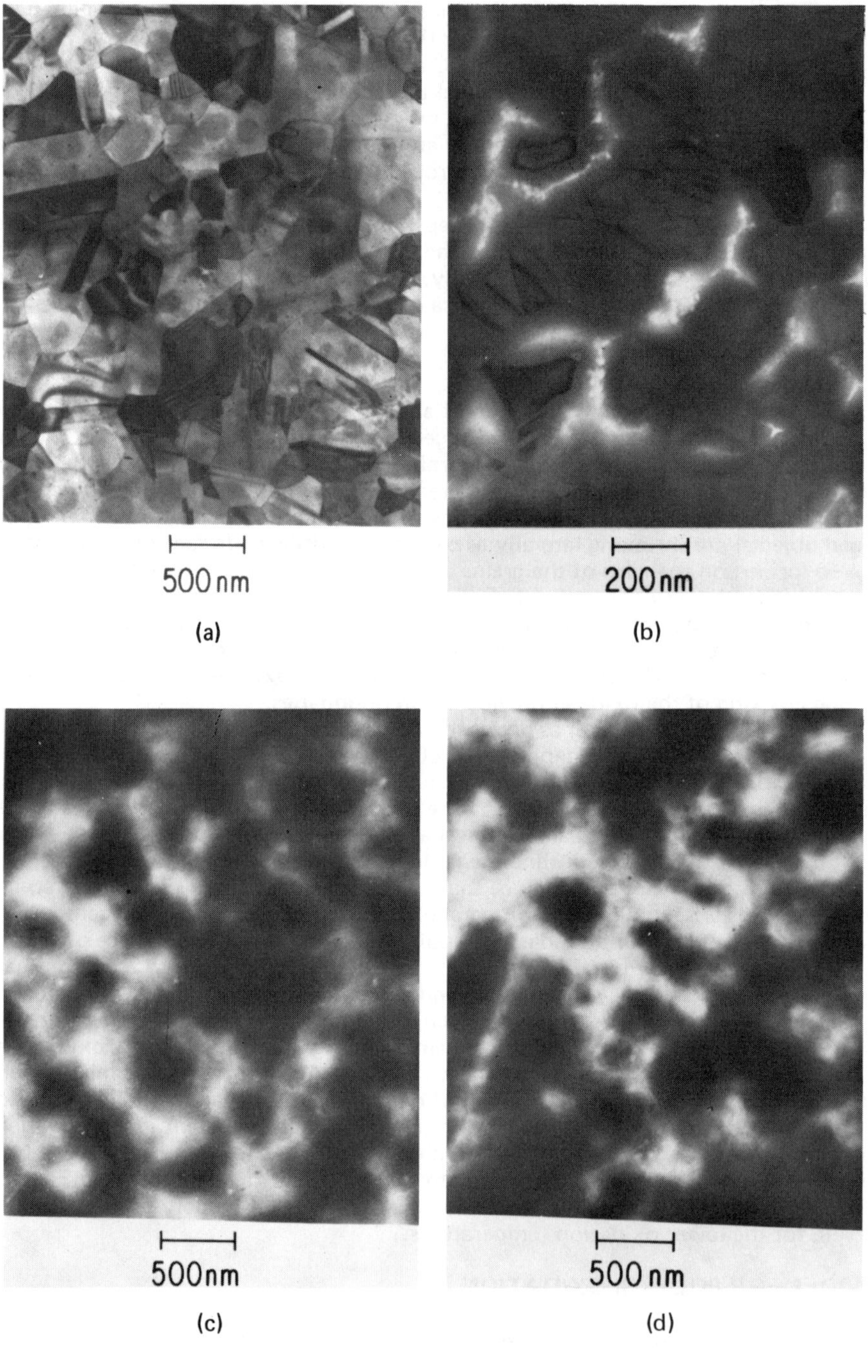

Fig. 1. CVD polycrystalline Si films at various stages of oxidation (a) unoxidized, (b) 50 nm surface oxide, (c) 150 nm surface oxide, (d) completely oxidized.

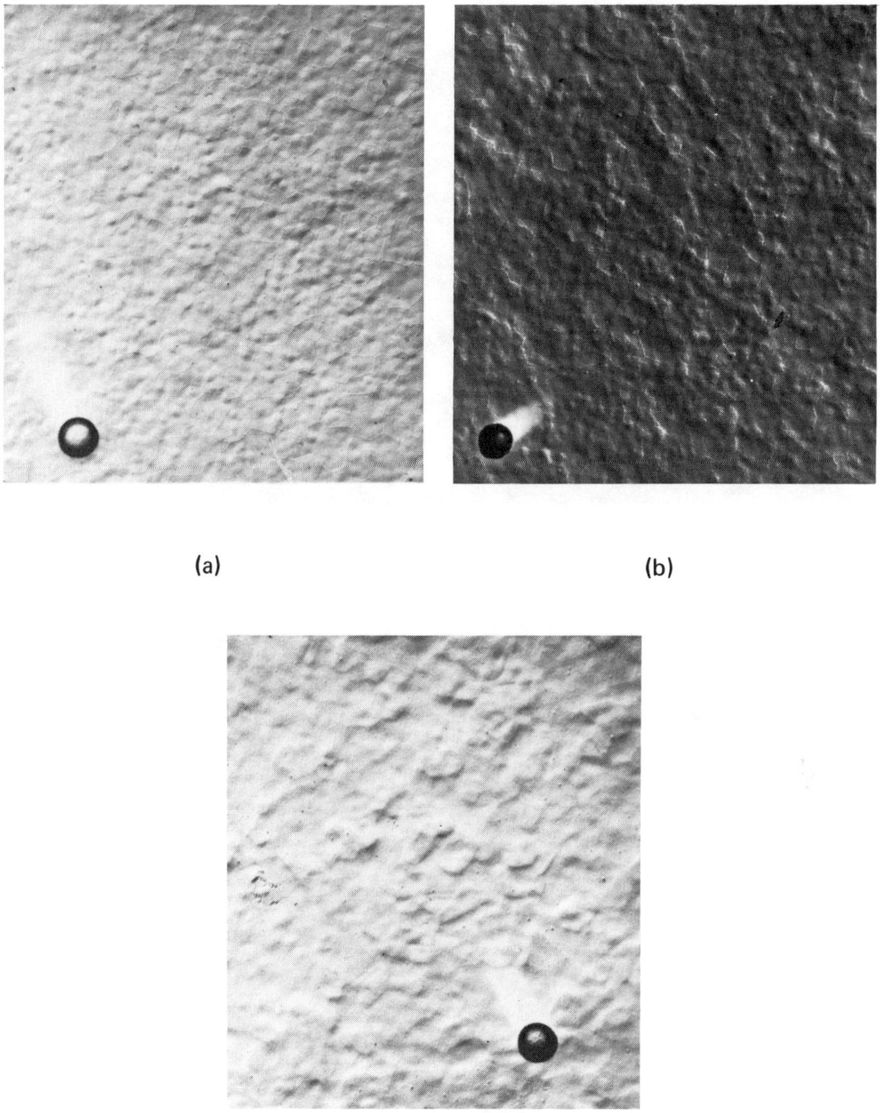

(a) (b)

(c)

Fig. 2. Replicas of poly-Si and poly-ox surfaces (a) poly-Si, CVD preparation at 800°C + $POCl_3$ doping (b) after oxidation at 1000°C for 1 hr., (c) after removal of oxide by HF etching. The magnifications are identical and the latex spheres are ~500 nm in diameter.

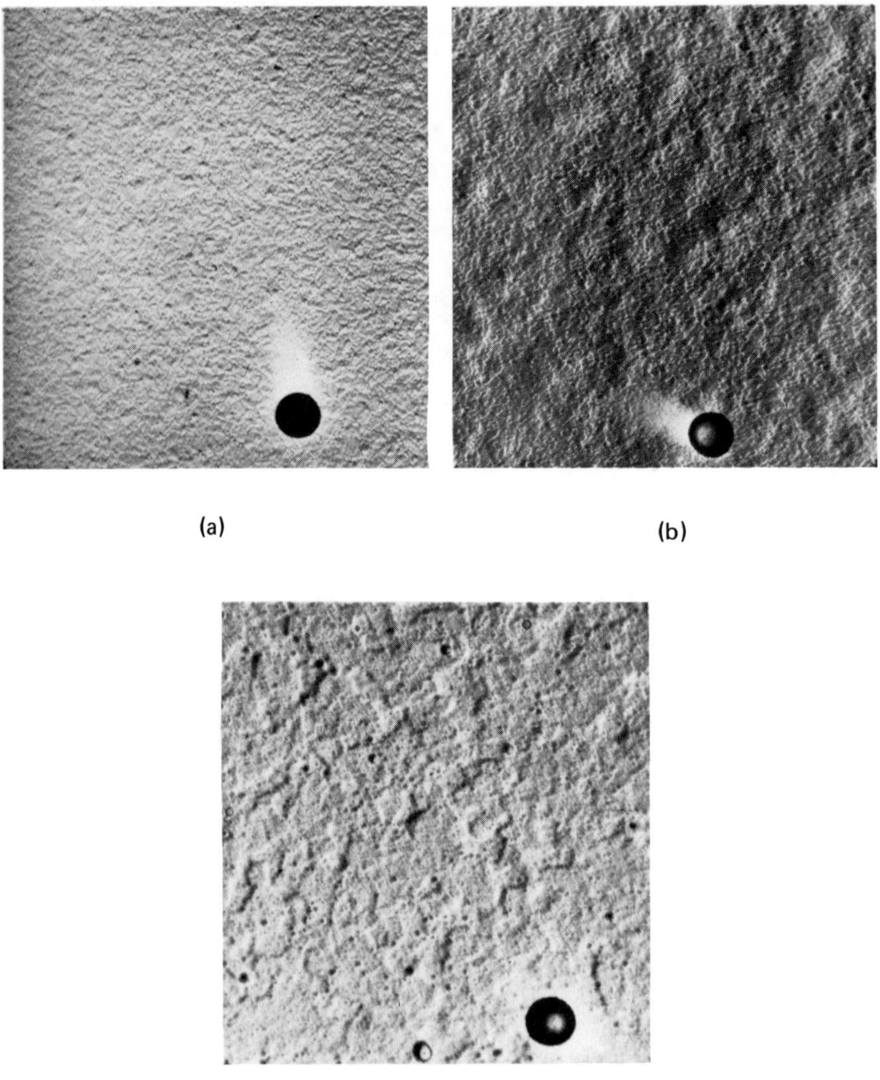

(a) (b) (c)

Fig. 3. Replicas of poly-Si and poly-ox surfaces (a) poly-Si, CVD preparation at 650°C + $POCl_3$ doping, (b) after oxidation of 1000°C for 1 hr., (c) after removal of oxide by HF etching. The magnifications are identical and the latex spheres are ~500 nm in diameter.

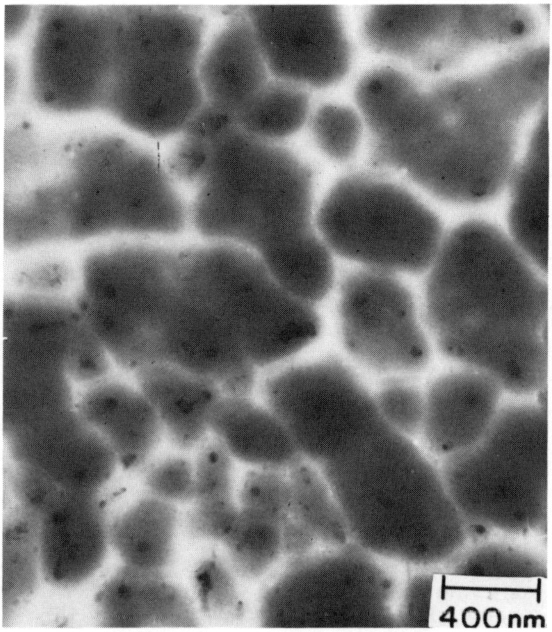

(a)

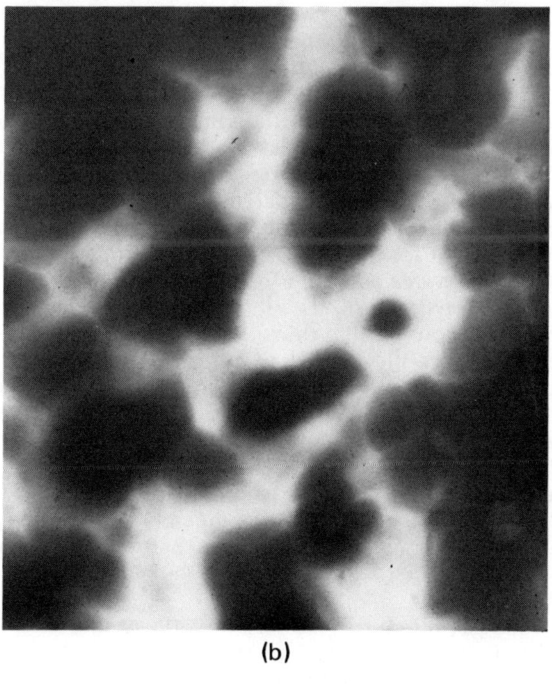

(b)

Fig. 4. Poly-Si sample as in Fig. 2, oxidized to the same extent in dry O_2 at (a) 800°C and (b) 1150°C.

The conversion of Si to SiO_2 involves an increase in volume by a factor of ~2.2. For the case of intergranular oxidation, as is observed for poly-Si, this increase of volume within the grain boundaries means that a large compressive stress is anticipated to develop during oxidation unless the required volume is obtained. For the oxidation of a free Si surface, the oxide expands outward as it forms. However, deep within a grain boundary this free expansion is not possible. It is also observed that poly-Si films can be totally oxidized and the films do not peel or delaminate or otherwise fail. Stress measurements on poly-Si films before and after oxidation showed only a slightly increased (< 10%) compressive stress[5]. Therefore either Si and/or SiO_2 must flow during oxidation of poly-Si to create the required volume. If it is assumed that SiO_2 flows, then the oxidizing grain boundaries would serve as sources of SiO_2 and the oxide would flow out of the grain boundaries as it is produced. This would result in thicker oxide near underlying grain boundaries than over mid-grain regions. If Si flows, then the Si would migrate from the boundaries to the mid-grain regions leaving less Si near boundaries and more near the mid-grain regions to fuel the oxidation. This would result in thinner poly-ox over grain boundaries as is in fact observed in Figs. 1 and 4. Therefore, it is likely that Si is the predominant migrating species under oxidation conditions. This mechanism can be understood with the aid of Fig. 5. This figure shows that as an idealized grain morphology (Fig. 5a) is oxidized (5b, c,d) a compressive stress develops due to the molar volume increase from Si to SiO_2. When the stress, τ, exceeds a critical value under specific oxidation conditions, Si flows in the direction of the arrows away from the oxidizing grain boundaries. In this way the necessary volume is produced to keep τ at the critical value. This type of plastic deformation is referred to as Nabarro-Herring creep[8,9].

STRUCTURE–RELIABILITY RELATIONSHIPS

As mentioned previously there are three poly-ox reliability implications to be discussed: high leakage currents in poly-ox, premature failure related to the high leakage currents and initial electrical shorts between poly-Si lines and the substrate.

From the oxidation results shown in Figs. 1, 2 and 3 it is clear that the oxidation of poly-Si causes roughness to develop both in the poly-ox as thickness undulations in the oxide itself and on the poly-Si surface via the intergranular oxidation.

It is seen in Fig. 1 b, c and d for incompletely oxidized poly-Si that the area of the thinnest poly-ox regions is small compared to the thicker regions of oxide. With the use of conventional optical (interference, ellipsometry) or mechanical (step height) film thickness measurement techniques, the measured oxide film thickness will be weighted toward the larger area and therefore thicker poly-ox regions. The electric field values reported in the literature[1,2,3] are calculated from the applied voltage divided by the measured poly-ox film thickness. Therefore, these electric field values will be erroneously low since they are based primarily on the maximum film thickness while the observed oxide currents are primarily related to the minimum thickness or maximum field conditions. As estimated from the TEM micrographs there is about 30% thickness fluctuation in the oxide on poly-Si. An overestimate of the film thickness of 30% would yield a calculated electric field 1.4 times lower than the maximum field which determines the conductivity. From DiMaria and Kerr's data[2] a factor of 1.5–2 is required to bring the poly-ox conductivity data in agreement with oxide grown on single crystal silicon. Therefore the observed poly-ox thickness fluctuations can also be responsible for the high poly-ox currents in addition to the poly-Si asperities[3].

It was shown that the poly-Si surface became roughened during oxidation. The intergranular mode of oxidation causes the grains to shrink laterally hence the height to width of the Si protrusions increases. Lewis[4] has reported that cathode surface protuberances can cause local field enhancement effects and that the height to width ratio and proximity of the protuberances are important factors that determine the

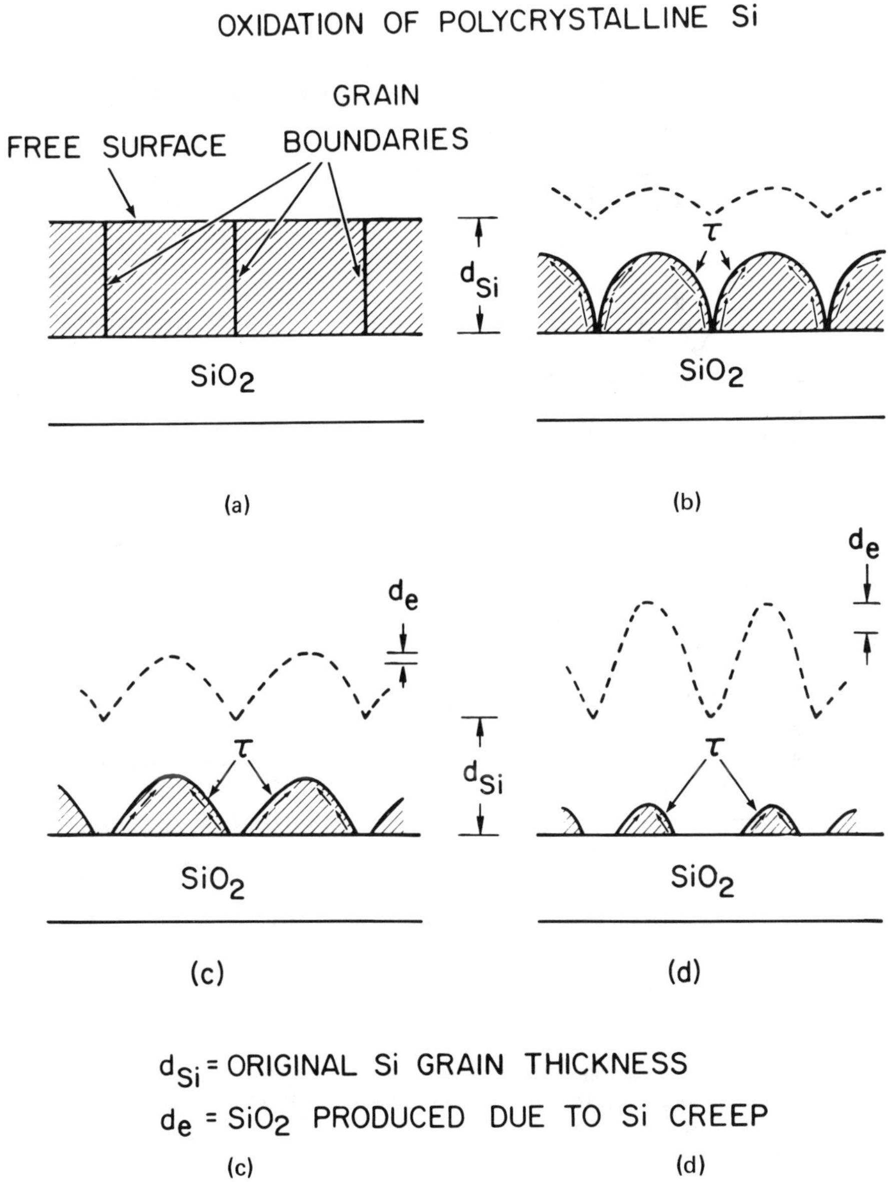

Fig. 5. Model for the intergranular oxidation of poly-Si, (a) unoxidized ideal grain structure, (b) (c) (d) show progressive stages of oxidation.

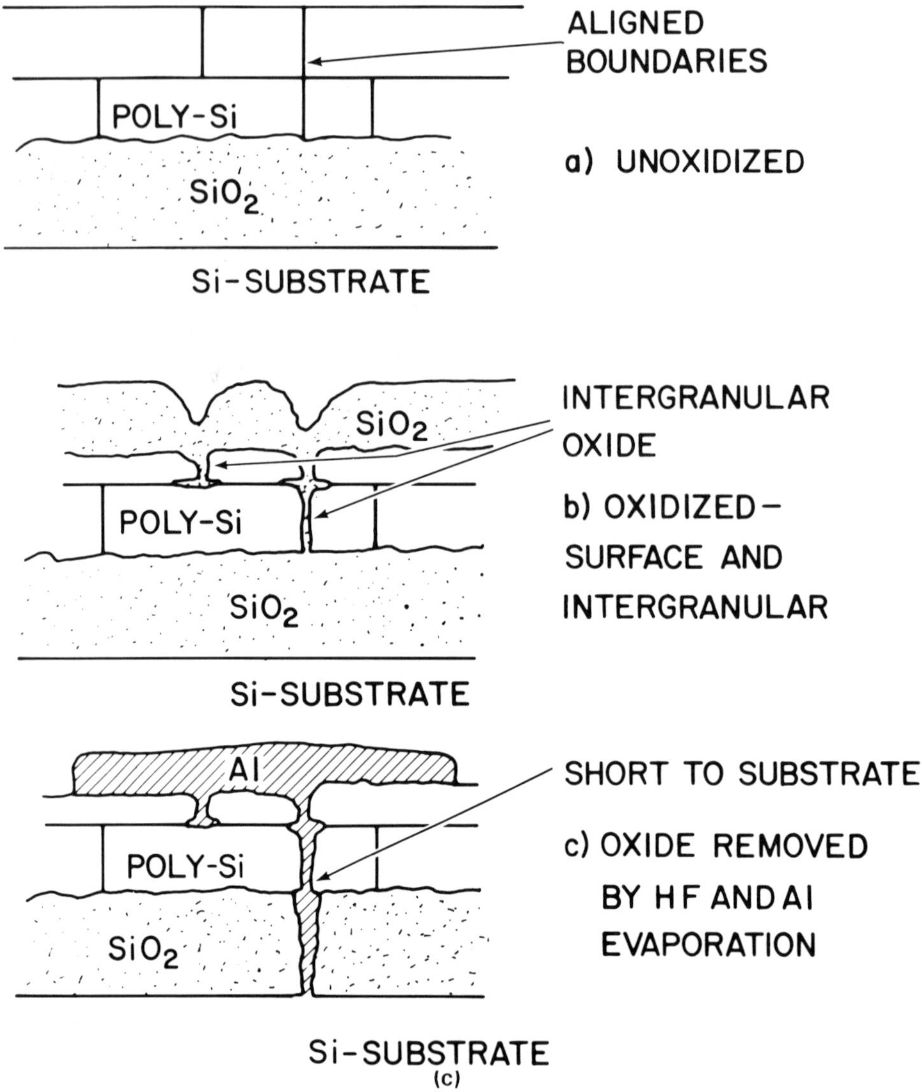

Fig. 6. Poly-Si shorting problem (a) idealized unoxidized aligned poly-Si grains (b) after intergranular oxidation (c) after HF removal of the intergranular oxide and Al metallization.

actual amount of the field enhancement. According to Anderson and Kerr[3] poly-Si surface asperities with a height to width ratio of ~2 are found and with an areal density of ~$5x10^9$ cm^{-2}. These numbers yield a mean field enhancement factor of less than 2 according to the calculations of Lewis[4]. Anderson and Kerr[3] report a field enhancement of 10 from Lewis' calculations but it appears that they only considered isolated surface asperities while their SEM micrographs show closely spaced surface irregularities. Therefore, poly-Si surface asperities and thickness undulations in the poly-ox probably contribute nearly equally to the higher than expected currents in the oxide. Anderson and Kerr[3] also report that poly-ox grown at higher temperatures yield relatively lower conductivities but never as low as the thermal oxide grown from single crystal silicon. This high temperature effect may be due to the fact that the thickness undulations are minimized because of viscous flow of SiO_2 in addition to Si creep at the higher temperatures. Figure 4 substantiates this idea by showing a smoother poly-ox grown at the higher temperature.

The shorting problem is easily understood by reference to Fig. 6. Figure 6a shows aligned or nearly aligned grain boundaries which when exposed to oxidizing conditions can cause oxide penetration deep within the poly-Si layer and even through the poly-Si. When the poly-ox is removed in HF based etchants to expose the poly-Si for Al contact metallization, a hole is also etched through the integranular oxide and through the underlying oxide. The Al metallization will cause metal to short from the top contact to the substrate through the etched hole as shown in Fig. 6c.

CONCLUSIONS

The electrical reliability problems associated with poly-ox have been shown to be related to the morphology of the poly-Si as surface roughness, to the morphology of poly-ox through oxide thickness undulations and to intergranular oxidation. Therefore, certain electrical properties of poly-Si and poly-ox have been explained as a result of understanding the mode of oxidation of poly-Si, specifically the resulting poly-Si and poly-ox structures.

ACKNOWLEDGEMENTS

This research was partially supported by the Defense Advanced Research Projects Agency and monitored by the Deputy for Electronic Technology, RADC, under Contract F19628–76–c–0249.

REFERENCES

1. S.A. Abbas and C.A. Barile, 13th Annual Proceedings, Reliability Physics Symposium, IEEE, Las Vegas, (1975).
2. D.J. DiMaria, and D.R. Kerr, *Appl. Phys. Lett.*, 27, 507 (1975).
3. R.M. Anderson and D.R. Kerr, *J. Appl. Phys.*, 48, 4834 (1977).
4. T.J. Lewis, *J. Appl. Phys.*, 26, 1405 (1955).
5. E.A. Irene, D. Dong, and E. Tierney, *J. Electrochem. Soc.*, submitted for publication 1979.
6. E.A. Irene, *J. Electrochem. Soc.*, 121, 1613 (1974).
7. R.M. Finne and D.L. Klein, *J. Electrochem. Soc.*, 114, 965 (1967).
8. F.R.N. Nabarro, *Bristol Conf. on Strength of Solids*, Physical Soc., London p. 75 (1948).
9. C. Herring, *J. Appl. Phys.*, 21, 437 (1950).

Stevens, Vander Voort, McCall, eds. Microstructural Science, Volume 8

A STUDY OF GRAIN BOUNDARY EMBRITTLEMENT IN Cu–3Al–1 Si USING AES*

C.L. White**, T.R. Odom***, and R.E. Clausing**

ABSTRACT

Single-phase, FCC Al–Si–bronze alloys are used in a variety of applications requiring good corrosion resistance, ductility, and strength. They normally exhibit tensile elongations between 20 and 60%, depending on annealing conditions, and are not usually prone to intergranular failure. However, samples of Al–Si–bronze (Cu–3 wt % Al–1 wt % Si) were found to fail intergranularly, with unusually low tensile elongation. Bulk chemical analysis of these bars indicated higher than normal concentrations of lead in the alloys. Auger electron spectroscopy (AES) of intergranular fracture surfaces revealed the grain boundaries to be highly enriched in lead. There is little doubt that this lead enrichment is responsible for the abnormally low ductility and intergranular failure in this Al–Si–bronze.

EXPERIMENTAL PROCEDURE AND RESULTS

The bar stock examined in this study was obtained from a single commercial vendor. The alloys had been induction melted using a mixture of virgin and recycled melt stock. Ingots were hot extruded to bar form, air cooled, cold drawn, straightened and stress relief annealed for 1 h at approximately 425°C. Bulk chemical analyses for these bars are given in Table I.

Table I.

Bulk chemical analysis of Al-Si-Bronze samples.
Values are wt%.

Bar	Al	Si	Pb	Fe	Sn	Zn
1	3.2	0.89	0.004	<0.01	0.002	0.12
2	3.48	0.98	0.006	0.024	0.003	0.041
3	3.4	1.0	0.020	0.014	0.007	0.11
4	3.5	1.03	0.022	0.03	<0.01	0.13

*Research sponsored by the Materials Sciences and Enriching Operations Divisions, U.S. Department of Energy under contract W–7405–eng–26 with the Union Carbide Corporation.
**Metals and Ceramics Division, Oak Ridge National Laboratory, Oak Ridge, Tennessee 37830 USA.
***Technical Services Division, Paducah Gaseous Diffusion Plant, Paducah, Kentucky 42001 USA.

Subsize-transverse tensile specimens, having a rectangular gage cross section (0.89 x 2.54 mm) and a 19.05 mm gage length, were tested at room temperature. Results of these tensile tests are summarized in Table II. Bar #1 exhibited good tensile properties, typical of Cu-3 Al-1 Si, while bars #2, 3, and 4 had low strength and elongation, and failed intergranularly. The failure modes reported in Table II were determined using scanning electron microscopy. Scanning electron micrographs of failed tensile specimens from bar #1 and bar #3 are shown in Figs. 1 and 2, respectively.

Table II.

Room temperature tensile test results for Al-Si-Bronze specimens

Bar	0.2% Offset Yield Stress (MPa)	Ultimate Tensile Strength (MPa)	Elongation (%)	Fracture Mode	Number of Samples
1	282	384	22	D	4
2	209	260	5	GBF	3
3	209	220	3	GBF	3
4	201	218	4	GBF	3

Optical metallography of as-received bar stock revealed evidence of plastic deformation and annealing twins, but the grain boundaries in all samples appeared to be free of second-phase films that might induce intergranular fracture (Fig. 3). As-received samples from bars #2, 3 and 4 exhibited occasional grain boundary cracks, usually less than one grain diameter in length, indicating that some cracking occurred in these bars during processing. Metallography of failed tensile specimens indicated that some additional secondary cracking took place during tensile testing. However, none of these cracks appeared to be significantly more than one grain diameter in length.

AES has been used extensively during the past ten years to study solute segregation to grain boundaries in alloys that can be fractured intergranularly. Typically, samples are fractured inside the AES ultra-high vacuum analysis chamber at pressures on the order of 2×10^{-8} Pa in order to minimize contamination of the exposed grain boundaries. AES on the fracture surface allows the elemental composition of the top 2—3 atom layers on that fracture surface to be determined. It is these same 2—3 atom layers along a grain boundary that are most likely to be enriched in trace impurity elements. Once an analysis of the grain boundary fracture surface has been obtained, it can be sputter ion etched to remove atoms, and reanalyzed to determine the composition of the underlying material. By alternately analyzing and sputter ion etching the surface, information about the composition versus distance profile going into the grains can be obtained.

AES is sensitive to only the top few atom layers on a sample because it involves detection of low-energy Auger electrons. Auger electrons (often with characteristic energies in the range between 100 and 1000 eV) occur for all elements except hydrogen and helium. Auger electrons originating more than 2—3 atom layers below the surface have a very large probability of being inelastically scattered and hence losing their characteristic energy, so that only those originating within the top few atom layers contribute to peaks in the observed Auger spectrum. AES and its use to study grain boundary composition has been reviewed extensively in the literature,[1,2] and the reader is referred to these reviews for discussions of the physics and experimental application of this technique. Only a brief description of the experimental details pertinent to the present study will be presented here.

Because tensile specimens from bars #2, 3 and 4 failed intergranularly, AES analysis was performed on transverse samples from these bars. Samples approximately 0.9 x 2.5 x 10.0 mm were notched to a cross section of 0.9 x 2.0 mm. Each sample was fractured at room temperature by bending inside the AES analysis chamber at an ambient pressure of 1–2 x 10^{-8} Pa. Immediately following fracture, the sample was positioned in front of the AES cylindrical mirror analyzer (CMA) for analysis of the fresh fracture surface.

AES analysis was performed using a lock-in-amplifier detection scheme to provide a dN(E)/dE vs E spectrum. Typical analysis conditions included use of a 4 keV primary electron beam, 1.11 eV/s pass energy sweep rate, 2 eV_{pp} pass energy modulation and a 300 ms lock-in-amplifier time constant. The primary beam current varied betwen 1 and 10 μA depending upon lens settings in the electron gun.

The CMA used in this study had a coaxial scanning electron gun, with a minimum beam diameter of ~5 μm. The sample current was used to modulate the image intensity on a video TV monitor in an SEM mode, providing a sample current image of the rastered region. Auger spectra obtained while rastering the beam at TV scan rates provide an average surface composition over the rastered region. Individual spots were also selected for analysis, providing an average surface composition over the area covered by the static electron beam.

Fig. 1. A scanning electron micrograph of the fracture surface of a subsize transverse tensile specimen from bar #1.

Fig. 2. A scanning electron micrograph of the fracture surface of a subsize transverse tensile specimen from bar #3.

Fig. 3. An optical micrograph showing the microstructure of bar #3.

An Auger spectrum from the fresh fracture surface of a sample of bar #4 is shown in Fig. 4. This spectrum was obtained while rastering the primary electron beam over a 600 x 600 μm region in the center of the fracture surface. The Auger peaks were identified by comparison with published spectra of pure elements and compounds.[3] The major elements indicated in this spectrum are copper and lead. Aluminum has Auger peaks at 68 and 1396 eV, while silicon has peaks at 92 and 1619 eV. The 68 eV aluminum peak is barely detectable in Fig. 4, while the 92 eV silicon peak is obscured by lead peaks at 90 and 94 eV. The higher energy aluminum and silicon peaks were detectable using a 6 keV primary beam and 5 eV_{pp} pass energy modulation.

The peak-to-peak intensities of the Auger peaks are approximately proportional to the atomic concentrations of the elements giving rise to those peaks.[4] In order to correct for differences in sensitivity between different peaks and different elements, sensitivity factors obtained from published spectra of pure elements and compounds have been used.[3] Because the sensitivity factor for the lead doublet 90–94 eV is much larger than the sensitivity factor for the 105 eV copper peak, the lead (90–94 eV)/copper (105 eV) peak height ratio must be multiplied by 0.08 to get an estimate of the lead/copper atomic ratio. A calculation for the spectrum in Fig. 4 yields a Pb/Cu ratio of 0.17. The rastered spectrum from the fresh fracture surface of a sample from bar #3 is very similar to the one for bar #4, having a Pb/Cu ratio of 0.18.

The rastered spectrum from bar #2 indicates an average Pb/Cu ratio of 0.06 at the grain boundaries, which is significantly lower than for bars #3 and 4. In addition to lead, however, tellurium peaks at 483 and 491 eV were also observed. The Te/Cu ratio was small, approximately 0.006, as determined using a method similar to the one used to obtain the Pb/Cu ratios.

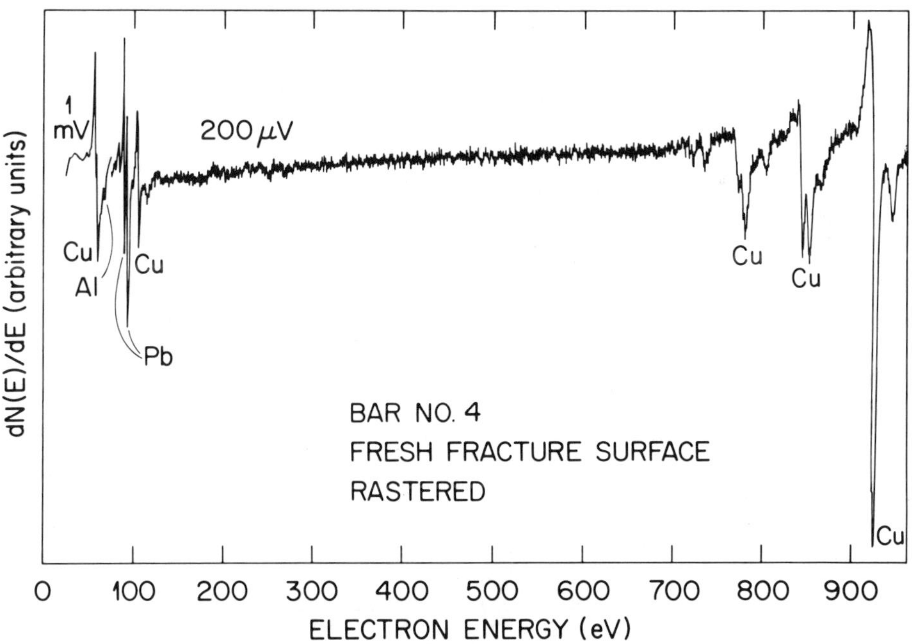

Fig. 4. An Auger spectrum from the fresh fracture surface of a sample from bar #4. This spectrum was taken while rastering the primary electron beam over a 600 x 600 μm area in the center of the fracture surface. The Pb/Cu ratio for this area is 0.17.

Following the initial analyses of the fresh fracture surface, sample current image and elemental lead maps of the surface were obtained. Elemental lead mapping was performed using a 5 eV_{pp} pass energy modulation and setting the CMA pass energy at the extreme negative deviation for the 90/94 eV lead doublet. The lock-in- amplifier output was then used to modulate image brightness on an oscilloscope as the primary electron beam was rastered over the fracture surface. Variations in the size of the lead doublet (which is detected as a single peak because of the larger pass energy modulation) show up as brightness variations in the elemental map. Unfortunately, other factors such as variations in surface orientation and crystallography can also affect signal intensity and background level in an Auger spectrum, which will in turn affect lock-in-amplifier output. Therefore, the variations indicated by these maps should always be verified by obtaining Auger spectra from the locations of interest.

Both spot analyses and analyses obtained while rastering the primary beam over small regions of the fracture surfaces indicated significant area to area variations in Pb/Cu ratios. Partial Auger spectra from three regions on the fractured sample from bar #3 are shown in Fig. 5. These partial spectra indicate the range of Pb/Cu ratios observed on this sample, Pb/Cu = 0.36 for area A, 0.08 for area B, and 0.25 for area C. The locations of these areas are indicated in the scanning electron micrographs of Fig. 6 (taken after the Auger analysis was completed).

Scanning electron microscopy of the fracture surfaces analyzed using AES indicated that they were all intergranular. Figure 6(a) shows a scanning electron micrograph of the Auger sample from bar #3. Figures 6(b), (c), and (d) shows high magnification scanning electron micrographs over areas A, B, and C (respectively) referred to in Fig. 5. Area A was quite smooth with no indication of localized dimpling, while area B was extensively dimpled and area C had a mixture of smooth and dimpled regions.

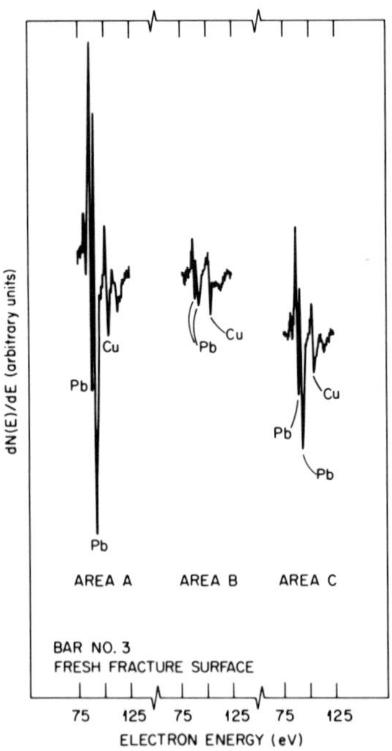

Fig. 5. Partial Auger spectra (75—125 eV) from three areas on the fresh fracture surface of the sample from bar #3. The Pb/Cu ratio is 0.36 for area A, 0.08 for area B, and 0.25 for area C.

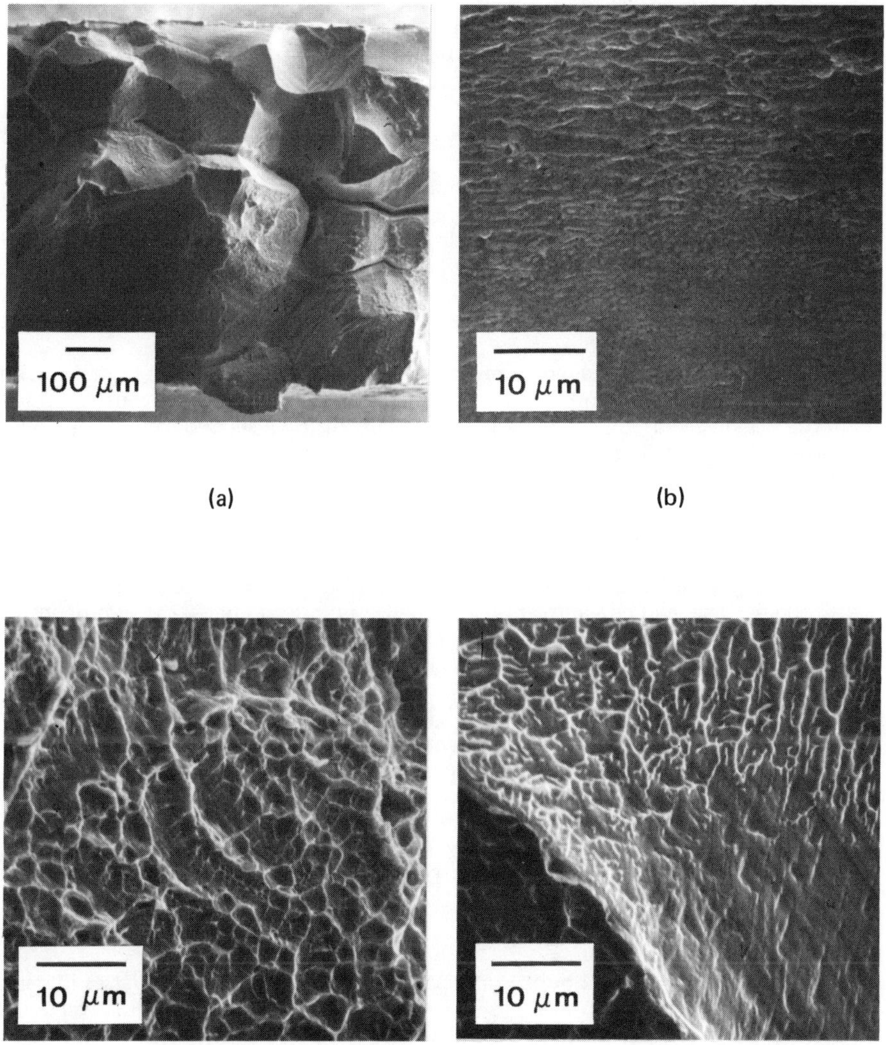

(a) (b)

(c) (d)

Fig. 6. Scanning electron micrographs of the Auger sample from bar #3; (a) low magnification micrograph showing locations of areas A, B, and C when partial spectra in Fig. 5 were taken, (b) higher magnification micrograph of area A, (c) higher magnification micrograph of area B, and (d) higher magnification micrograph of area C.

Following AES analysis of the as-fractured surface, the samples were positioned in front of a normal incidence inert ion sputter gun. Sputter etching of the surface with 1 keV argon ions was carried out in order to remove small quantities of material from the fracture surface. By alternately analyzing the fracture surface, and sputter etching it, the compositional variation in a direction normal to the fracture surface could be studied. Typical "sputter profiles", showing the variation in Pb/Cu atomic ratio as a function of sputtering time are shown in Fig. 7. Lead/copper ratios in Fig. 7 were obtained from rastered spectra to provide an average fracture surface composition.

Sputtering rates were calibrated by sputtering Ta_2O_5 films under identical conditions and determining the change in film thickness as a function of sputtering time. The sputtering conditions used in this study removed 2.5 Å of Ta_2O_5 per minute of sputtering. The exact relationship between the sputtering rate for Ta_2O_5 and that for copper and lead is not known, however, the available information suggests that they probably do not differ by more than a factor of two or three.[5]

The principal experimental results obtained in this study can be summarized as follows:

1. Bar #1 exhibited good ductility, while bars #2, 3 and 4 did not.
2. Both the yield and tensile strengths of bar #1 were greater than for bars #2, 3 and 4.
3. Scanning electron microscopy indicated that the bars which exhibited low ductility (#2, 3 and 4) failed intergranularly.
4. Optical metallography failed to reveal any second-phase film at the grain boundaries that might induce embrittlement. Small intergranular cracks were observed in the "as received" bars #2, 3 and 4. Some additional secondary cracking occurred during tensile testing. However, these secondary cracks seldom exceeded one grain diameter in length.
5. Bulk chemical analysis indicated that the lead concentration in bars #3 and 4 is four to five times that for bar #1. The lead concentration in bar #2 is only 50% greater than for bar #1.
6. AES analysis indicated extensive segregation of lead to grain boundaries in bars #3 and 4. The lead concentration at grain boundaries in bar #2 is significantly lower than for bars #3 and 4 but still greatly above the bulk level. Trace levels of tellurium were also observed at the grain boundaries of bar #2.
7. Large variations in lead concentration from point to point on a given fracture surface were observed. In areas where localized analysis and scanning micrographs were obtained, the highest lead concentrations were associated with smooth grain boundary facets, while lower lead concentrations were associated with dimpled facets.
8. Sputter profiling indicated that most of the lead enrichment at the grain boundaries was confined to within a few atom layers of the grain boundary, but significant levels of lead were still observed after removing more than 40 atom layers.

DISCUSSION

These results leave little doubt that lead segregation to grain boundaries in bars #2, 3 and 4 is directly responsible for their brittle, intergranular failure. The occurrence of small grain boundary cracks in the as-received bars #2, 3 and 4, indicates that crack nucleation was probably not a limiting factor in the failure mechanism. The limited amount of secondary cracking observed in failed tensile specimens indicates that failure probably involved the propagation of a single super-critical crack, as opposed to linking up of numerous independently nucleated cracks. The lower yield strengths in bars #2, 3, and 4 could be caused by reductions in net load bearing cross section and stress concentrations resulting from these internal cracks.

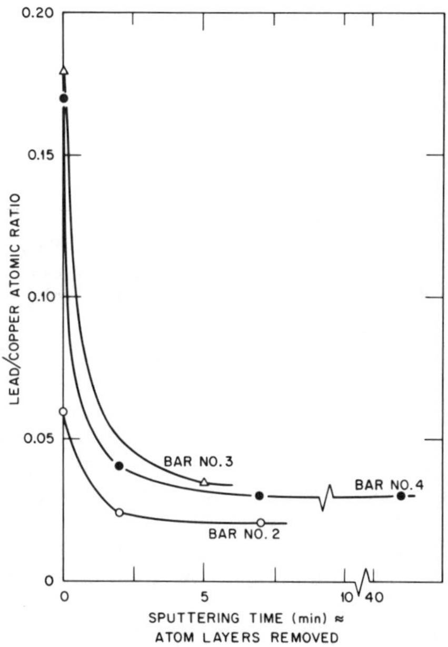

Fig. 7. Sputter profiles, showing the variation of Pb/Cu ratios with sputtering time for bars #2, 3, and 4. The Pb/Cu ratios were obtained from Auger spectra taken while rastering the primary electron beam over a 600 X 600 μm area near the center of the fracture surface. The sputtering rate was approximately one atom layer per minute.

Similar grain boundary embrittlement phenomena in copper or copper-base alloys have been attributed to very small concentrations of bismuth,[6–9] antimony,[10] and tellurium.[11] In the case of bismuth embrittlement of copper, the strong segregation of bismuth to the grain boundaries has been repeatedly verified using AES. Tellurium segregation has also been observed to coincide with embrittlement by that species. While segregation of antimony in copper has not been directly observed, there seems to be little doubt that it also occurs in association with embrittlement due to that element.

The segregation of lead to grain boundaries in Cu–3 Al–1 Si is consistent with its low solid solubility in copper ($<$ 0.29 wt% at 600°C) and its large atomic radius (1.75 Å) relative to copper (1.28 Å). Hondros[12,13] has noted an empirical correlation between solute enrichment at grain boundaries and the reciprocal atomic solid solubility of the solute in the solvent. The maximum lead enrichment observed at Al-Si bronze grain boundaries in this study is in reasonable agreement with Hondros' correlation.

The mechanism by which segregated lead, bismuth, tellurium, or antimony might embrittle copper or Cu–3 Al–1 Si is not clear. Hondros and McLean[9] have studied the embrittlement of copper by bismuth in some detail. They conclude that the bismuth segregation alters the cohesive energy of the grain boundary, allowing slip bands intersecting at grain boundaries to create sufficient stress concentrations to nucleate grain boundary cracks. Seah[14] has suggested an alternative explanation for segregation-induced embrittlement, postulating a drop in cohesive ***stress*** rather than cohesive ***energy*** as the result of oversize solute atoms segregating to grain boundaries. Seah's treatment was applied to temper embrittlement in low-alloy steels, and was found to be consistent with published experimental results. Contrary to this, White, Clausing, and Heatherly[15] observed that the effects of solute segregation to grain boundaries in a number of other solute-solvent systems (including phosphorus-tungsten and thorium-iridium) do not appear to support Seah's conclusions.

Lead is certainly an oversize solute in Al–Si–bronze, making the present results consistent with Seah's treatment. Other explanations, such as those based on reductions in cohesive energy, cannot be dismissed because grain boundary and surface energy data for the copper-lead (or the Cu-Al-Si-Pb) system, which would allow the effect of lead segregation on cohesive energy to be evaluated, are not available. Without such data and without atomistic models for intergranular crack nucleation and propagation, detailed understanding of grain boundary embrittlement phenomena is not possible.

In general, one might expect solute segregation to vary from boundary to boundary because the structural details of the boundaries will vary.[16] Low energy grain boundaries would be expected to have less solute enrichment than high energy (highly disordered) grain boundaries because the driving force for segregation is the lowering of grain boundary energy. Coupled with the expected variation in solute segregation to different boundaries is the likelihood that intergranular fracture will preferentially expose the most embrittled boundaries for analysis.[12] In those areas where the fracture path is forced to follow less embrittled boundaries, one might expect more ductility (as evidenced by dimpling of the grain boundary facets) or localized regions of transgranular failure.

Boundary to boundary variations in bismuth segregation to copper grain boundaries has been reported by Powell and Mykura[7] and Powell and Woodruff.[8] Powell and Mykura reported that the lowest bismuth concentrations were observed on the roughest grain boundary facets. Powell and Woodruff, however, after a more extensive investigation of boundary to boundary variations concluded that there was no apparent correlation between the roughness of the grain boundary facets and the bismuth concentration observed on them.

The results of the present study clearly indicate a boundary to boundary variation in Pb/Cu ratio. Comparison of the partial Auger spectra in Fig. 5 with scanning electron micrographs in Fig. 7 suggest that the highest Pb/Cu ratios are associated with smooth grain boundary facets (area A) and low ones with extensive dimpling (area B). Unfortunately, the number of individual grain boundary facets examined was too small to generalize this observation.

Solute-enriched regions at grain boundaries are normally observed to extend only a few atom layers to either side of the boundary center plane.[16] This is expected because the highly distorted core of the boundary is confined to that region. The sputter profiles in Fig. 6 all show large decreases in Pb/Cu ratio after approximately two atom layers have been removed from the intergranular fracture surfaces, indicating the highly lead enriched region is very thin.

Longer sputtering times, up to 41 min, failed to reduce the Pb/Cu ratios to levels that would be indicative of the bulk. In each case, the sputter profile leveled off at a Pb/Cu ratio of ~ 0.03. Similar observations have been reported in several studies of solute segregation to grain boundaries where sputter profiling was performed.[17-21] Unequal sputtering rates for copper and lead, surface roughness, shadowing of the ion beam, and redeposition of sputtered material from lead-rich precipitates could all contribute to this effect. While none of these possibilities can be absolutely ruled out, shadowing of the ion beam should have been minimized by use of the normal incidence sputter gun. Spot analysis on both smooth and rough areas of the fracture surface failed to reveal any significant difference in Pb/Cu ratio after sputtering, suggesting that surface roughness is not a critical factor. Scanning and optical microscopy failed to reveal any evidence of precipitates, suggesting that if they are present, they are probably only a few tens of nanometers in diameter. By process of elimination, these observations suggest that either unequal sputtering rates for lead and copper or redeposited lead are probably responsible for the failure to approach bulk Pb/Cu ratios with sputtering.

CONCLUSIONS

The experimental results presented in this paper clearly indicate a connection between the low ductility, intergranular failure, and lead enrichment at the grain boundaries in Al–Si bronze. In one case (bar #2), it seems likely that tellurium also played a role in the embrittlement. Grain boundary Pb/Cu ratios varied from area to area, the largest being associated with smooth grain boundary facets, and the lowest with dimpled facets.

REFERENCES

1. D.F. Stein, W.C. Johnson, and C.L. White, "Study of Grain Boundary Segregation Using Auger Electron Spectroscopy," *Grain Boundary Structure and Properties,* G.A. Chadwick and D.A. Smith, (eds.) Academic Press, pp. 301–51 (1976).
2. A. Joshi, "Experimental Techniques to Study Segregation", *Interfacial Segregation,* W.C. Johnson and J.M. Blakely (eds.), ASM, pp. 39–109 (1979).
3. L.E. Davis, et al., *Handbook of Auger Electron Spectroscopy,* Second Edition, Physical Electronics Industries (1976).
4. R.E. Weber and A.L. Johnson, "Determination of Surface Structures Using LEED and Energy Analysis of Scattered Electrons," *J. Appl. Phys.* 40, 314–318 (1969).
5. G.K. Wehner, "The Aspects of Sputtering in Surface Analysis Methods", *Methods of Surface Analysis,* A.W. Czanderna (ed.), Vol. 1, Elsevier, pp. 5–37 (1975).
6. A. Joshi and D.F. Stein, "An Auger Spectroscopic Analysis of Bismuth Segregated to Grain Boundaries in Copper", *J. Inst. Met.,* 99, 178–81 (1971).
7. B.D. Powell and H. Mykura, "The Segregation of Bismuth to Grain Boundaries in Copper-Bismuth Alloys", *Acta Met.* 21, 1151–56 (1973).
8. B.D. Powell and D.P. Woodruff, "Anisotropy in Grain Boundary Segregation in Copper-Bismuth Alloys", *Phil. Mag.,* 34, 169–76 (1976).
9. E.D. Hondros and D. McLean, "Cohesion Margin of Copper", *Phil. Mag.,* 29, 771–95 (1974).
10. L.M.T. Hopkins, "The Intergranular Brittleness of Single-Phase Copper-Antimony Alloys," *J. Inst. Met.,* 84, 102–08 (1955-56).
11. H.L. Marcus and N.E. Paton, "Metastable Grain Boundary Segregation in Copper Containing Tellurium", *Met. Trans.,* 5, 2135–38 (1974).
12. E.D. Hondros, "Grain Boundary Segregation, The Current Situation and Future Requirements", *J. de Physique,* 36(10), 117–35 (1975).
13. D. Gupta, "Influence of Solute Segregation on Grain-Boundary Energy and Self Diffusion," *Met. Trans. A,* 8A, 1431–38 (1977).
14. M.P. Seah, "Segregation and the Strength of Grain Boundaries", *Proc. Roy. Soc. (London) A.,* 349, 535–54 (1976).
15. C.L. White, R.E. Clausing, and L. Heatherly, "The Effect of Trace Element Additions on the Grain Boundary Composition of Ir + 0.3 Pct W Alloys," *Met. Trans. A,* 10A, 683–91 (1979).
16. R.W. Balluffi, "Grain Boundary Structure and Segregation", *Interfacial Segregation,* W.C. Johnson and J.M. Blakely (eds.) ASM, pp. 193–236 (1979).
17. H.L. Marcus and M.E. Fine, "Grain Boundary Segregation in MgO-Doped Al_2O_3," *J. Amer. Cer. Soc.,* 55(11), 568–70 (1972).
18. R. Viswanathan, "Temper Embrittlement in a Ni-Cr Steel Containing Phosphorus as Impurity", *Met. Trans.,* 2, 809–15 (1971).
19. R. Viswanathan and T.P. Sherlock, "Long-Time Isothermal Temper Embrittlement in Ni-Cr-Mo-V Steels", *Met. Trans.,* 3, 459–68 (1972).

20. P.V. Ramasubramanian and D.F. Stein, "An Investigation of Grain-Boundary Embrittlement in Fe-P, Fe-P-S, and Fe-Sb-S Alloys," *Met. Trans.* 4, 1735–42 (1973).
21. M.P. Seah and E.D. Hondros, "Grain Boundary Segregation", *Proc. Roy. Soc. (London) A,* 335, 191–212 (1973).

Stevens, Vander Voort, McCall, eds. Microstructural Science, Volume 8

FIELD ION MICROSCOPE OBSERVATIONS OF OXYGEN ADSORPTION ON TUNGSTEN AT 5° AND 300°K

Young-Won Kim*

ABSTRACT

The adsorption and desorption of oxygen on tungsten has been observed using a field ion microscope at 5° and 300°K with a background pressure of <2x10-10torr. At both temperatures the tungsten specimen tip revealed surface disorder for the adsorption of oxygen. Exposure to a dose of 2x10-5torr-sec at 5°K resulted in disordering the top surface atomic layer: any further exposure did not extend appreciably the disorder below the first layer. Careful observations suggest that physical adsorption is the predominant process at 5°K and even the strongly-bound oxygen atoms hardly penetrate into the substrate sites. The effect of 300°K adsorption was about 3 atomic layers deep, indicating that the adsorbed oxygen molecules may be dissociative and penetrative.

INTRODUCTION

Oxygen adsorption on tungsten, leading to oxidation, has been subjected to a great deal of studies[1] because of its interesting nature, both chemical and physical. It seems to be agreed that oxygen is chemically adsorbed on tungsten surface at or above room temperature (RT). As temperature is lowered, however, the physical adsorption may become dominant, probably because of condensation and decreased chemisorption heat[2] and/or decreased surface binding energy[3]. In effect, some field ion microscope (FIM) observations[4,5] show that oxygen is not strongly bound at 78°K although adsorption may be dissociative depending on its adsorption site. However, field emission observations have demonstrated that a certain degree of somewhat strong binding may occur on open planes at temperatures even lower than 20°K[6]. Therefore, it was decided in this experiment to determine whether strong binding occurs between oxygen and tungsten surface at lower temperatures, and the first FIM observations of tungsten surface oxygen adsorbed at 5°K are reported here.

EXPERIMENTAL PROCEDURE

The field ion microscope (FIM) specimens were prepared from a 99.98% pure tungsten wire after decarburizing and then degassing in the UHV furnace attached to the FIM, Fig. 1. The all-metal demountable FIM, originally built for defect studies[7], was modified for the present experiment. The background pressure was about 1x10-10torr and the FIM specimen temperature was maintained at ~5°K during observation.

Before oxygen adsorption, the specimens were shaped in the FIM by the field evaporation technique to obtain the best image voltage (BIV) of about 10–12kV.

*Structural Metals Branch, Air Force Materials Laboratory, WPAFB, Dayton, Ohio 45433 USA.

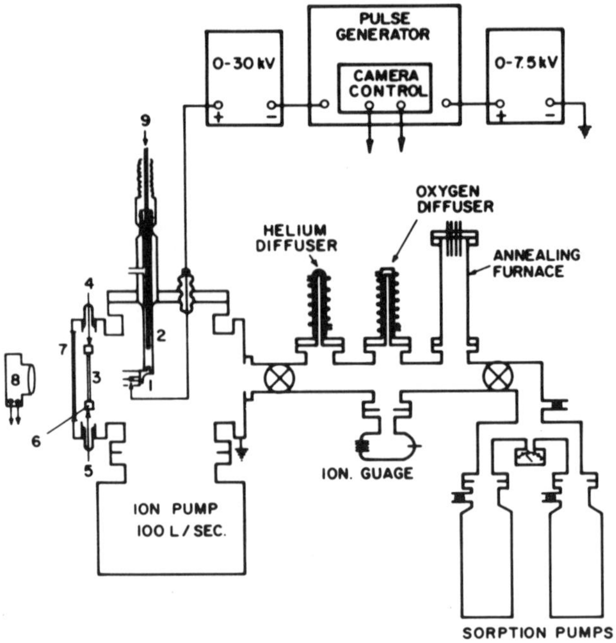

Fig. 1. All-metal demountable field ion microscope. 1) specimen holder, 2) helium liquefier, 3) screen-channel plate set, 4&5) high voltage feed throughs, 6) screen set holder ring, 7) viewing window, 8) motor drive camera, 9) cold helium flow.

The images of clean surfaces were then recorded as a function of the tip voltage up to the BIV. A typical FIM micrograph of a [110]–oriented tungsten tip is shown at the best image voltage, Fig. 2.

For the oxygen adsorption at 5°K, the imaging gas (He) was evacuated with the specimen held at the BIV, and oxygen was introduced into the system through the heated silver thimble (oxygen diffuser, Fig. 1) to the desired pressure. Then the adsorption of oxygen on the specimen tip was attained by simply dropping the specimen voltage to zero. After a predetermined exposure the oxygen was pumped out to $<5\times10^{-10}$ torr to make further adsorption practically negligible. The adsorption procedure at RT was, in principle, the same except the specimen was shaped at 5°K, warmed up to RT while maintaining the BIV, then allowing absorption to occur. The specimen was then cooled rapidly down to 5°K.

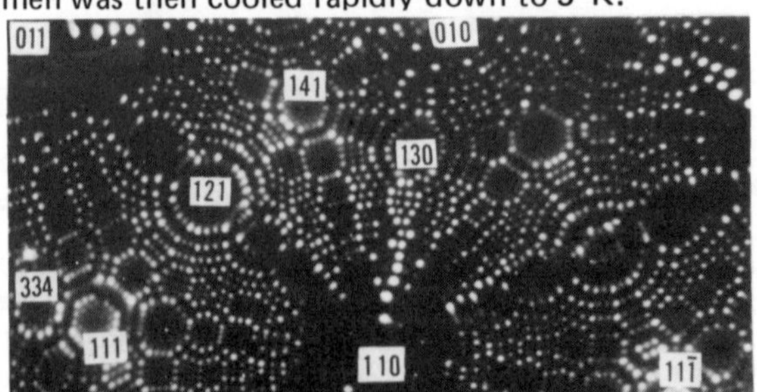

Fig. 2. FIM micrograph of clean, [110]-oriented tungsten at BIV (10.0kV).

Adsorption effects and desorption phenomena were observed at 5°K at an imaging gas pressure of ~3x10^{-5} torr. Images were recorded directly on Kodak Tri-X film at f3.5 with exposures less than 10 sec.

RESULTS AND DISCUSSIONS

Overall Observations

The specimens for 5°K adsorption were exposed in 1x10^{-6} torr oxygen to total doses ranging from 3x10^{-6} to 8x10^{-4} torr-sec. These exposures are sufficient to form many layers of oxygen adsorbates on the tungsten surface. The effects of adsorption on the surface could be visualized in the FIM in the form of the surface disorder caused by an electric field applied to the specimen. The disorder increased as exposure increased. At exposures > ~2x10^{-5} torr-sec, however, the degree of disorder remained the same within the dose limit used.

Development of surface disorder during increasing specimen voltage for a specimen exposed at 6x10^{-4} torr-sec. is shown in Figs. 3b–d. The oxygen-bearing surface revealed some large bright spots at ~0.2BIV. The bright spots were unstable and many of them disappeared or desorbed immediately at the indicated voltage or higher. This process continued up to ~0.4BIV but the number of bright spots increased. As the voltage was raised further, the bright spots became smaller in size but greater in number. The bright spots were randomly distributed mainly along <011> zones, Fig. 3b. This feature is in striking contrast to the clean surface at the same voltage, Fig. 3a, where the crystallographic planes along <011> zones are outlined by bright plane edges. The number of bright spots increased continuously, over wider area, with increasing voltage, Fig. 3c. At this point (0.78BIV) some of the spots were still unstable, flickering and changing sites. The observations so far described are more or less similar to those [4,5] on the tungsten surface oxygen-adsorbed at ~80°K. At 0.88BIV, all the spots were almost stable and represented the regular tungsten tip surface, Fig. 3d.

The seemingly clean surface seen in Fig. 3a was far from clean. This was demonstrated by lowering the voltage and observing the existence of previous features, for example, those of Fig. 3c. This indicates that many of the adsorbed oxygen atoms are strongly bound and secure themselves at strong binding sites such as ledges. In fact, careful observations showed that all the bright spots were present on the plane edges, with some exceptions, of open planes like {111}. However, their presence from the effect of adsorbed oxygen were confined to the first atomic layer of the specimen. A perfectly clean surface image was obtained after one monolayer was field evaporated.

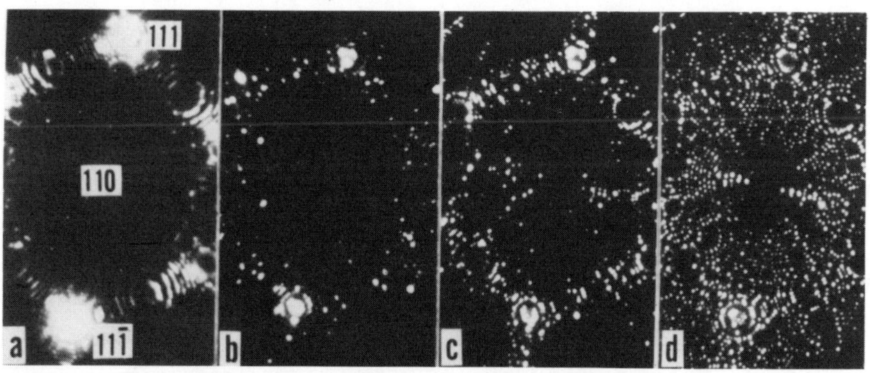

Fig. 3. Clean surface (a) and surfaces exposed at 5°K to 6x10^{-4} torr-sec. (b–d). (a) 0.49BIV, (b) 0.49BIV, (c) 0.78BIV, (d) 0.88BIV.

The specimen surface exposed at RT behaved somewhat differently from that at 5°K. The major difference was the depth of disorder; it increased gradually with increasing exposure, down to a few atomic layers deep. Figs. 4a–c show the response of the specimen surface exposed to $6x10^{-6}$ torr-sec. to increasing voltage. The behavior and motion of bright spots were similar to that at 5°K. However, more spots were present over a wider area and the surface disorder was two layers deep. Holding the tip voltage at BIV for 30 sec. denuded most of the first surface layer. The resulting image is shown at 0.71BIV in Fig. 4c where edges of open planes are now visible. Denuding the next disordered layer required higher applied voltage but still less than that (1.25BIV) for the evaporation of self-tungsten atoms. The clean surface was obtained after evaporation of about two atomic layers. At higher exposures, for example, at $3x10^{-3}$ torr-sec., the bright spots were present randomly all over the surface, Fig. 4d. It was not easy to tell how many surface layers were disturbed by this exposure. It is estimated, however, that the surface disorder was about 3–4 atomic layers deep. The depth of disorder or oxygen penetration did not increase appreciably any further with increasing exposure up to 0.1 torr-sec., the maximum dose used in this experiment.

Observations on {111} Area

This section described the behavior of individual bright spots in loosely packed {111} and some adjacent planes.

Some typical motions of bright spots in {111} areas are shown in Figs. 5 and 6 for 5°K exposures at $6x10^{-5}$ torr-sec. and $7.2x10^{-4}$ torr-sec., respectively. In both cases, the degree of surface disorder was similar, disappearing after desorbing one monolayer. Some interesting evaporation features were also observed. In a {334} plane just above {111}, Fig. 5, the evaporation of atoms did not occur normally. Rather, one of seven atoms in the plane disappeared from the edge, Fig. 5b, at 0.85 BIV and this was followed by the evaporation of an atom from the inside, Fig. 5c. The case shown in Fig. 5b is different from that of a clean surface which normally shows the last evaporations to consist of inside atoms. Sometimes bright spots developed 'inside a plane. This is shown for {111} in Figs. 5d and 5e where a gradually brightening spot finally disappeared at the BIV. Similar behavior was observed in Fig. 6, but the spot was mobile and migrated to the edge of the plane. Also seen in Fig. 6 is the development of a pair of bright spots. A spot in {332} at low voltage gained a neighbor at higher voltage to form a complex, as seen in Fig. 6c. This pair was quite stable up to the BIV, Fig. 6d. Lastly, the clean surface obtained by field evaporation shows occasionally missing atom(s), as shown in {543} on Fig. 5f, suggesting that oxygen binding to the surface was very strong in this case.

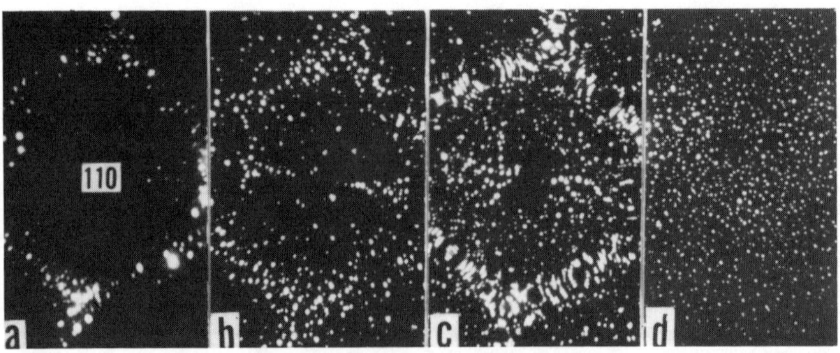

Fig. 4. Surfaces exposed at RT to $6x10^{-6}$ torr-sec. (a,b,c) and $3x10^{-3}$ torr-sec. (d). (a) 0.50BIV, (b) 0.62BIV, (c) 0.71BIV, (d) BIV.

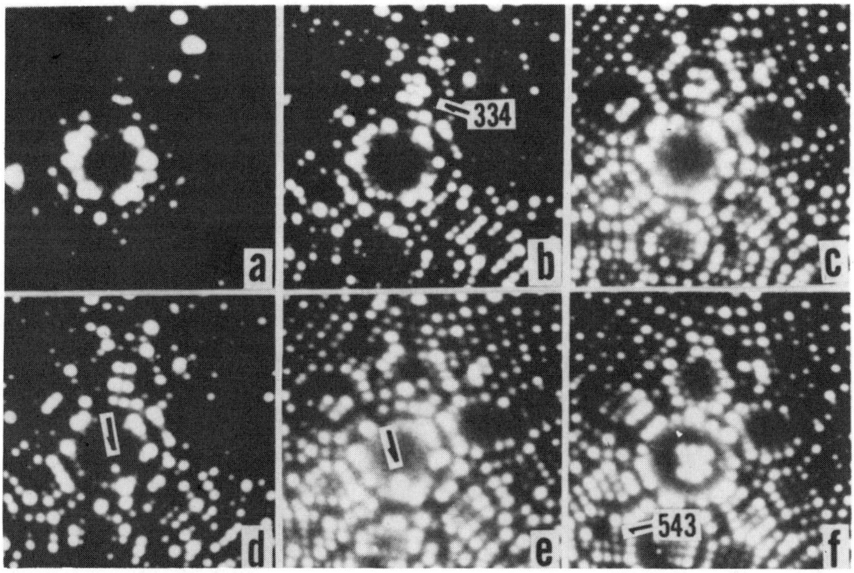

Fig. 5. (111) area exposed to 6x10^{-5} torr-sec. at 5°K. (a) 0.68BIV, (b) 0.85BIV, (c) 0.94BIV, (d) c → 0.85BIV, (e) 0.98BIV, (f) e → 1.15BIV → 0.98BIV.

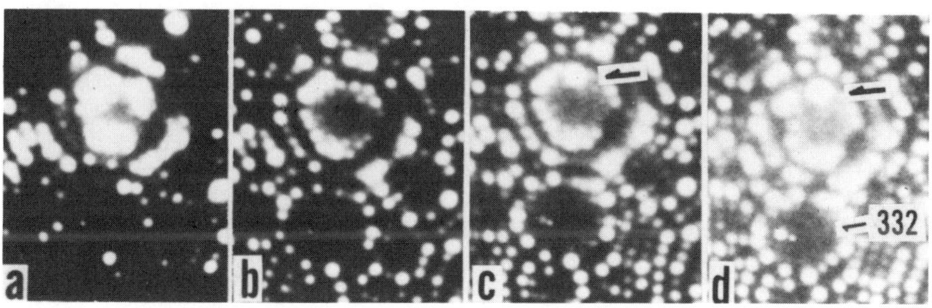

Fig. 6. (111) area exposed to 6x10^{-4} torr-sec. at 5°K. (a) 0.78BIV, (b) 0.84BIV, (c) 0.92BIV, (d) 0.98BIV.

The above observations suggest that some adsorbed oxygen is in a quite strongly bound state up to the BIV. The presence or the after-effect of adsorbed oxygen, however, is confined to the plane edges of the first atomic layer. The presence of bright spots in planes may not be the rule but rather, the exception. This is comparable to the observations {4} for oxygen adsorbed on tungsten at 78°K. Also, bright spots in planes, perhaps as well as on edges, may not be direct images of oxygen but rather, tungsten-oxygen complexes. Furthermore, some of the spots in planes might be wandering tungsten atoms. Finally, it is noted that oxygen adsorption at 5°K did not show evidence of being dissociated.

CONCLUSIONS

1. Surface disorder due to 5°K oxygen adsorption on tungsten is confined to the first monolayer.

2. Oxygen is generally bound to the surface rather weakly, but some strong binding occurs, especially on the edges.
3. Oxygen adsorbed at RT tends to penetrate the substrate. The resulting disorder may be ~3–4 atomic layers deep within the doses used.
4. Binding of oxygen to the surface is much stronger at RT than at 5°K, but adsorbed oxygen shows similar behavior at both temperatures.

REFERENCES

1. B. McCarrel, *J. Chem. Phys.* 46, 863–69 (1967).
2. D.O. Hayward and B.M.W. Trapnell, *Chemisorption*, Butterworths, London, 194–8 (1964).
3. T. Engel and R. Gomer, *J. Chem. Phys.*, 52, 1832–41 (1970).
4. R.T. Lewis and R. Gomer, *Surf. Sci.*, 26, 197–229 (1971).
5. G.K.L. Cranstoun and J.S. Anderson, *Surf. Sci.*, 32, 397–421 (1972).
6. R. Gomer and J.K. Hulm, *J. Chem. Phys.* 27, 1353–76 (1957).
7. Y.W. Kim, Ph. D. Thesis, University of Connecticut, (1976).

STRUCTURE–PROPERTY–BEHAVIOR CORRELATIONS: I, INFLUENCE OF STRUCTURE

Chairpersons: I. LeMay and L.E. Samuels

Stevens, Vander Voort, McCall, eds. Microstructural Science, Volume 8

USE OF THE GLOBAL AND TOPOLOGICAL PARAMETERS IN MICROSTRUCTURE–PROPERTY CORRELATION

F.N. Rhines*

ABSTRACT

It is a fundamental property of the microstructure of materials that its features are irregular in both size and shape. Quantitative measurements of the one, two and three-dimensional properties, the one- and two-dimensional curvature and the topological properties can be made, however, by the global parameters, which are totals of each of the geometric properties. These can be evaluated statistically by well-known methods of quantitative microscopy. Those physical properties, which are themselves additive, correlate directly with global parameters. Physical processes, such as grain growth, solution heat treatment and sintering also occur inhomogeneously. Their description and kinetics commonly involve the use of ratios of the global parameters, which are meaningful when the distribution of values making up the ratio is either constant, or can be defined. The resulting relationships have been found to be very simple, having the directness of natural laws. Most of the physical processes operate as multiply connected systems, the structure of which can be described quantitatively by the use of the topological parameters: number and connectedness.

INTRODUCTION

The existence of a direct relationship between the internal structure of a material and its properties is one of the oldest and most fundamental ideas in the field of material science and technology. Yet progress with its realization, in a quantitative sense, has been slow and frustrating. Until a few years ago we had uncovered little more than qualitative generalizations, such as the greater hardness associated with a finer grain structure. Recent successes and advances in the development of quantitative relationships are an outcome of the nearly coincidental evolution of three kinds of geometric concepts that have made it possible to characterize microstructure with a degree of reality and exactness not previously attainable. The first of these is the now familiar *quantitative microscopy,* initiated more than a century ago by Delesse[1] and brought to a state of comprehensive utility during the post World War II era. Second has been the application of *descriptive topology* to the characterization of the connective nature of microstructure, promoted largely by C.S. Smith[2], during the same period. Third has been the recognition of the *global parameters* and their utilization in formulating structure-property relationships. The latter has been a special interest of the present writer[3] in recent years. Taken together, these concepts have provided a tool that is turning the once qualitative field of structure-property relationships into an exact science, by making it possible to characterize microstructure both fully and quantitatively.

A principal barrier to progress in the past seems to have been our failure to take microstructure at its face value. Instead, we have endowed it with geometric properties that it does not really possess. We have referred to the microstructures of materials as

*Department of Materials Science and Engineering, University of Florida, Gainesville, Florida 32611 USA.

being plate-like, spheroidal, acicular and equiaxed. We have described processes in which we have postulated spherical grains and cubic precipitates, sometimes of uniform size. In so doing, we have created a dream-world of microstructure. Thinking to simplify, we have actually excluded any possibility of achieving a true expression of the nature and magnitudes of microstructure. Recognizing that microstructures can display traits that distinguish, for example, pearlite from Martensite, it is essential to appreciate also that, in reality, these are irregular structures. No two carbide plates in pearlite are ever precisely alike in detail. No two grains in a polycrystalline aggregate are ever exactly the same. The microstructures of materials are not repeating patterns, but rather exist in an infinity of shapes. This difference from feature to feature is a non-trivial intrinsic property that results from the space-filling nature of the structure of materials. Without such fitting together of the parts of the structure, materials would lack the coherence that gives them their physical properties. In characterizing microstructure for correlation with physical properties, therefore, it is essential to provide for its natural irregularity, not to ignore it, or play tricks with it.

The *global parameters* of microstructure are those geometric properties which are additive for the entire sample of the material. They are, therefore, unaffected by shape, size and irregularity of the features of the microstructure. The total volume of the α-phase in an $\alpha + \beta$ alloy, for example, can be measured and expressed exactly, even though no two particles of the phase are of the same size, or shape. The total surface area of the grains in a polycrystalline aggregate is measurable and is independent of the sizes and shapes of the individual grains. Lineal features, such as the total length of triple line, or of dislocation edge, are measurable without reference to shape. Less generally appreciated is the mathematical truth that curvature is additive. Accordingly, total curvature of line, or of surface, is a global property that can be expressed without reference to its distribution along the line, or upon the surface. The same is true of torsion of line, which may be thought of as spiral curvature.

These are also the geometric properties of microstructure which have been found amenable to measurement by methods of *quantitative microscopy.* The relationships by which they are evaluated are summarized for two-dimensional subjects in Table I and for three-dimensional subjects in Table II. Although it may appear strange to describe a structure in these terms, it can be seen that the global parameters have a logic of their own. The first two parameters in Table I and the first three in Table II are magnitudes of the first, second and third dimensions. The next three in Table II are curvatures of the first and second dimension. The last two parameters in each table are topological and relate to the manner in which structure is put together. This group of global parameters, taken together, makes it possible to characterize microstructure in the terms in which the physical properties respond to structure and, at the same time, to do so quantitatively. This, as will be suggested by what follows, results in significant simplification of the relationships of physical metallurgy and related fields.

Table I

Global parameters of areal measurement

Geometric Property in unit area	Symbol	Measuring Relationship	Nature of Sampling Device	Units
Length of line	L_A	$L_A = \frac{1}{2}\pi N_L$	N_L = line intercepts with unit length	cm^{-1}
Area fraction	A_A	$A_A = L_L = P_P$	L_L = length in unit length P_P = points per point	cm°
Curvature of line	C_A	$C_A = \pi T_A$	T_A = lineal tangensies in unit area	cm^{-2}
Number	N_A	Direct count	N_A = number in unit area	cm^{-2}
Connectedness	G_A	$G_A = B_A - K_A + 1$	B_A = branches in unit area K_A nodes in unit area	cm^{-2}

Table II

Global parameters of volume measurement

Geometric Property in unit volume	Symbol	Measuring Relationship	Nature of Sampling Device	Units
Length of line	L_V	$L_V = 2\ N_A$	N_A = point intercepts in unit area	cm^{-2}
Area of surface	A_V	$A_V = 2\ N_L$	N_L = line intercepts with unit length	cm^{-1}
Volume fraction	V_V	$V_V = A_A = L_L = P_P$	A_A = area in unit area L_L = length in unit length P_P = points per point	cm°
Curvature of line	C_V	$C_V = \pi T_V$	T_V = planar tangencies in unit volume	cm^{-3}
Torsion of line	R_V	$R_V = \pi I_V$	I_V = inflections in unit volume	cm^{-3}
Curvature of surface	M_V	$M_V = \pi T_A$	T_A = lineal tangencies in unit area	cm^{-2}
Number	N_V	Direct counting	Serial sections	cm^{-3}
Connectedness	G_V	$G_V = B_V - K_V + 1$	B_V = branches in unit volume K_V = nodes in unit volume.	cm^{-3}

As they undergo such processes as plastic deformation, grain growth, phase transformation, solution heat treatment and precipitation, materials also behave inhomogeneously on a fine scale. Their over-all behavior is, however, that of a coordinated system, because, in these processes no change can occur locally without affecting the process elsewhere in the body of material. The behavior is that of a connected system and the *topological parameters* number and connectivity are suitable for the expression of such associations.

Number is here used in its most general sense, typically: number of particles of a designated phase, number of grains, number of grain faces, number of grain edges, number of pores. These are always whole numbers, never fractions. Number is, of course, independent of the sizes and shapes of the entities counted. It must, however, be evaluated in the dimension to which it refers, or in some higher dimension, but never in a lower dimension. Thus, the number of points along a line can be measured by lineal sampling, or by counting in a surface, or in a volume. But the number of points on a surface cannot be determined by lineal sampling. It must be counted by areal sampling, or in three dimensions. What is most important is that number in space must be counted in space. This requirement sometimes necessitates arduous procedures and makes topological characterization time-consuming.

Connectivity is a fundamental property of networks, composed of links, or *branches,* joined at *nodes.* The numerical value of the connectivity of a network may be visualized as the maximum number of branches that could be severed without separating the network into parts. At the limit of the severing, the network becomes a single chain, or a tree with limbs and twigs that form no loops. In order to evaluate connectivity in a microstructure, use is made of the relationship:

$$\text{Connectivity} = \text{Branches} - \text{Nodes} + 1 \qquad (1)$$

It is necessary, accordingly, to make two kinds of number counts, i.e. the number of branches and the number of nodes. Depending upon the kind of network under scrutiny, this could mean the numbers of grain edges and of grain corners, or the numbers of channel pores and of junction pores, or, equivalently, the numbers of any kind of linkage feature and of its joints. Connectivity is always expressed in whole numbers and is independent of the shapes and sizes of the features making up the network.

CORRELATIONS WITH MICROSTRUCTURE

When the global parameters are applied to structure-property problems, both old and new, one is inevitably impressed by the beautiful simplicity of the relationships that emerge. It is like coming out of a forest into a formal garden. Property relationships stand before one in pristine purity. Processes are seen as systems of exquisite organization. Dramatically, the structure-property aspects of physical metallurgy and ceramics take on the appearance of an exact science.

Direct Relationships

Simplest among the structure-property relationships are those between individual global parameters and properties of the additive type. Typically, the density (δ) of a two-phase alloy is proportional to the volume fraction (V_V) of one of the phases, Fig. 1:

$$\delta = a + k\,V_V \qquad (2)$$

It is easy to perceive in this case that the density should be an additive property and should correlate with the total volume of the added phase.

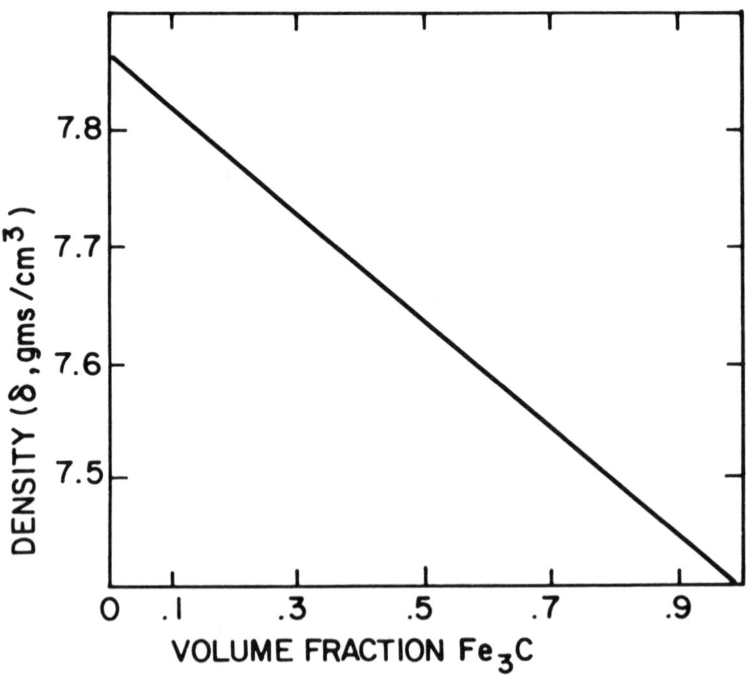

Fig. 1. Density of an α + Fe_3C alloy as a function of the volume fraction (V_V) of Fe_3C.

Another rather obvious case is to be found in the effects of grain and phase boundary on a mechanical property such as Brinell Hardness (BHN)[4]. Clearly the magnitude of any such effect should be proportional to the total area of the interface (S_V), thus:

$$BHN = a + k\ S_V \tag{3}$$

This relationship has been found to apply generally to the grain boundary area of annealed polycrystalline metals, Fig. 2, and also to such two-phase alloys as pearlitic steels[5], Fig. 3, where the interface is between two different phases, ferrite and cementite. The full power of this relationship is yet to be realized in the characterization of the properties of polyphase aggregates in general. Its applicability is not limited, of course, to Brinell Hardness. The yield strength, if properly measured, follows the same law, but the ultimate tensile strength and the engineering elongation appear to correlate better with other kinds of microstructural properties.

Another direct relationship that is now well established is that between the total length (L_V) of dislocation edge and the square of the stress (σ) to initiate plastic flow,[6] Fig. 4:

$$\sigma^2 = a + k\ L_V \tag{4}$$

Because grain boundaries, as well as dislocations within the grains, provide resistance to plastic flow (Equation 3), this relation appears unambiguous only when comparing the behavior of materials having the same grain size, but different degrees of cold work.

The ultimate tensile strength (UTS) of a ductile porous body, such as a sintered metal, has been found to correlate well with the projected area of ductile rupture surface in a tensile fracture,[7] Fig. 5:

$$UTS = a + k\ A_A \tag{5}$$

This measurement is best made by the use of a scanning electron microscope,

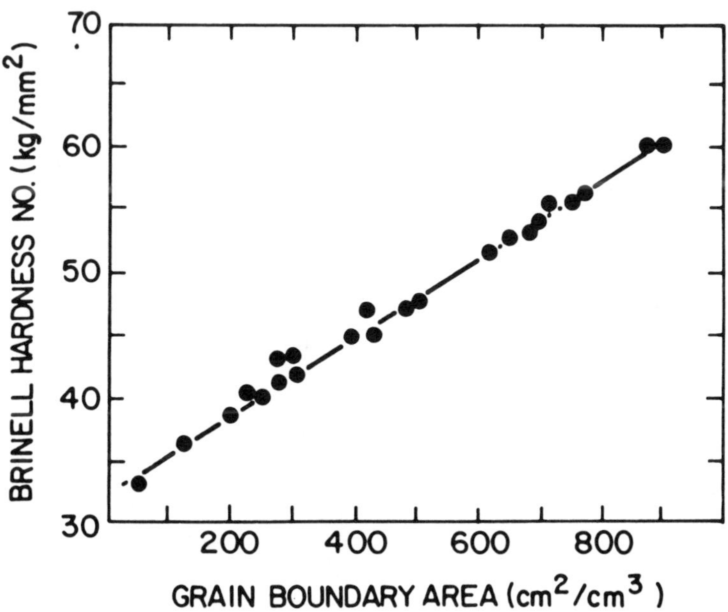

Fig. 2. Brinell Hardness Number of fully annealed 70–30 Cu-Zn brass as a function of the total area per unit volume (S_V) of grain and twin boundary area. From data of Babyak and Rhines[4].

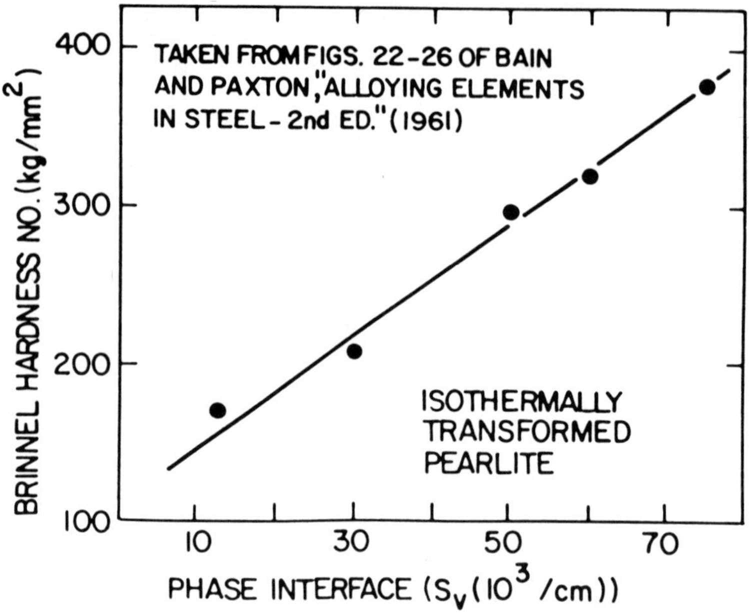

Fig. 3. Brinell Hardness Number of isothermally transformed eutectoid steel as a function of the total area per unit volume (S_V) of phase interface between ferrite and cementite. From data and photomicrographs in Alloying Elements in Steel – 2nd Ed., Bain and Paxton [5].

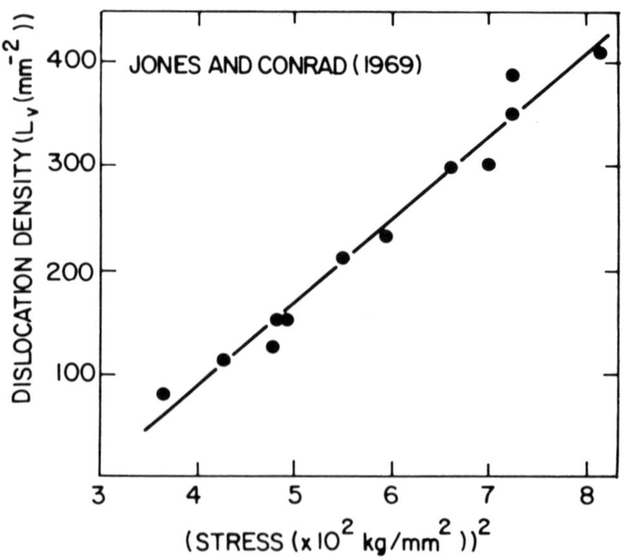

Fig. 4. Square of the stress to initiate plastic deformation as a function of the total length (L_V) of dislocation edge in unit volume. From data of Jones and Conrad [6].

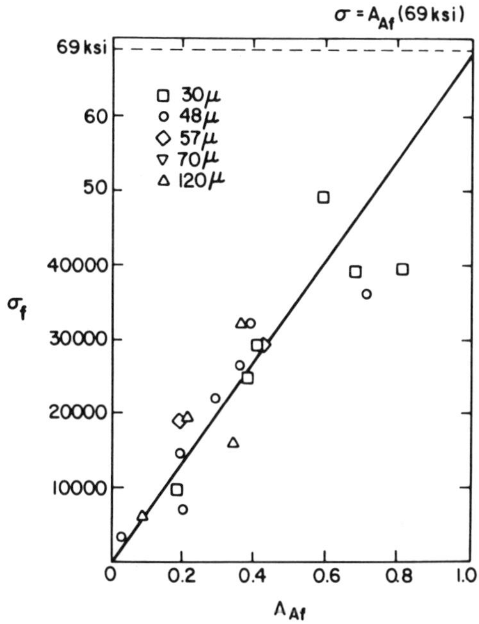

Fig. 5. Ultimate Tensile Strength (σ_f) versus area fraction of fractured metal (A_{A_f}) in the fracture of a sintered nickel powder as density increases. From Rhines, DeHoff and Whitney[7].

This measurement is best made by the use of a scanning electron microscope, which fulfills the geometric requirements of the area fraction measurement. In the case of brittle materials, which fracture conchoidally, it has been found[8] that the ultimate tensile strength is proportional to the length per unit area (L_V) of edge of the rupture areas according to the first relationship in Table I, Fig. 6:

$$UTS = a + k\ L_A \qquad (6)$$

Frequently, the conchoidal fracture surface is crossed by markings, called rivers, that correspond to the meeting of two crack systems. When these are present, they must be included with the total of fracture edge in the metallographic measurement and the calculation of tensile strength.

Relationships Involving Multiplicative Combinations of Global Parameters

Metallurgical processes that are sensitive to microstructure usually occur at many localities where they respond to the local geometry in more or less complex ways. In order to express this situation for the system as a whole, it may be necessary to describe what happens on the average. This involves using the global parameters in the form of ratios, as reciprocals, or in more complex combinations. The averages, or ratios that result will normally represent a range of values that are not simply additive, because they are products of a multiplicative relationship. In order to apply them to a property correlation, it becomes necessary either to know the distribution of the values represented by the ratio, or average, or to be so fortunate as to have the values all alike. This is a general problem which may be more clearly appreciated by the consideration of a simple two-dimensional example.

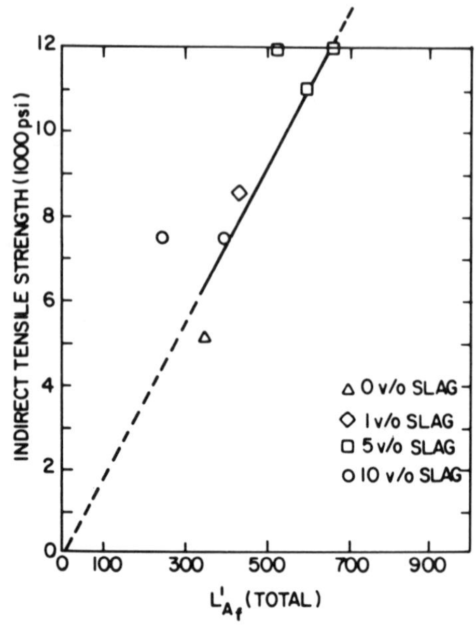

Fig. 6. Ultimate Tensile Strength of a sintered mixture of hematite and slag as a function of the length (L_A) per unit area of fracture edge. From a dissertation by Paulo Pinheiro de Neto Silva, Univ. Fla [8].

Suppose that I have a roll of ribbon on which I desire to inscribe a maximum number of circles, of a specified diameter, and I wish to relate the number of circles per unit area to the average width of the ribbon. The average is readily obtained by dividing the area of the ribbon by its length, but the result may not be useful for determining the number of circles that can be drawn in unit area, because the ribbon may not be of uniform width, as illustrated in Fig. 7. If the edge of the ribbon is regularly serrated, so that its pattern is fully definable, a correlation with the number of circles can be made, although the number will differ from that for the case of a ribbon of uniform width. Should the edge be irregular, however, no prediction of the number of inscribed circles is possible. It is thus with products of any of the global parameters. The distribution of values contained in the product must be regular, or otherwise definable, for products of global parameters to correlate quantitatively with physical behavior. Fortunately, nature appears to be generous in providing regular cases. Some examples follow.

Steady State Grain Growth [9]

Steady state grain growth is an increase in the average grain size that occurs spontaneously in single phase materials during protracted heating. Since every grain differs from every other grain in shape and size, the meaning of average size requires careful consideration. Involvement with shape can be avoided by expressing the size in terms of the volume of the grain. The *average grain volume* ($\bar{V}$) can then be measured by dividing the volume of the specimen by the global parameter *total number of grains* (N), whence:

$$\bar{V} = 1/N_V \tag{7}$$

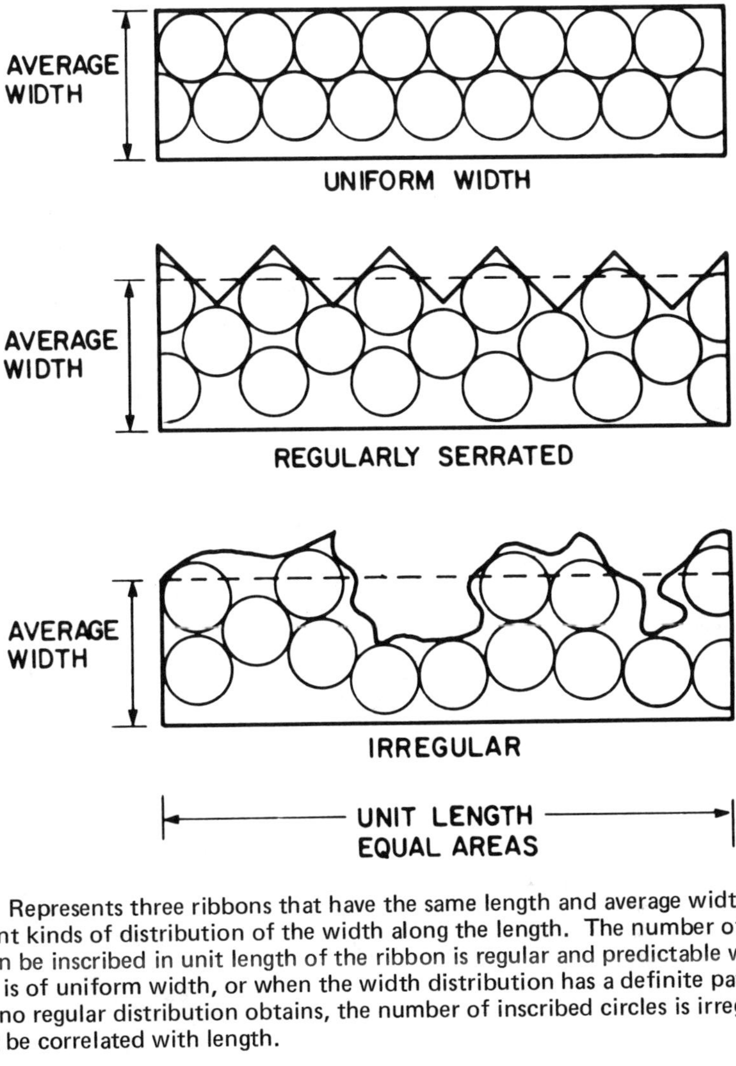

Fig. 7. Represents three ribbons that have the same length and average width, but different kinds of distribution of the width along the length. The number of circles that can be inscribed in unit length of the ribbon is regular and predictable when the ribbon is of uniform width, or when the width distribution has a definite pattern. Where no regular distribution obtains, the number of inscribed circles is irregular and cannot be correlated with length.

This does not tell anything about the sizes of individual grains, however, and grain growth might be expected to be grossly affected by the distribution of sizes in the mixture. Experimentally, it has been found that, in steady state grain growth, the average grain volume is proportional to time (t), Fig. 8:

$$\overline{V} = a + k\,t \tag{8}$$

Studies by Okazaki and Conrad [10], as well as by Patterson [11], have shown that the grain size distribution is log normal, Fig. 9, and that the slope of the log normal plot ($\ln \sigma_V$) remains constant during grain growth. Thus, a knowledge of ($\overline{V}$), together with the corresponding ($\ln \sigma_V$) tells exactly how many grains of what sizes are contained in the sample at a specified moment in time. Increase in ($\overline{V}$) occurs as a simple magnification of the structure. The analogy between this case and that of the two-dimensional ribbon, cited above, should be apparent.

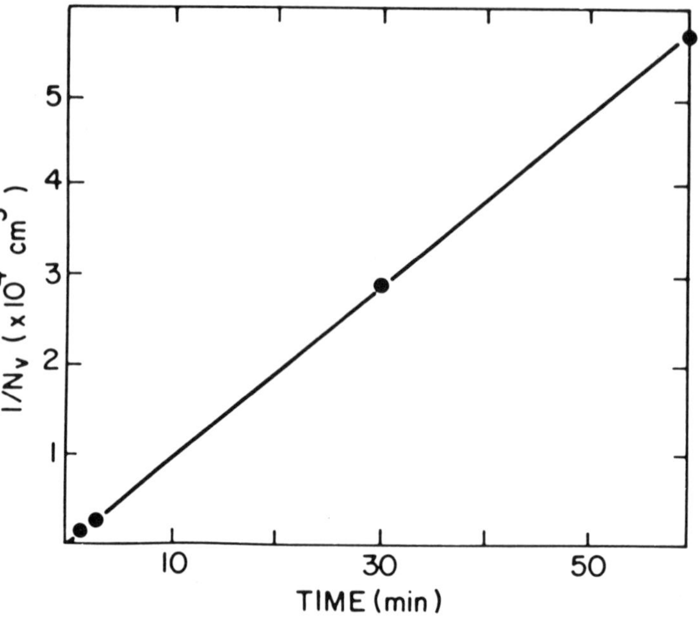

Fig. 8. Average grain volume ($1/N_V$) as a function of time in steady state grain growth of aluminum. From Rhines, Craig and DeHoff[9].

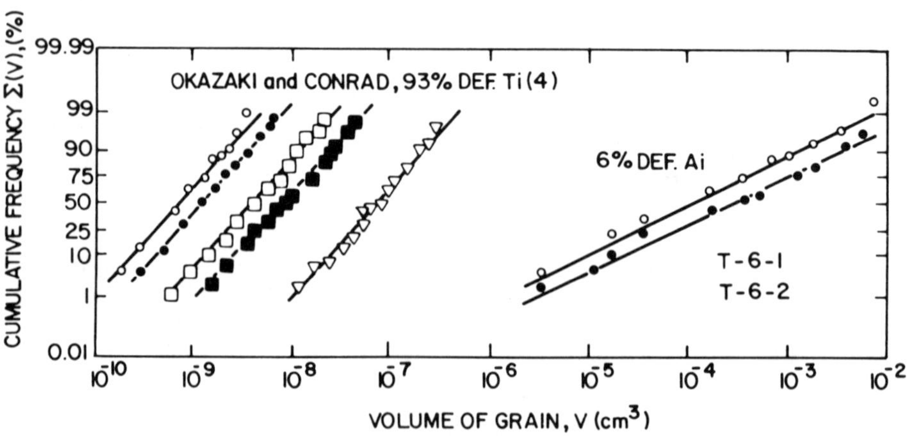

Fig. 9. Log-normal plots of grain volume data of Okazaki and Conrad[10], and of Patterson[11]. The slopes remain constant within each group as the average grain volume increases in steady state grain growth.

In order to derive the kinetics of the change that takes place with increase in $(\bar{V})$, it is necessary to be able to describe what happens in the system step by step, i.e., it is necessary to describe the *path of grain growth.* The grains do not simply increase in volume, because the total volume of the system remains constant. Rather, growth occurs by the disappearance of some grains, their volume being transferred to other grains and distributed through the system. No grain can collapse without every other grain in the system being affected to some degree. Grain growth is accordingly a multiply connected feed-back process. As such, it is a topological process that can be described only through the use of topological parameters. The appropriate global topological parameter is number, by which the grain boundary network can be characterized quantitatively.

According to network law, in any unbounded system of contiguous cells, the sum of the cells (N_V) and the edges (E_V) is equal to the sum of the faces (F_V) and corners (C_V):

$$N_V + E_V = F_V + C_V \tag{9}$$

Where the form of the network is controlled by surface tension forces, as it is in the case of grain growth, three grains meet along each edge and four grains meet at each corner, so that there are twice as many edges as corners in the system. Under these circumstances, equation (9) becomes:

$$N_V + C_V = F_V \tag{10}$$

Each grain is characterized, in a topological sense, if one knows either its number of corners, or its number of faces. Thus, the polycrystalline aggregate is composed of grains topologically characterized by the numbers of their faces and existing in a distribution made up of different numbers of grains in each topological class from one face to fifty, or more. Williams and Smith[12] have shown that the volumes of the individual grains in a size distribution are proportional to the numbers of their faces. Accordingly, it appears that not only is the grain volume distribution log normal, but the distribution of grains according to their numbers of faces is log normal as well. Furthermore, Williams and Smith showed that the average grain in any distribution has seven sides in the network, equivalent to a tetrakaidecahedral figure. With this information, it becomes possible to add to the description of the grains in a distribution the numbers of faces on each grain.

Grain growth may now be visualized as a process in which the grains are individually demoted from one topological class to the next lower class by the loss of a face. When seven faces have been lost to the system, the equivalent of one average grain will have vanished and (N_V) will be decreased by one, thereby increasing the average grain volume. The faces that are lost in this way are always three-edged faces, Fig. 10, which are capable of collapsing as their three corners come together and the pair of grains that had shared the face draw apart. This topological exchange results, at the same time, in the creation of a new three-edged face, so that the process can continue indefinitely.

The driving force for grain growth has its origin in the surface tension (γ) of the grain boundary. Surface tension acting upon a curved boundary exerts a pressure (P) that is proportional to the curvature:

$$P = \gamma \tfrac{1}{2}\left(\frac{1}{r_1} + \frac{1}{r_2}\right) \tag{11}$$

Migration of the grain boundary occurs when this will result in a loss of grain boundary area in the system, i.e., energy dissipation. In general, such motion has the effect of balancing the grain boundary forces in the system and of improving the equiaxedness of the grains. This leads to the occasional closure of a three-edged face and, ultimately, to the collapse of a grain. This, of course, upsets the balance of forces and transmits a

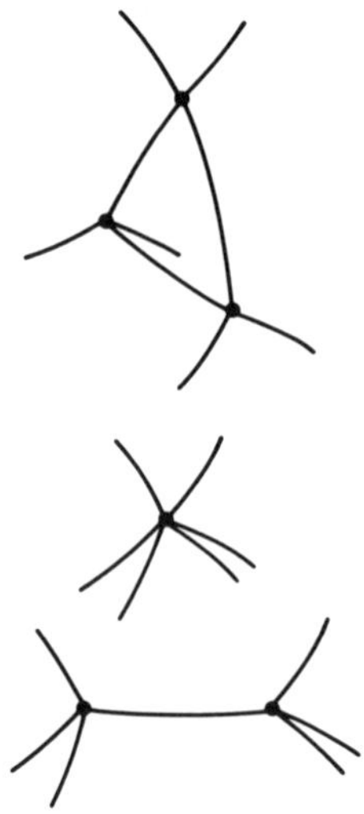

IN THE NETWORK	C	F	E
BEFORE	3	I	3
AFTER	2	0	I
LOST	I	I	2

Fig. 10. As a pair of neighboring grains moves apart, their mutual triangular face (top sketch) closes (center sketch) to produce a new triple line joining two new corners (bottom sketch). This occurs with a net loss of one corner, one face and two edges[9].

wave of grain boundary motion throughout the system. The frequency with which these events occur is proportional to the frequency of three-edged faces in the network, which, in turn, is determined by the grain size distribution (1n σ_V).

It has been demonstrated[9] that there is a direct relation between the volume swept out by grain boundary, as it migrates, and the number of grains that are caused to collapse. Each time that grain boundary sweeps through the equivalent of the entire volume of the specimen, four grains are collapsed in the system. Using this number, together with the total surface curvature, (M_V), the total surface area of grain boundary (S_V), the number of grains (N_V, the surface tension (γ) and the grain boundary mobility (μ), the rate constant for equation (9) has been evaluated (see Appendix I):

$$k = 4\,\mu\,\gamma\,M_V \cdot S_V/N_V \tag{12}$$

Here the product ($M_V \cdot S_V/N_V$) is found experimentally to be a constant, as it should be if (1n σ_V) is constant. The rate equation for grain growth then becomes:

$$(1/N_V)_t = (1/N_V)_0 \left\{ 1 + 4\,\mu\,\gamma\,M_V \cdot S_V/N_V\,t \right\} \tag{13}$$

This relationship is especially remarkable for the fact that it includes no adjustment constant, as have the grain growth expressions of the past. Every factor in the equation is either an established physical constant, or can be evaluated precisely by quantitative microscopy. This is possible because the global parameters have provided, for the first time, a true expression of the geometric factors.

Even more important, the global parameters and particularly the topological parameters have provided a vocabulary by means of which we can not only express complex behavior in a multiply connected feed-back process, but can visualize the process itself.

The exquisite organization of a natural process which has thereby been revealed is truly astounding. It surpasses, in its coordination of a variety of factors, the inventiveness of the most talented of theoreticians. Yet, it is in no way built upon theory, but is simply a mathematical expression of direct measurements. Such is the power of the global parameters.

Solution Heat Treatment

Some progress has been made in the analysis of several other processes of physical metallurgy by the use of the global parameters. Each such case has brought with it surprises. While it would serve no purpose to describe these in detail at this time, a brief look at two of them will show that the grain growth case is fairly typical of what is being found.

The process of solution heat treatment, in which particles of a minor phase are being taken into solid solution by heat treatment, has usually been assumed to be controlled by the rate of diffusion of the dissolving component in the matrix. This would require that the particles become smaller, until they ultimately disappear. When the solution of $CuAl_2$ in aluminum was followed by the use of the global parameter measurements[13], however, it was found that the reciprocal of the volume fraction (V_v) remaining out of solution is proportional to the time (t) of heating:

$$1/V_v = a + k\,t \qquad (14)$$

and at the same time the reciprocal of the particle surface area (S_v) and the reciprocal of the total surface curvature (M_v) are also proportional to time, Fig. 11. In other words, the volume fraction, the total surface area and the total curvature remain in constant proportion to one another, Fig. 12. This can mean only that the particle size distribution remains constant and that the solution process proceeds by a reduction in the numbers of particles of all sizes in proportion to their frequency in the population.

Such regular behavior implies, once more, that the particle size distribution is regular. Indeed, Hren et al[14], found a log normal distribution of Gunier-Preston zones in an alloy of identical composition. Whatever distribution of particle sizes prevails, however, it is clear that it remains constant. The rate-controlling mechanism is, therefore, associated with the rate at which each particle evaporates into the matrix. It is most probable that this depends upon the surface curvature of the particle and is faster the sharper the curvature. Solution then proceeds as a cascading topological process. Further study will be needed to fully describe this case. It appears simpler than grain growth, but is again a highly organized behavior.

Sintering

The process of sintering is of special interest, because its path can be described by the use of the topological parameter connectedness (G_v)[15]. Three stages of sintering are recognized. During the first stage, the connectedness of the system that has been established by the stacking of the powder remains constant while the surface contour changes under the driving force of surface tension so as to minimize the surface area. In the second stage the connectedness decreases progressively to zero while the ratio of pore volume to pore surface (V_v/S_v) remains constant. In the third stage the number of isolated pores (N_v) diminishes as the pores spheroidize to become fewer but larger, the connectedness meanwhile remaining at zero.

Thus far, only the second stage has yielded to detailed characterization. Its regular behavior is indicated by the constancy of the ratio (V_v/S_v) which says that the mean free path across the porosity stays the same while connectivity decreases. Under these conditions it has been found possible to deduce the sintering force per unit area

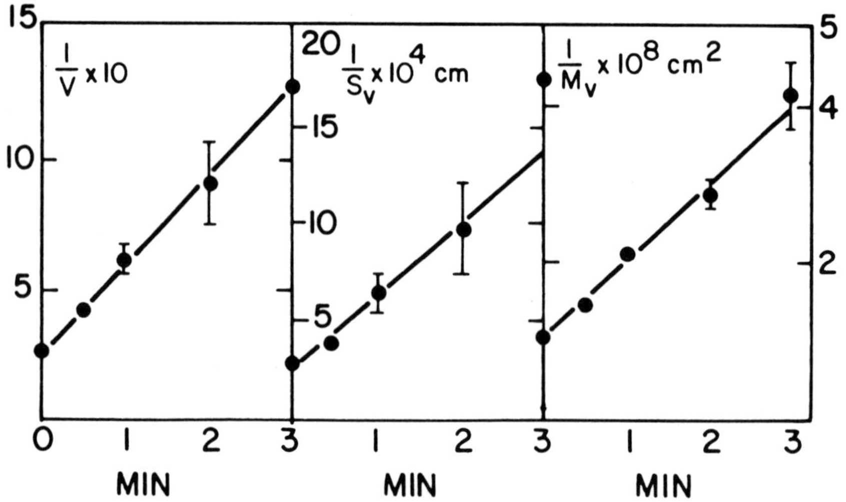

Fig. 11. Inverse of the volume fraction (V_V), total surface area in unit volume (S_V) and total curvature of surface in unit volume (M_V) of $CuAl_2$ particles in a matrix of aluminum solid solution as a function of time at 500°C. From Rhines and Hernu [13].

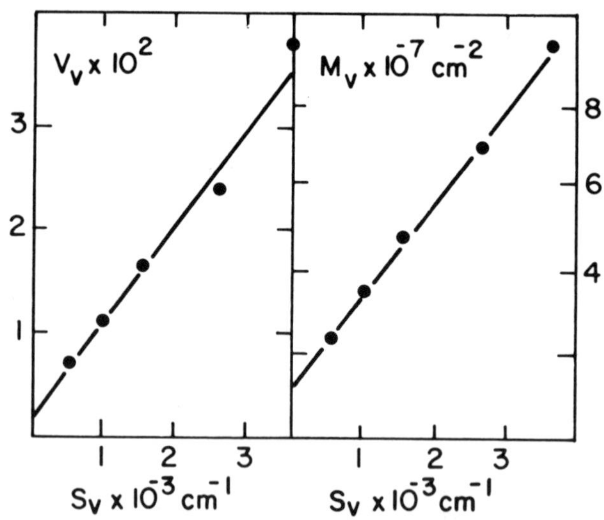

Fig. 12. Ratio of volume fraction (V_V) and also of total curvature of surface (M_V) to total surface area (S_V) of $CuAl_2$ particles in an alloy of Al + 4% Cu during solution heat treatment at 500°C. From Rhines and Hernu [13].

(F/a) from the surface tension (γ) and the global parameters volume fraction of porosity (V_V), total curvature of pore surface (M_V) and total pore surface area (S_V) [16], Fig. 13:

$$F/a = \gamma \, V_V \cdot M_V/S_V \tag{15}$$

Again, this relationship is remarkable for the absence of any arbitrarily adjustable

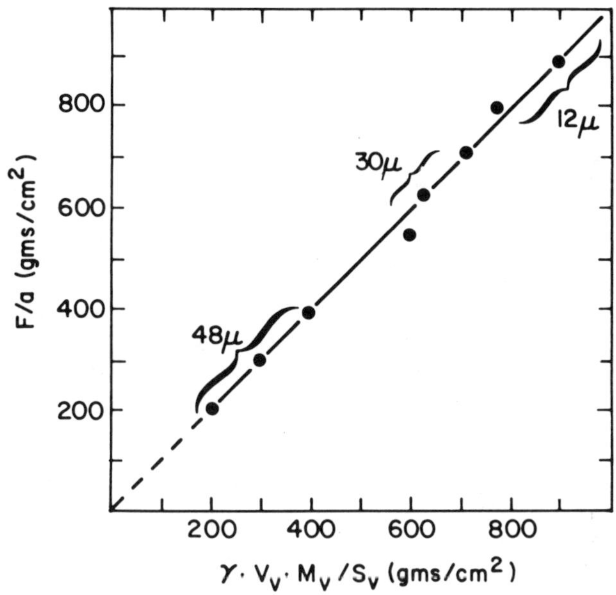

Fig. 13. Measured sintering force (F/a) versus the force derived from the surface tension (γ) and the global parameter ratio ($V_V \cdot M_V/S_V$). From data of Gregg and Rhines [16].

constant, all factors being fully defined, as in a natural law. This relation can in fact be used for the direct measurement of the surface tension of the metal, simply by measuring the sintering force in a weighing experiment.

SUMMARY

In summary, the features of the microstructures of materials are irregular in shape and size, but their additive geometric properties can be correlated with additive physical properties through the use of the global parameters, which are all measurable by the use of well-known methods of quantitative microscopy.

Physical processes, such as grain growth, solution heat treatment, sintering and others require for their description the use of global parameters in multiplicative combination, as averages, reciprocals, ratios and more complex forms. These products can yield unique expressions of the processes only when the distribution of values that make up the average, or other product, are regular, or somehow definable.

The structures of materials differ fundamentally from isolated objects in that they are interconnected. This characteristic requires the use of one or more topological parameters (number and connectedness) for its exact description. Physical processes often proceed by change in the topological state of the system.

Correlations of global properties with physical properties are generally very simple. Processes described in terms of the global parameters, including the topological parameters, may be exact, having the characteristics of natural laws.

The global parameters, including the topological parameters, provide not only a mathematics for the precise expression of structure-property relationships, but also a language through which such relationships can be visualized and discussed.

BIBLIOGRAPHY

1. Delesse, A., "Procede Mechanique pour Determiner la Composition des Roches", *Ann. Mines* 13 (Fourth Series), 379–388 (1848).
2. Smith, C.S., "Grains, Phases and Interfaces, an Interpretation of Microstructure", *Trans. AIME* 175, 15–51 (1948).
3. Rhines, F.N., "Microstructure Property Relationships in Materials", *Met. Trans.* 8A, 127–133 (1977).
4. Babyak, W.J. and Rhines, F.N., "The Relation Between the Boundary Area and Hardness of Recrystallized Cartridge Brass," *Trans. AIME* 218, 21–33 (1960).
5. Bain, E.C. and Paxton, H.W., *Alloying Elements in Steel* (book), Amer. Soc. for Metals, Metals Park, Ohio, 2nd Ed. (1961), Figs. 22–26.
6. Jones, R.L. and Conrad, H., "The Effect of Grain Size on the Strength of α-titanium at Room Temperature", *Trans. TMS-AIME* 245, 779 (1969).
7. Rhines, F.N., DeHoff, R.T. and Whitney, E.D., "Quantitative Determination of the Structure-property Relationships in Nuclear Fuel Materials", Final Rep., U.S. AEC No. AT–(40-1)4212, (1975).
8. Pinheiro, Paulo, de Neto Silva, "Cold Strength of Synthetic Ferric Oxide Slag Briquettes in Relation to their Microstructural Evolution During Sintering," Dissertation, University of Florida, (1973).
9. Rhines, F.N., Craig, K.R. and DeHoff, R.T., "Mechanism of Steady-state Grain Growth in Aluminum," *Met. Trans.* 5, 413–425 (1974).
10. Okazaki, K. and Conrad, H., "Recrystallization and Grain Growth in Titanium: I. Characterization of the Structure", *Met. Trans.* 3, 2411–2421 (1972).
11. Patterson, B.R., "The Grain Size Distribution of Aluminum," Dissertation, University of Florida, (1978).
12. Williams, W.M. and Smith, C.S., "A Study of Grain Shape in an Aluminum Alloy and Other Applications of Stereoscopic Microradiography," *Trans. AIME* 194, 755–765 (1952).
13. Rhines, F.N. and Hernu, Jean-Brice, "Topological Evolution of $CuAl_2$ Particles During Solution Heat Treatment," *Scripta Met.* 11, 155–157 (1977).
14. Harkness, S.D., Gould, R.W. and Hren, J.J., "A Critical Evaluation of X-ray Small Angle Scattering Parameters by Transmission Electron Microscopy: GP Zones in Al Alloys," *Phil. Mag.* 19 No. 157, 115–128 (1969).
15. Rhines, F.N., "A New Viewpoint on Sintering," *Trans. MPA* 91–101 (1958): Plansee Seminar 3rd (1958).
16. Gregg, R.A. and Rhines, F.N., "Surface Tension and the Sintering Force in Copper", *Met. Trans.* 4, 1365–1374 (1973).

APPENDIX

The derivation of the rate equation for steady state grain growth, as presented in reference (9) had a dimensional defect amounting to a factor of an extra cm^3. It is now believed that this difference arose through the use of the mobility (μ) in the form for the unlimited grain boundary, instead of modifying it for the case of the connected network. Since the migration of any grain boundary must be transmitted throughout the system, much as an electrical impulse is transmitted throughout an electrical network, the motion is proportional to the connectivity of the network. Accordingly, the mobility that applies in the case of grain growth is believed to be the product of the connectivity (G_v) and (μ). But, in a surface tension controlled network the connectivity is equal to the number of corners (C_v) and the network mobility can be represented by the product ($\mu \, C_v$).

The steps in the derivation given in reference (9) then become: The average pressure exerted upon a point on the grain boundary:

$$P = \gamma M_v/S_v \qquad (\text{dynes/cm}^2)$$

The total force on all of the grain boundary is:

$$F = \gamma M_v \qquad (\text{dynes/cm}^3)$$

The number of sweeps made by all of the grain boundary through unit volume is:

$$\mu C_v \gamma M_v S_v \qquad (\text{cm}^{-6}\ \text{sec}^{-1})$$

The number of grains lost per cm^3 in one second is:

$$4 \mu C_v \gamma M_v S_v \qquad (\text{cm}^{-6}\ \text{sec}^{-1})$$

The total volume transferred from lost grains to remaining grains is:

$$4 \mu C_v \gamma M_v S_v/N_v \qquad (\text{cm}^{-3}\ \text{sec}^{-1})$$

The average growth per grain is:

$$4 \mu C_v \gamma M_v S_v/N_v^2 \qquad (\text{sec}^{-1})$$

For the case of the surface tension controlled network C_v/N_v is equal to 6, whereby the volume increase per grain per second becomes:

$$4 \cdot 6 \mu\ \gamma M_v S_v/N_v \qquad (\text{cm}^3\ \text{sec}^{-1})$$

And in time (t):

$$24 \mu \gamma M_v \cdot S_v/N_v\ t \qquad (\text{cm}^3)$$

The rate equation then is:

$$1/(N_v)_t = 1/(N_v)_0 \quad \{1 + 24 \mu \gamma M_v \cdot S_v/N_v\ \ t\}$$

Stevens, Vander Voort, McCall, eds. Microstructural Science, Volume 8

WEAR OF ALUMINUM ALLOYS

T.S. Eyre*

ABSTRACT

Aluminum alloys containing silicon, copper, iron and nickel have been rubbed against steels of different hardness and also a grey cast iron. In all cases both oxidative and metallic wear occurs. The transition load between oxidative and metallic wear is increased as the silicon content increases and the addition of copper also brings about significant increases. The presence of copper is beneficial irrespective of its effect on the mechanical properties produced by solution treatment and ageing. In oxidative wear, transfer of iron from the counterface to the aluminum alloy occurs and a very complex surface containing AlFe and $\alpha\, Al_2O_3$ is produced. Wear in the oxidative regime occurs by fracture and spalling of the oxide from the aluminum alloy surface. Metallic wear occurs by plastic deformation, fracture and metal transfer of aluminum to the surface of the steel counterface. Wear behavior of these alloys is not related in any way to their bulk mechanical properties or microstructure. Considerable further effort is required before our knowledge of surface properties is sufficient to design improved wear resistant alloys for engine components.

INTRODUCTION

In attempting to save energy the kerb weight of automobiles is being reduced. Aluminum casting alloys are, therefore, being constantly reviewed in an attempt to replace cast iron components[1] and a substantial change is likely to occur in the next few years. Cylinder blocks have been used in many production automobiles from as early as 1960 onwards. The development of the Reynolds 390 alloy for the Chevrolet Vega and the use of an etch treatment to leave the silicon particles standing proud of the load bearing surface was the first success in omitting a cast iron liner. In Europe cylinder blocks without cast iron liners have been increasingly used, for example the Reynolds 390 alloy is used in the Porsch 928, the Daimler Benz 450 SLC and the Rolls Royce Camargue.

The investigation reported here has been carried out within a framework of past and ongoing investigations into the wear of grey cast iron piston rings and cylinder liners and the application of surface treatments and therefore some of the more important metallurgical aspects of wear are already well established.[2] It seems logical, therefore, to extend these studies to alloys which are likely to find increasing use in both the automobile and general engineering industries. Considerable economic savings are possible if the aluminum alloy surface itself can be used directly, or with some simple treatment as the bearing surface.

LITERATURE SURVEY

The first part of the experimental work was reported by Beesley and Eyre[3] in which they demonstrated that the addition of copper was beneficial and the addition of zinc was detrimental to the seizure resistance of pure aluminum. Arguments based

*Brunel University, Uxbridge, Middlesex UB8 3PH, ENGLAND.

on experimental evidence were put forward for the formation of an AlFe compound as well as copper segregation to the wear surface. These effects promoted greater protection of the aluminum by the complex surface layer built up during mild (oxidative) wear.

Shivanath et al.[4] and Clegg and Das[5] have considered the wear of aluminum silicon casting alloys which are already widely used in the automobile industry. Shivanath concluded that the information available was often contradictory, probably because of the complex composition of the different commercial alloys used, and also the variation in the experimental procedures. Shivanath, therefore, based his work on binary alloys of aluminum containing up to 20% silicon. These alloys were produced by both sand and chill casting and the 11% Al/Si alloy was also modified. It was concluded that these alloys exhibited both mild (oxidative) and severe (metallic) wear, the transition being dependent upon load. The transition load increased linearly as the silicon content increased. The oxidative and metallic wear rates were not significantly affected by the silicon particle size and the oxidative wear rate was not affected by the silicon content. Only the transition load was affected by silicon content.

The work of Clegg and Das[5] which was confined to commercial alloys also showed that wear rate was not affected by modification treatments that varied the silicon particle size over a wide range. Shivanath[4] and Clegg and Das[5] agreed that the wear behavior was independent of the bulk mechanical properties of the alloys.

The effect of the counterface material is very important in tribological studies and although Shivanath et al. and Clegg and Das describe precisely the steels they used, they did not vary the counterface in any way. Sarkar[6] however, concluded that there was no difference in wear of Al/Si alloys when worn dry against steel and grey cast iron, a rather surprising conclusion in view of work carried out on other metallic systems which show the general superiority of grey cast iron.

In this investigation, it was therefore, decided to extend the studies of Shivanath et al. in an attempt to upgrade the information on the binary alloys, gradually to equivalent commercial alloys, similar to those investigated by Clegg and Das. To do this, the investigation has been broken down into three parts to examine:

1) The effect of nickel, iron and copper in amounts usually present in commercial alloys.
2) The effect of heat treatment on the alloys containing different amounts of copper.
3) The effect of different counterface materials.

EXPERIMENTAL DETAILS

The conditions used to manufacture the alloys and to carry out the wear tests and subsequent examination procedures are all substantially the same as those used by Shivanath et al.[4]

RESULTS

All of these alloys exhibited two types of wear curves. For oxidative wear, sometimes immediately, but certainly after a short interval of rubbing, a steady state condition was established with both surfaces showing a smooth oxidized appearance. Some tests were continued for longer periods to establish that no further changes occurred, but most tests were terminated after one hour. For metallic wear, usually after a short period of time which decreased as the load was increased, transfer of aluminum to the disc surface occurred, the wear rate increased rapidly and the wear pin was very quickly worn away (Fig. 1).

It was possible to plot the oxidative wear rates and the metallic wear rates against the applied load to obtain a transition diagram for each material. The effect of different alloys are shown together in Fig. 2, and it will be observed that:

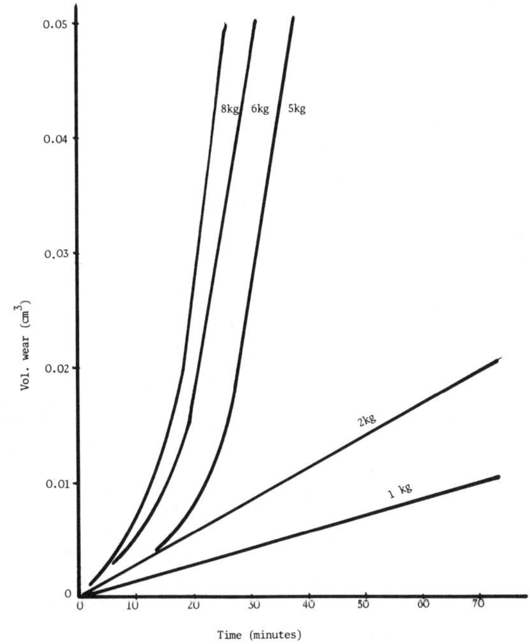

Fig. 1. Wear of Al/16% Si alloy.

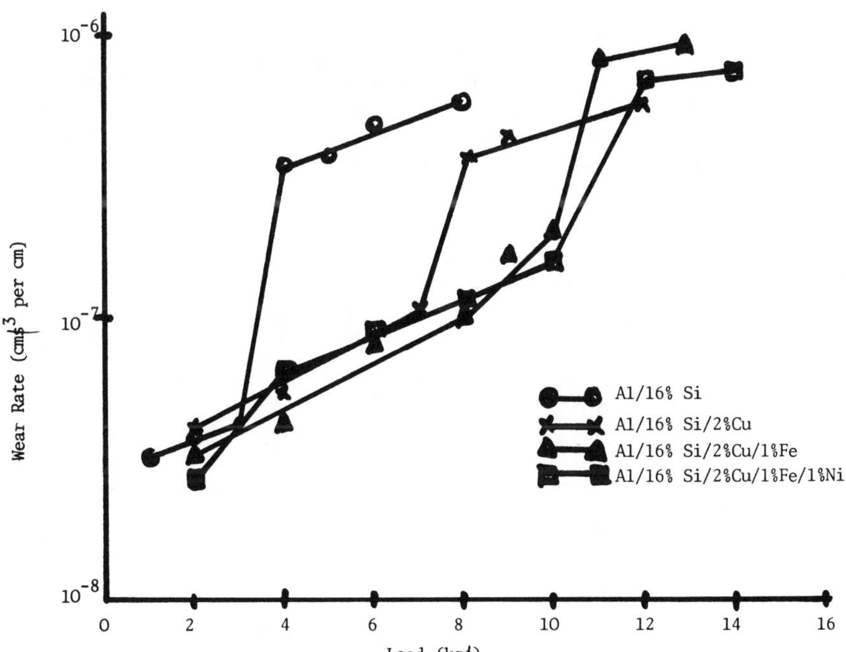

Fig. 2. Wear of different alloys against steel disc A.

1) The oxidative wear rates for all alloys fall in a narrow band on a straight line gradually increasing with the load.
2) The metallic wear rates for all alloys fall in a narrow band on a straight line increasing with load. The wear rate is an order of magnitude higher than oxidative wear rates.
3) There is a fairly sharp transition between oxidative and metallic wear.
4) Copper and to a lesser extent iron have a marked effect on increasing the transition load and nickel has no significant effect.

The alloys containing copper were produced both in the as-cast, solution-treated and fully aged conditions with 1, 2 and 4% copper. In all conditions the wear properties were not significantly altered by the actual copper content.

One of the main departures in this investigation from the experimental procedure used by Shivanath et al.[4] was to use a different steel disc (Table 1). It became obvious in this investigation that the transition loads were much lower than those previously reported by Shivanath et al.[4]. In the earlier work, the transition load for the Al/16% Si alloys was approximately 16 kg whereas in the present investigation a load of 4 kg was obtained. Some of the pins and discs used by Shivanath were still available and further experiments were carried out (Fig. 3) which verified Shivanath's results of a transition load of 16 kg. Wear pins produced in the present investigation were also worn against the disc material used by Shivanath and verified that the transition load increased from 4 to 16 kg. Further experiments were carried out to determine the factor that controlled the behavior of the disc material. Essentially the difference in wear behavior obtained against different counterfaces could only be caused by 1) variation in surface finish, 2) variation in composition of the steel, 3) variation in hardness of the steel.

In all cases the variation in the surface finish was insignificant, so it was decided to temper the harder steel (750 H_V) down to the hardness of the softer steel (350 H_V). When this was carried out the transition load was reduced for the harder steel used by Shivanath to that of the softer steel used in this investigation (Fig. 3).

Flake graphitic cast iron is widely used in wear applications and it was expected that the wear of aluminum alloys against this metal would be markedly different. Although it was not possible to obtain grey discs of 350 H_V, in fact they were approximately 250 H_V. It will be seen from Fig. 4 that the transition load for the Al/Si alloy was markedly increased when worn against the grey cast iron.

Table 1. Composition of disc materials

Composition	C	Si	Mn	Ni	Cr	Mo	P	S	H_V
Steel A	0.35	0.10	0.45	1.3	0.9	0.20	-	-	
	0.45	0.35	0.70	1.8	1.4	0.35	0.05M	0.05M	350
Steel B	0.90	0.10	0.30	-	1.0	-	-	-	750 and
	1.20	0.35	0.75	-	1.	-	0.05M	0.05M	350
Grey C.I.	3.5	2.1	0.8	-	-	-	0.10	0.10	250

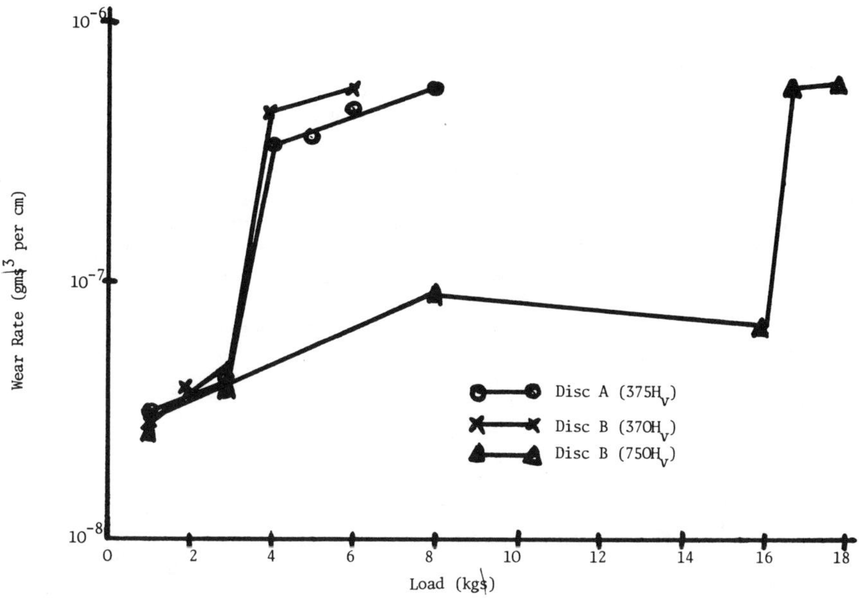

Fig. 3. Wear of Al/16% Si alloy against different disc steels.

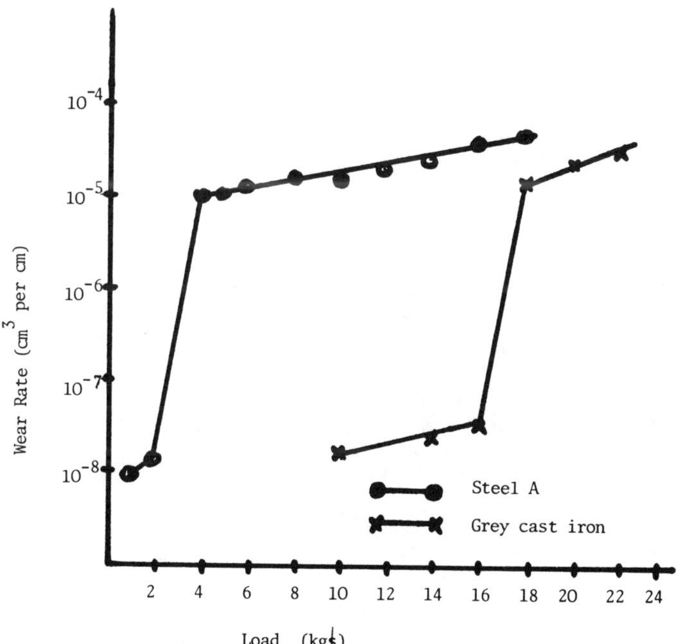

Fig. 4. Wear of Al/16% Si alloy against steel and grey cast iron.

METALLOGRAPHIC OBSERVATIONS

Figure 5 shows the macroscopic appearance of the two basic wear mechanisms observed in all the experiments carried out. Oxidative wear produced smooth, black wear tracks and black powdery wear debris. Metallic wear produced adhesion of aluminum to the steel disc with very rough surfaces and metallic debris. Examination of the wear surfaces in greater detail showed that the oxide layer built up in grooves formed in the early stages of wear and appears to be consolidated into quite discrete layers (Fig. 6). Fracture of the oxide occurs by the tensile stresses developed by sliding, and a plate-like debris separates from the surface (Fig. 7). The oxide is often thicker than 10μm which suggests either very high temperatures have been encountered or that oxidation rates are enhanced by friction. The oxide layer was examined using EPMA and in all cases iron has been transferred to the aluminum surface, and its presence through the total thickness of the oxide indicates the complicated nature of the transfer process (Fig. 6). This is further evidence to support the earlier conclusion of Beesley and Eyre[3] who stated that these oxidative surfaces contained complex AlFe compounds. It is possible that the presence of iron stimulates oxidation through an exothermic reaction:

$$3Fe_3O_4 + 8Al \rightarrow 9Fe + 4Al_2O_3$$

However, tests with thermocouples embedded in wear pins in all cases indicated that the temperatures were limited to a maximum of 150°C in the oxidative regime and 250°C in the metallic regime.

Electron diffraction confirmed the presence of AlFe and also $\alpha - Al_2O_3$ and in view of this evidence it is very difficult to accept the view of those[7] who try to discuss oxidative wear on the basis of static oxidation theory. The ploughing and consolidation mechanism observed by Stott et al.[8] would appear to be more appropriate to these observations. With the analytical techniques available during this investigation it was not possible to examine the precise role of copper in oxidative wear and it is difficult on the basis of information in the literature to speculate on this matter. However, it may well be that local enrichment of copper to the surface as well as *in situ* ageing processes could strengthen the complex surface layer and also increase its cohesion to provide a higher transition load.

Examination of metallic wear shows that it is controlled by surface deformation, fracture and metal transfer without any significant oxidation occurring. Once metallic wear begins, the process operates too quickly for any regeneration of the oxide to play a significant part. The actual wear rates are 1) oxidative wear rate 10^{-8} to 10^{-7} $cm^3 .cm^{-1}$, and 2) metallic wear rate 10^{-5} to 10^{-4} $cm^3 .cm^{-1}$.

A particular feature noted during the examination of the metallography of surface layers exhibiting oxidative and metallic wear was that irrespective of the original size of the silicon particles they were all broken up to about the same size (2 to 5μm) (Fig. 8). This observation explains why the original particle size has no effect on wear behavior, it is the surface characteristics produced dynamically during wear that conditions the subsequent behavior of the metal surface. Any increase in the hardness of the steel counterface has the effect of reducing adhesion, the oxidative wear regime does not break down as readily and the transition load is markedly increased. The presence of graphite in the cast iron counterface prevents adhesion of aluminum by acting as a solid lubricant and the metal transfer that would otherwise occur is inhibited.

Figure 9 shows the plate shape of the debris from oxidative wear and this can be directly compared to Fig. 7 which shows its separation from the wear surface.

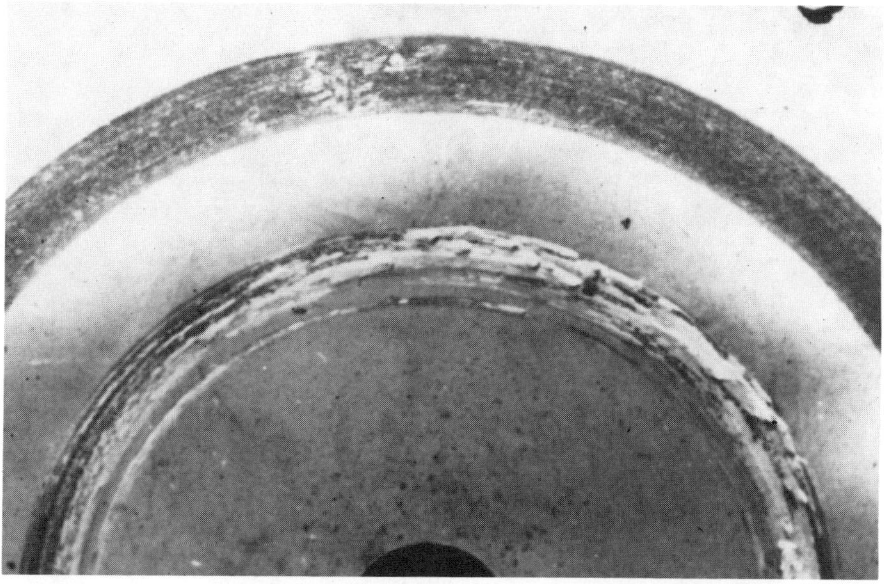

Fig. 5. Oxidative wear track (outer) and metallic wear track with transfer particles (inner). (2X)

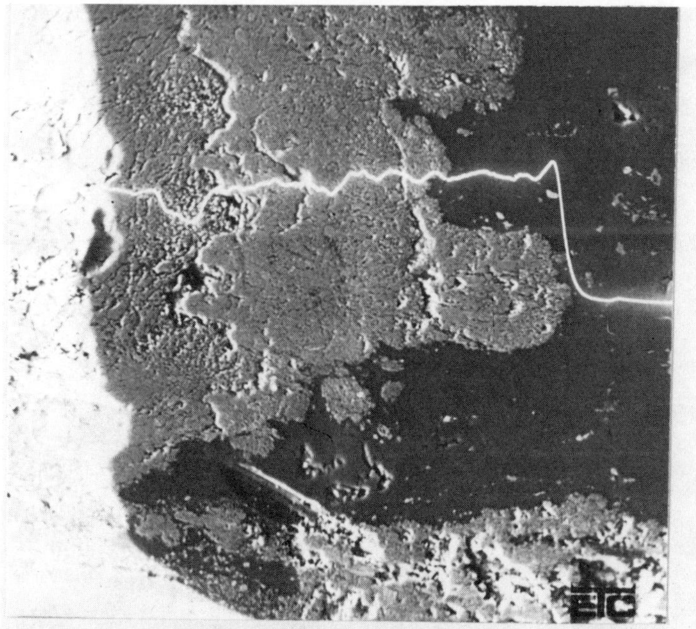

Fig. 6. Taper section showing layered build up of oxide in wear grooves. EPMA line trace for iron showing higher concentration in the surface layer. (300X).

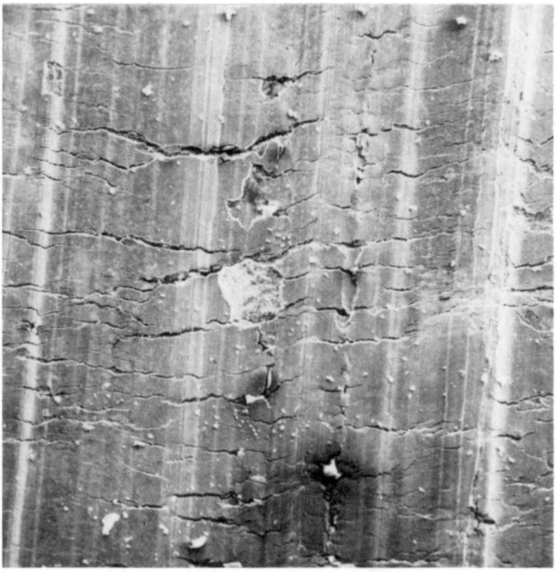

(a)

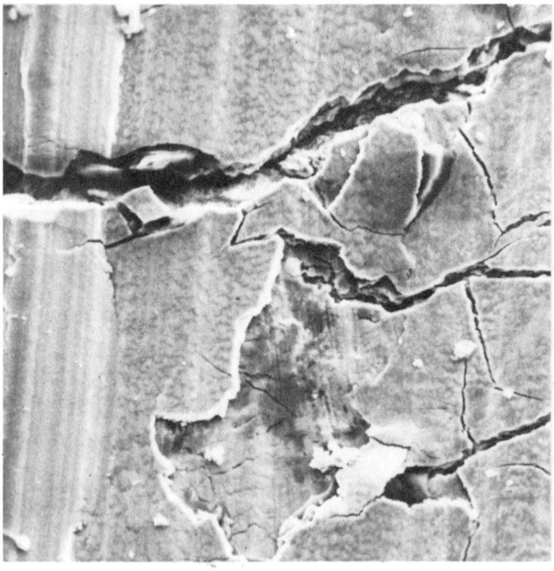

(b)

Fig. 7. Surface of Al/16% Si alloy after wear at 1 kg, viewed in the SEM. (a) 200X, (b) 1000X.

Fig. 8. Taper section through Al/16% Si alloy after wear at 1kg, showing break up of Si particles. (200X).

DISCUSSION

This investigation illustrates how important it is to appreciate that wear behavior is affected by a large number of factors, some of the more important of which are applied load, sliding speed and the counterface material. The first two factors are related to the temperature developed during rubbing which is then largely instrumental in influencing the nature of the protective films formed and the deformation of the substrate metal. The counterface material is mainly related to its susceptibility to promote adhesion with aluminum and it has been shown that both the hardness and composition are important. Wherever possible, aluminum alloys should be used against a graphitic cast iron surface rather than steel. If steel must be used for other reasons it should be used in the hardened condition.

All of the alloys examined exhibit the same pattern of behavior, oxidative wear breaking down to metallic wear with an increase in wear rate by at least an order of magnitude. The load at which the breakdown occurs (transition load) is controlled by silicon content and not by the silicon particle size. Copper has a marked effect on increasing the transition load, however, increasing it beyond 1 percent is not advantageous. The condition of the copper containing alloys does not influence the transition load.

The transfer film formed during oxidative wear is very complex and both AlFe and αAl_2O_3 have been found on the aluminum alloy wear surface. It may be very misleading to think in terms of static oxidation behavior. The observations from metallographic examination indicate the layered build up of the surface film by a compaction process. Oxidative wear proceeds by tensile fracture and spalling of a plate like debris and metallic wear by plastic deformation and fracture followed by transfer of the aluminum alloy to the counterface.

This investigation demonstrates that there is a clear relationship between the characteristics of the debris, the surfaces from which it has separated and thus the wear

(a) (b) (c)

Fig. 9. Debris produced in oxidative wear (a) EPMA for Al (b) and Fe (c). (200X).

mechanism producing it. This suggests therefore that it will be possible to produce a "Wear Atlas" for given combinations of materials.

There is no relationship between the mechanical properties and wear behavior of these alloys. At the present time our knowledge of surface properties is insufficient to explain all of the results obtained in this investigation. However, the overall result, is not too disappointing particularly if this work stimulates further metallurgical studies of wear behavior of both these and other materials of construction. Above all else there is an opportunity to design and produce surfaces which may be supported by cheaper and more readily available bulk materials.

ACKNOWLEDGEMENTS

A number of investigators working in the author's laboratory have contributed to the results reported here and these include Li Fah Wong, F. Chadorshabchi and S. Molina. Some of this work was supported by the Science Research Council. British Aluminum Co. Ltd. supplied some of the metals used.

REFERENCES

1. Owen W.S., "Can a simple heat treatment save Detroit?", *Mets. Tech,* 7, 1, 1–13 (1979).
2. Nadal J., Eyre T.S., "Cylinder liner wear in low speed diesel engines," *Trib. Int.* 11, 5, 267–271 (1978).
3. Beesely C., Eyre T.S., "Friction and wear of aluminum alloys containing copper and zinc," *Trib. Int.,* 9, 2, 63–69, (1976).
4. Shivanath, R., Sengupta P.K., Eyre T.S., "Wear of aluminum – silicon alloys", *Brit. Foundryman,* 70, 12, 349–356 (1977).
5. Clegg, A.J., Das A.A., "The influence of structural modification on the wear resistance of a hypereutectic aluminum – silicon alloy", *Brit. Foundryman,* 70, 11, 333–339 (1977).
6. Sarkar A.D., "Wear of aluminum alloys", *Wear,* 31, 331–337 (1975).
7. Quinn T.F.J., "The effect of "hot spot" temperatures on the unlubricated wear of steel", *ASLE Trans.,* 10, 158–168 (1967).
8. Stott F.H., Wood G.C., "Influence of oxide on friction and wear of alloys", *Trib. Int.,* 11, 211–218, (1978).

Stevens, Vander Voort, McCall, eds. Microstructural Science, Volume 8

EFFECT OF PARTICLES ON THE TENSILE FRACTURE OF ALUMINUM ALLOYS

L.D. Kenny*, Larry R. Morris* and M. Ryvola*

ABSTRACT

The fracture behavior in tension of aluminum alloys containing micron sized intermetallic particles was studied as a function of particle volume fraction, particle spacing and solute in the aluminum matrix. The fracture mode changed from "chisel-point" to "double cup" to "cup and cone" as the particle density and solute content increased. The ratio of dimple diameter to particle spacing remained constant but the voids nucleated on particles before fracture account for only a small fraction of the dimples found on the fractured surface. Ductile fracture was initiated by the joining of only a small number of pre-existing voids in the center of the neck and the fracture spreads from this one site by the progressive nucleation of new voids immediately ahead of the crack front.

INTRODUCTION

After the point of plastic instability in a tensile test, strain is progressively localized due to the formation of a neck. In the necked region a second, but microscopic, localization of flow leads to fracture. Fracture is influenced by a wide variety of factors but in ductile materials the nucleation of voids on particles and void growth are usually considered critical. The mechanisms leading to ductile fracture have been examined in a number of alloys, including aluminum[1,2], steel[3,4], copper[5,6] and brass[7].

Three types or classes of fracture have been seen in F.C.C. metals[8,9]:

1. *Chisel-point fracture* is produced by a continuing shear deformation in metals of high purity. The regions on the sides of the chisel point consist of a highly slipped specimen surface rather than a genuine fracture surface.
2. *Double cup fracture* occurs in alloys containing a low volume fraction of particles. The central cavity produced by void growth and coalescence expands toward the specimen surface probably by a mechanism described by Rogers as "ductile cutting".
3. *Cup and cone fracture* occurs in alloys containing larger amounts of particles and matrix solute. The void growth and coalescence at the center of the neck leads to the formation of a central cavity or ductile crack. As the central cavity grows towards the surface, deformation is concentrated at the rim into zones of high shear strain at 45° to the tensile axis leading to shear fracture.

*Aluminum Company of Canada, Ltd., Research Centre, Kingston, Ontario, Canada.

The fracture strain has been related to microstructural features such as volume fraction of particles and particle spacing in both copper and steel alloys[3,6]. However, very little data are available in aluminum base alloys even though all commercial aluminum alloys contain various amounts of insoluble intermetallic particles. These intermetallic compounds are harder than the aluminum matrix, usually iron-rich, are about a micron in diameter, and originate from eutectic transformations during ingot solidification.

In this paper we examine the effect of the volume fraction of intermetallic particles on fracture strain and fracture morphology in two types of aluminum alloys.

EXPERIMENTS

The materials used and some relevant properties are listed in Table 1. Two basic types of alloy were used: high purity aluminum containing $FeAl_6$ or $NiAl_3$ particles, and aluminum plus 4% magnesium containing $FeAl_3$ particles. The magnesium, which remains in solid solution in the aluminum matrix, was added to increase the flow stress and work hardening rate. Less than 0.01% Fe or Ni remain in solid solution.

Test specimens were prepared by extruding 60 mm diameter unidirectionally solidified ingots, at 500°C, into 6.35 mm diameter rods. The extruded rods were, than annealed at 300°C. This technique produced specimens that contained a uniform dispersion of almost spherical particles in an annealed aluminum matrix, as is shown in Fig. 1.

The 6.35 mm diameter rods were pulled in tension at room temperature at a cross-head speed of 2.5mm/min. In many cases the test specimens were pulled to various stages prior to fracture and examined metallographically.

TABLE 1

Experimental Materials

Alloy (wt.%)	Particle Phase	Vol. Fract. of Particles (%)	Particle Dia. (μm)	Particle Spacing (μm)	0.2% Yield Stress (mPa)	Flow Stress at = 0.1 (mPa)	n Value
S.P. Al Base	-	0	-	-	16	42	.28
-0.15 Ni	$NiAl_3$	0.25	1.0	8.25	23	62	.25
-0.55 Ni	$NiAl_3$	0.90	0.69	3.71	24	71	.27
-1.02 Ni	$NiAl_3$	1.67	0.46	1.99	26	77	.25
-2.07 Ni	$NiAl_3$	3.41	0.41	1.40	56	117	.21
-4.12 Ni	$NiAl_3$	6.90	0.42	1.15	67	140	.20
-5.95 Ni	$NiAl_3$	10.10	0.30	0.69	91	189	.18
-0.48 Fe	$FeAl_6$	1.54	0.45	2.43	20	77	.28
-1.89 Fe	$FeAl_6$	5.86	0.38	1.09	73	143	.22
4 % Mg Base	-	0	-	-	76	201	.39
-0.13 Fe	$FeAl_3$	0.23	1.50	12.61	86	223	.38
-0.38 Fe	$FeAl_3$	0.67	1.45	8.59	87	239	.37
-0.90 Fe	$FeAl_3$	1.58	1.42	6.30	95	259	.36
-1.34 Fe	$FeAl_3$	2.36	1.20	4.66	94	275	.35
-2.13 Fe	$FeAl_3$	3.75	1.42	4.75	96	281	.35
-2.60 Fe	$FeAl_3$	4.58	1.50	4.66	103	279	.32

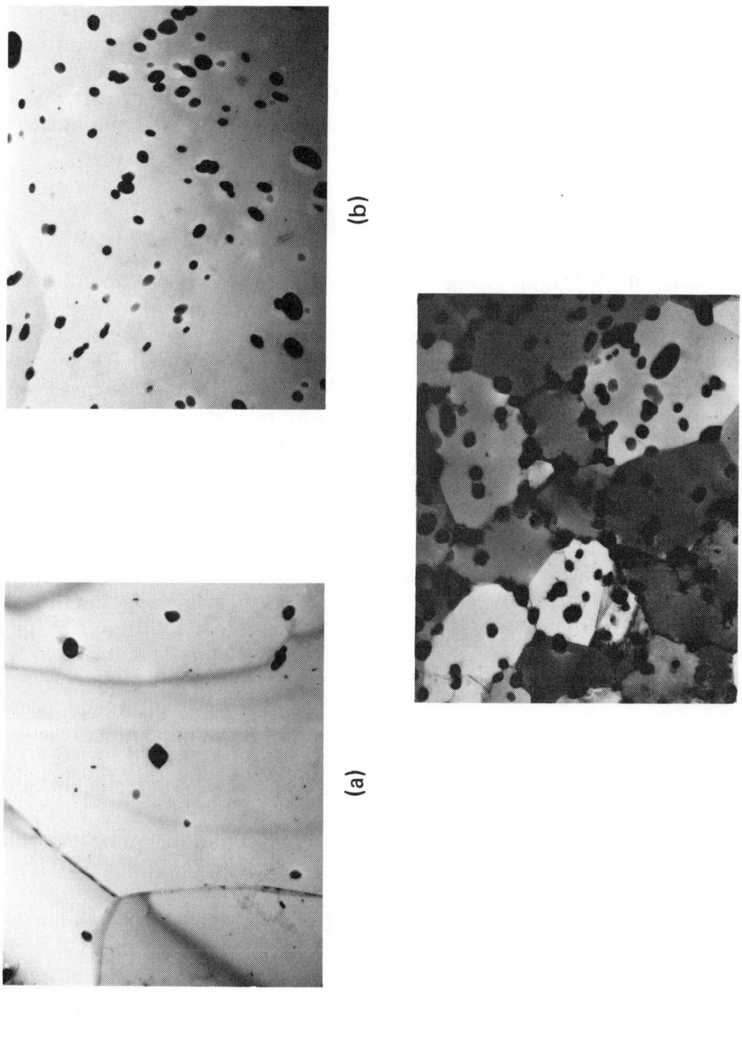

Fig. 1. Transmission electron micrographs from foils taken transverse to the direction of extrusion showing typical $NiAl_3$ particle dispersions. a) 0.25 volume % $NiAl_3$. b) 3.41 volume % $NiAl_3$. c) 10.10 volume % $NiAl_3$. 9200X.

Fracture strains were calculated as:

$\epsilon_f = 1n \frac{A_o}{A_f}$, where: "A_o" is the original cross-sectional area and "A_f" the average fracture area.

Particle spacings were determined by counting particles on polished sections and defining the spacing by the relation $L = N_p^{-1/2}$, where "L" is the particle spacing and N_p the number of particles per unit area. Dimple diameters were obtained by counting the number of dimples or voids on SEM fractographs and defining the average diameter as $D = N_D^{-1/2}$, where N_D is the number of dimples per unit area in the centre of the fracture.

Particle phases were identified by X-ray diffraction of extracted residues and particle volume fractions were calculated using the chemical analysis and known phase densities.

RESULTS

Void and crack formation at various stages of neck development are shown in Figs. 2 and 3 for an Al— 1.89% Fe specimen. Voids were always observed initially in a circular area at the center of the neck where the geometrically induced radial stresses were highest[10]. Void growth and linkage lead to the formation of a ductile crack which expands in the radial direction by void link-up, as shown in Figs. 2b and 3b. The fracture process is terminated by shear. Note that if the shear lips were to be fitted together in Fig. 2c a central cavity would still remain.

In Fig. 4, cross-sections of three Al—4% Mg—Fe fractured tensile specimens are shown. These specimens were etched after fracture to delineate areas of high strain using a decorating technique described previously[11]. In Fig. 5, a shear band near the specimen surface is shown in more detail. It is clear from these micrographs that both the average fracture strain and the relative proportions of central crack growth and terminal shear depend upon particle density.

The influence of $FeAl_3$ particles on the fracture type or class is also illustrated in Fig. 4. In the particle free Al—4% Mg alloy the fracture forms a double cup. With the addition of 0.23 vol. % of particles fracture changes to a cup and cone. In the alloys with a high purity aluminum base fracture was of the double cup type up to a particle volume fraction of 1.67%; beyond this concentration fracture changed to cup and cone. The exception to the above was the 99.99% Al specimen which failed by a slipping off process — a chisel-point fracture.

In the pure aluminum base alloys the voids appeared to nucleate at the particle-matrix interface. In the Al-4% Mg base alloys the voids also appeared to be initiated at cracks in the particles. In all cases the void density was one or two orders of magnitude smaller than the particle density. In other words, up to the point of fracture very few particles nucleate a void.

Fractographs of Al-Ni and Al-4% Mg-Fe alloys containing various amounts of particles are shown in Figs. 6 and 7. These areas, near the center of the fracture surface, show an equiaxed dimple structure. In double cup fractures the area near the fracture edge was relatively dimple free; for cup and cone fractures elongated dimples were found having approximately the same spacing as the dimples in the center.

The decrease in dimple diameter with increasing volume fraction of particles, evident in the fractographs, is plotted in Fig. 8. Dimple diameters for both types of alloy fall on the same curve; however, the ratio of dimple diameter to particle spacing, shown in Fig. 9, is higher for the pure Al base series than for the Al-4% Mg base series. Considering the difficulty in making accurate measurements on particle and dimple spacings, the measured ratio is considered to be a constant and does not change with particle volume fraction.

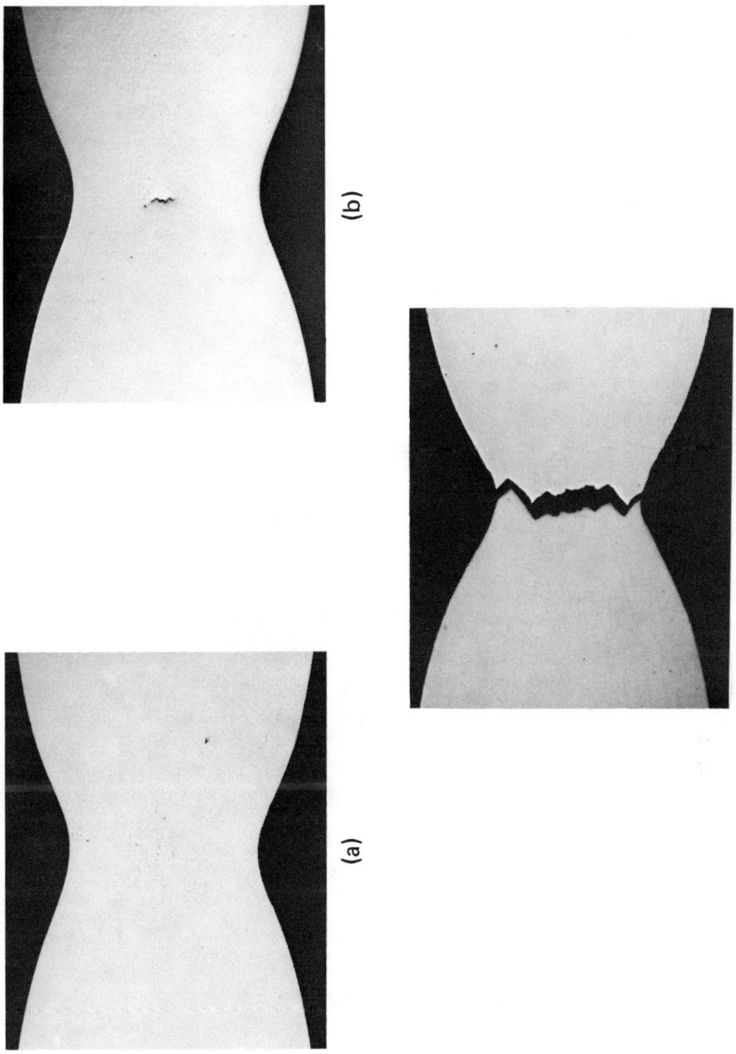

Fig. 2. Longitudinal cross-sections of neck areas showing void and crack formation in an Al-1.89 Fe (5.86 volume % $FeAl_6$) alloy. a) neck strain, $\epsilon_n = 1.39$. b) $\epsilon_n = 1.40$. c) $\epsilon_n = 1.53$. 15X.

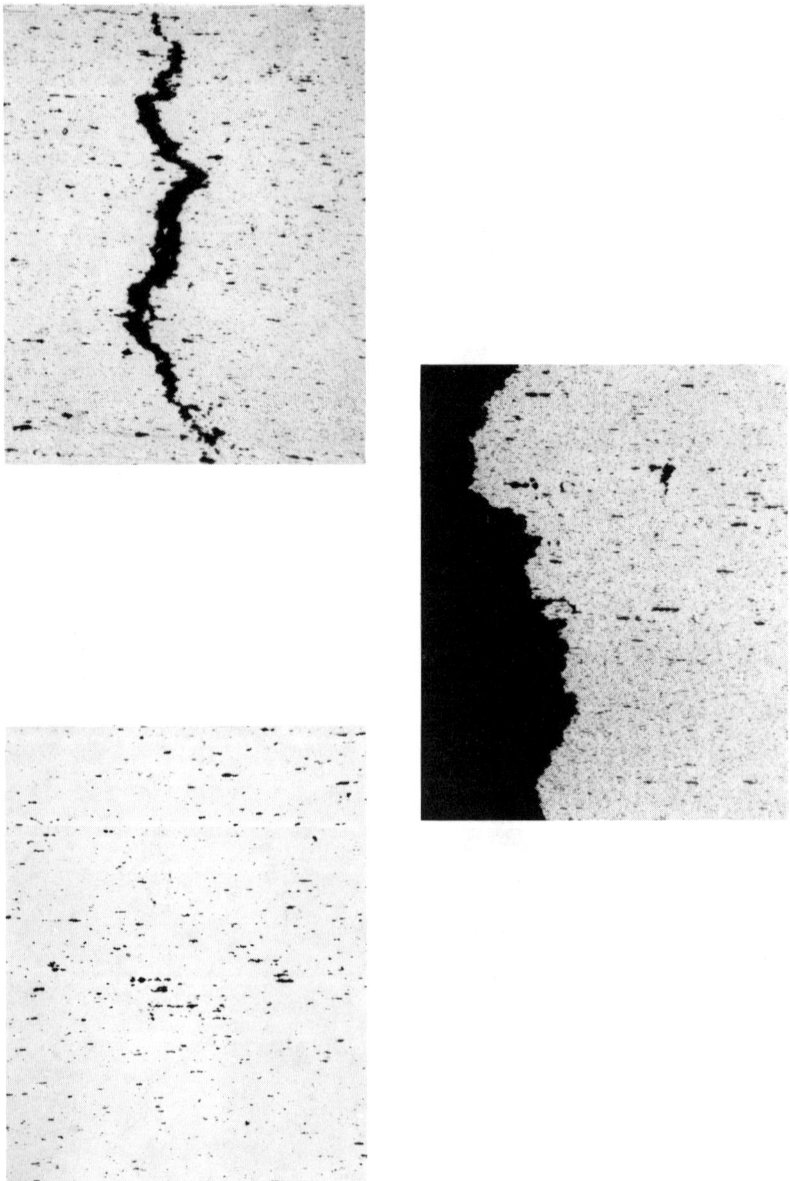

Fig. 3. Micrographs taken at the centre of the neck (or fracture) area corresponding to the samples shown in Fig. 2. 400X

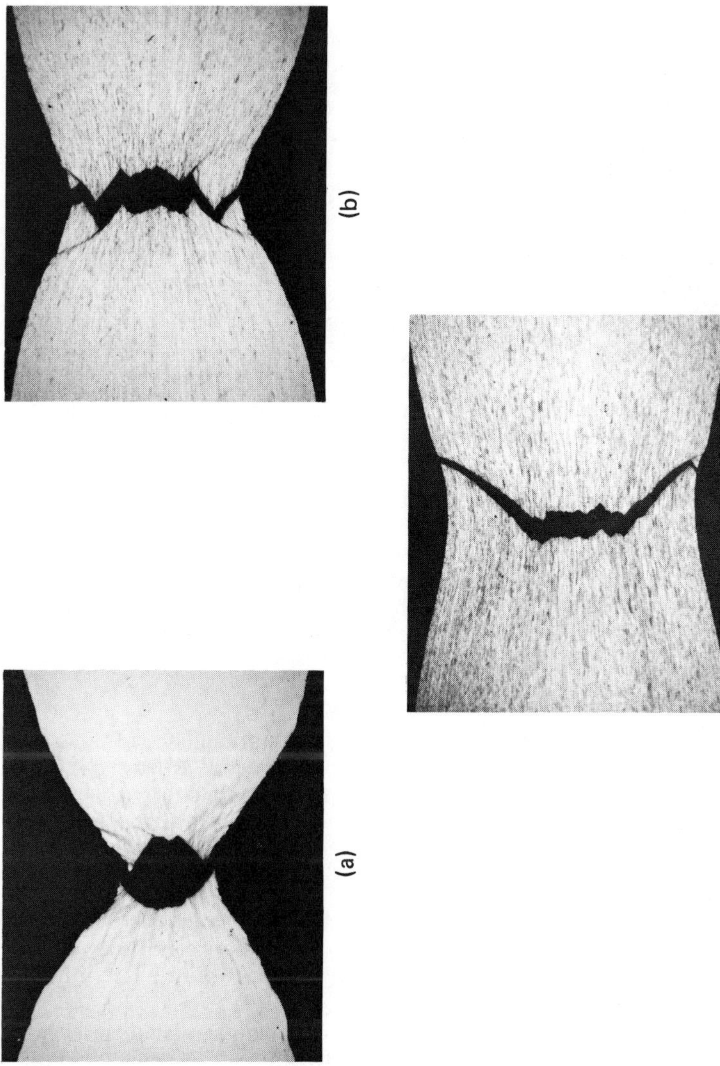

Fig. 4. Longitudinal cross-sections of Al–4% Mg–Fe fractured tensiles containing various volume fractions of $FeAl_3$ particles. a) $V_f = 0\%$. b) $V_f = 0.23\%$. c) $V_f = 2.36\%$. Specimens etched to delineate areas of high strain. 15X.

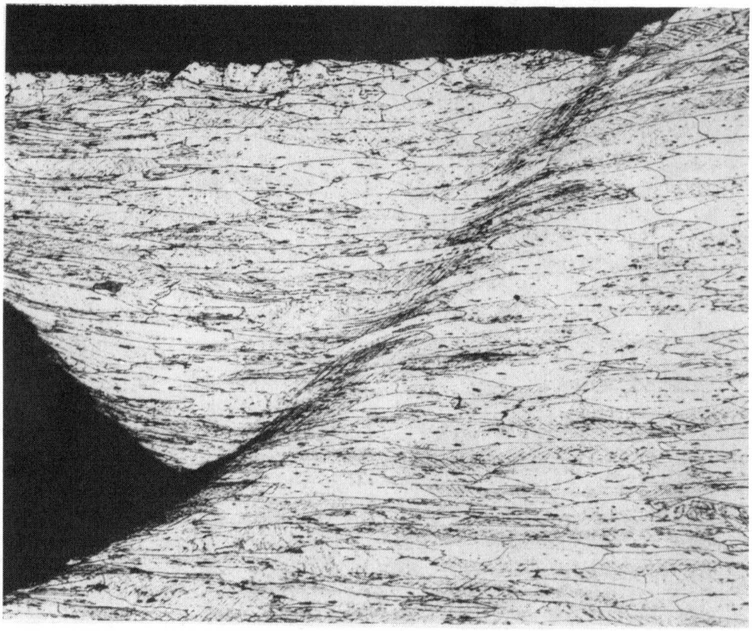

Fig. 5. Crack along a shear band at the outer lip in a Al—4% Mg — 0.13 Fe (V_f = o.23% $FeAl_3$) tensile specimen. 200X.

The influence of particle volume fraction on fracture strain is shown in Fig. 10. The trend shown in the graph — a rapid initial decrease in ductility with increasing particle content followed by a levelling out of the curve is similar to that found previously in spheroidized steel[3] and in sintered copper alloys[6]. It can be seen in Fig. 10 that strengthening the matrix with the solid solution of 4% Mg causes a large decrease in fracture strain for an equivalent volume fraction of particles.

DISCUSSION

The types of ductile fracture found in these aluminum alloys follow the trends observed in other FCC metals. The progression, as expected, is from chisel-point to double-cup to cup and cone as the particle density or solute content increases.

The decrease in dimple diameter with volume fraction of particles is a simple consequence of the fact that the ratio of dimple diameter to particle spacing is constant. The experimental results can be described by the relationship:

$$D = kdV_F^{-1/3}$$

where: D is the dimple diameter,
d is the particle diameter,
V_F is the volume fraction of particles, and,
k is a constant that varies with the fraction of particles that nucleate a void.

The addition of Mg reduced the ratio of dimple diameter to particle spacing, i.e., reduces "k", indicating that the increased flow stress increased the nucleation of voids at particles. The superposition of data points from both types of alloy, Fig. 8, is fortuitous as the decrease in "k" is compensated here by an increase in "d" (see Table 1).

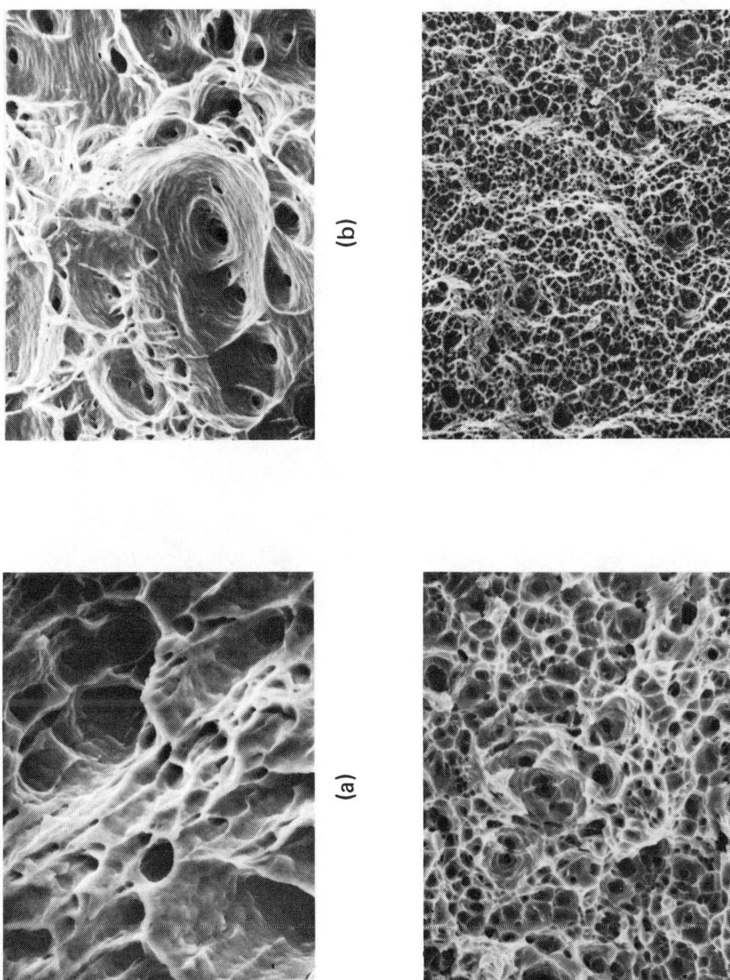

Fig. 6. SEM views of fracture surface for various volume fraction of $NiAl_3$ particles in 99.99% Al matrix. a) $V_f = 0\%$. b) $V_f = 0.25\%$. c) $V_f = 3.41\%$. d) $V_f = 10.10\%$. 640X.

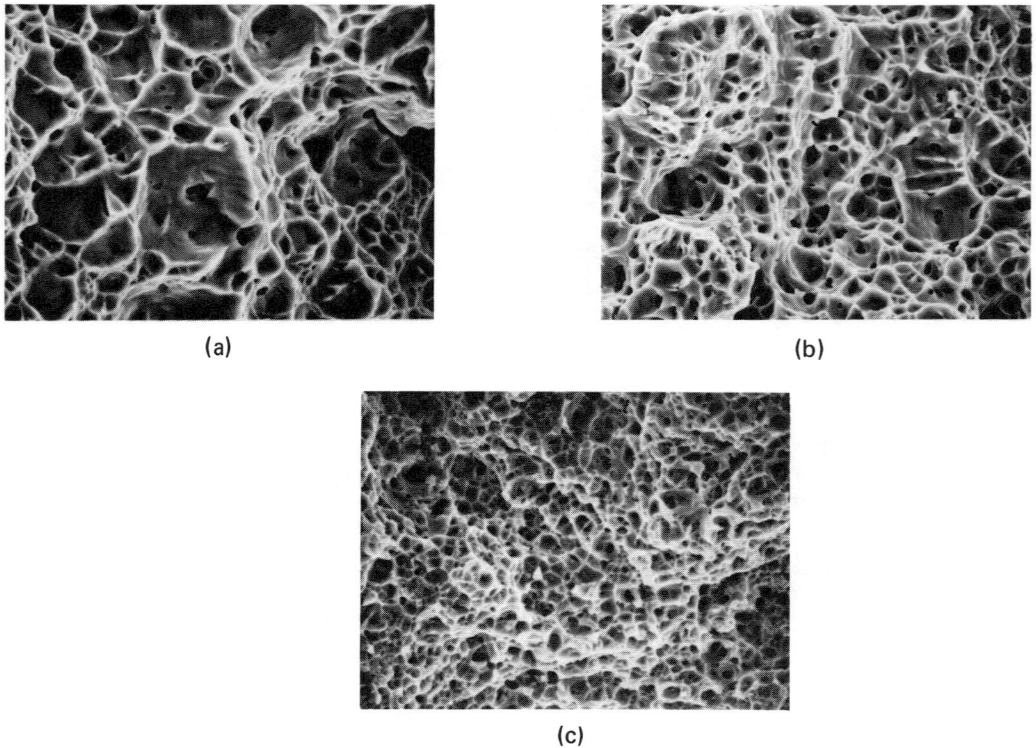

Fig. 7. SEM views of fracture surface for various volume fractions of $FeAl_3$ particles in Al—4% Mg matrix. a) $V_f = 0\%$. b) $V_f = 0.23\%$. c) $V_f = 2.36\%$. 640X.

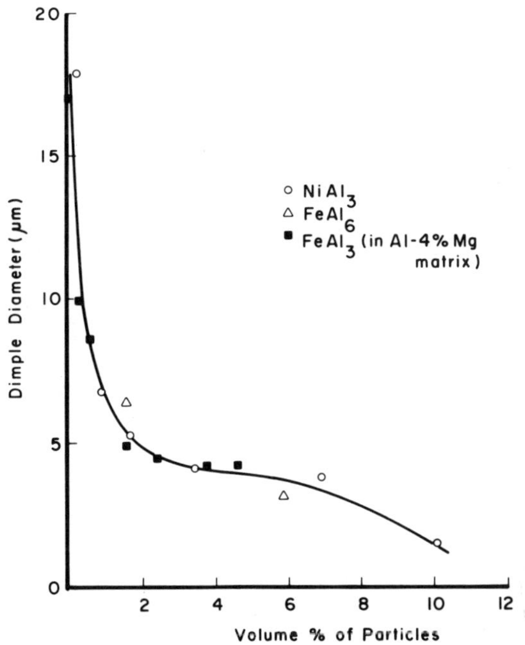

Fig. 8. Effect of volume fraction of particles on dimple diameter.

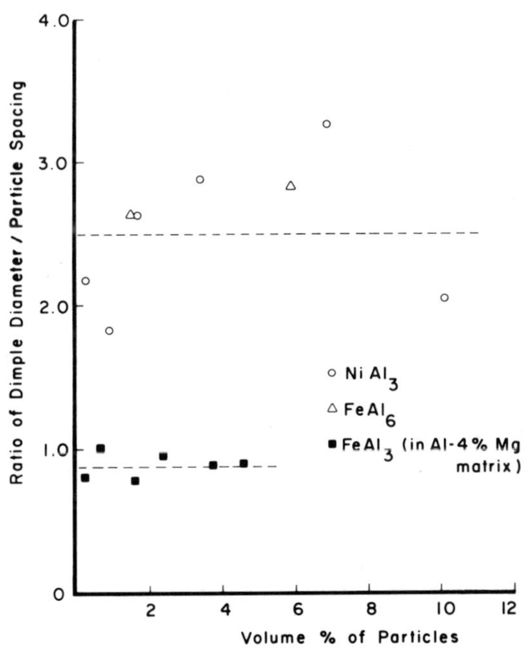

Fig. 9. Effect of volume fraction of particles on ratio of dimple diameter to particle spacing.

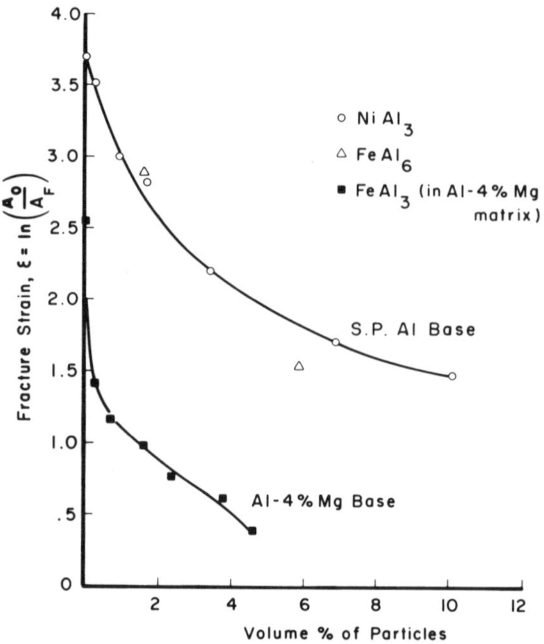

Fig. 10. Effect of volume fraction of particles and matrix composition on fracture strain in 6.35 mm diameter tensile specimens.

The general shape of the fracture strain vs volume fraction curves, Fig. 10, agrees with theoretical predictions [10,12], but the agreement is not good quantitatively — especially for the Al-Mg series of alloys. The reduction in fracture strain on adding Mg cannot be attributed to the increase in the proportion of shear failure since the events that follow the initiation of a central crack do not have a large influence on the measured final fracture area. The reduction in fracture strain must be attributed to the initiation of a central crack at lower strains. This could be due to the higher radial stresses resulting from the higher flow stresses in the Al-Mg alloys or to the non-uniformity of strain as evidenced by the "jerky flow" caused by Mg strain ageing.

The usual model of ductile fracture considers the nucleation, growth, and coalescence of voids in the whole of the cross-section of the neck leading to fracture of the specimen when the voids are all joined by necking or shear. Examination of the fracture surface and of the cross-section in stages up to fracture show clearly that the voids present just before fracture account for only a small fraction of the dimples found on the fracture surface. The central crack appeared to be initiated by the coalescence of only a small number of pre-existing voids — the fractographic evidence suggests 3 or 4 at most — and the ductile crack spreads from this one site by the progressive nucleation of new voids in the stress field immediately ahead of the crack front. A criterion for ductile failure must involve the conditions for the lateral growth of a crack from the center of the neck and not just consider the cooperative junction of holes across the diameter of the neck.

Some of the factors that lead to the initiation of fracture will be presented in a subsequent paper.

ACKNOWLEDGEMENT

To the Director of the Aluminum Company of Canada, Ltd: Research Centre, Kingston, for allowing publication of this work.

REFERENCES

1. G.Y. Chin, W.F. Hosford, Jr., and W.A. Backofen, "Ductile Fracture of Aluminum", *Trans AIME,* 230, pp. 437–449 (1964).
2. J. Gurland and J. Plateau, "The Mechanism of Ductile Rupture of Metals Containing Inclusions", *Trans ASM,* 56, pp. 442–454 (1963).
3. C.T. Liu and J. Gurland, "The Fracture Behaviour of Spheroidized Carbon Steels", *Trans ASM,* 61, pp. 156–167 (1968).
4. J.I. Bluhm and R.J. Morrisey, *Proc. of First Int. Conf. on Fracture,* Senadi, Japanese Society for Strength and Fracture of Materials, 3, pp. 1739–1780 (1966).
5. I.E. French and P.F. Weinrich, "The Shear Mode of Ductile Fracture in Commercial Copper", *Scripta Met.,* 11, pp. 965–968 (1977).
6. B.I. Edelson and W.M. Baldwin, Jr., "The Effect of Second Phases on Mechanical Properties of Alloys", *Trans ASM,* 55, pp. 230–250 (1962).
7. W. Rostoker and S.Y.K. Liu, "The Influence of Porosity on the Ductility of Sintered Brass", *Journal of Materials,* 5 (No. 3), pp. 605–617 (1970).
8. I.E. French and P.R. Weinrich, "The Tensile Fracture Mechanisms of F.C.C. Metals and Alloys – A Review of the Influence of Pressure", *J. Australasian Inst. of Metals,* 22 (No. 1), pp. 40–50 (1977).
9. H.C. Rogers, "The Effect of Material Variables of Ductility", in *Ductility,* ASM, Metals Park, Ohio, pp. 31–56 (1968).
10. F.A.A. McClintock, "A Criterion for Ductile Fracture by the Growth of Holes", *J. Appl. Mech.,* 35, pp. 363–371 (1978).
11. L.R. Morris, L.D. Kenny and M.R. Ryvola, "A Metallographic Examination of the Failure of Thin Sheet During Bending", in *Microstructural Science,* 7, Elsevier, pp. 59–67 (1979).
12. L.M. Brown and J.D. Embury, "The Initiation and Growth of Voids at Second Phase Particles", *Proc. Third Int. Conf. Strength of Metals and Alloys,* Cambridge, England, 1, pp. 164–169 (1973).

Stevens, Vander Voort, McCall, eds. Microstructural Science, Volume 8

METALLOGRAPHY AND KINETICS OF TRANSFORMATION AND PRECIPITATION OF BORON–CONTAINING HSLA (HIGH STRENGTH LOW ALLOY) STEELS

M.R. Krishnadev*, G.J. Sojka*, S.K. Banerji**, and T.B. Cameron***

ABSTRACT

The paper deals with the influence of boron on the transformation substructure and precipitation kinetics of HSLA steels with different microalloying (V, Cb, Ti) and low alloying (Cu) additives. The overall results indicate that (i) the age hardening response of the boron-containing HSLA steels is of the classical type, (ii) highly dislocated transformation substructure can be obtained with boron additions provided cooling rates are adequately controlled and (iii) there is considerable scope for developing very economical low carbon HSLA steels with good strength levels (450 to 700 MPa) with minute additions (0.009 percent) of boron.

INTRODUCTION

Boron is a very powerful hardenability-promoter, and is relatively inexpensive compared to other alloying elements (Mn, Cr, Mo, Ni) which improve hardenability[1]. However, its usage, particularly in HSLA steels, has been very limited. There are two reasons for this: the conflicting data on the effect of boron on toughness and weldability; the lack of understanding of the fundamental metallurgy of boron steels. Hence a project has been undertaken at Laval University to critically examine the influence of boron on the microstructure and mechanical properties of HSLA steels and to develop strong, tough boron HSLA steels for structural applications under rigorous climatic conditions (for example Arctic, Northern Quebec). The paper deals with the preliminary results.

EXPERIMENTAL[2]

The composition of the steels used in the study is shown in Table 1. Low carbon levels have been used to achieve good toughness and weldability. Ti, Zr, and Al are used to stabilize the nitrogen so that boron is free to act as a hardenability-promoter. The steels were cast as 45kg ingots and forged to 100mm square bars. These were soaked for one hour at 1250C, and then hot rolled to 12.7mm thick plates. The temperature of the final pass was around 1020C and the plates were air-cooled after the final pass.

Usually boron steels are used in the quenched and tempered condition; hence, the effect of quench and temper treatment on the microstructure and mechanical proper ties of the hot rolled boron-HSLA steels have been determined. To this end, samples from as-rolled plates have been re-austenitized at 950C for 15 min., water-quenched and then tempered for 1 hour at 600C.

*Department of Metallurgy, Universite Laval, Quebec GIK 7P4, Canada.
**Foote Minerals Company, Exton, Pennsylvania, USA.
***University of Connecticut, Storrs, Connecticut, USA.

Table I — Composition of Boron–HSLA Steels (weight percent)

STEEL	C	Si	Mn	Al	Ti	B	S	V	Cb	Cu	Zr	N	P
	%					ppm	%				ppm		
I	0.053	0.26	1.62	0.023	0.037	10	0.014	---	---	--	50	70	20
II	0.034	0.23	1.50	0.021	0.037	25	0.016	0.087	---	--	60	60	50
III	0.040	0.24	1.53	0.016	0.068	24	0.020	---	---	--	50	80	30
IV	0.035	0.22	1.54	0.017	0.035	9	0.021	---	0.065	--	70	70	50
V	0.039	0.25	1.52	0.018	0.035	9	0.019	---	---	1.79	20	60	40

Balance Fe

RESULTS AND DISCUSSION

Microstructural Studies

As-rolled microstructures.

Figures 1a to 1d show the optical microstructure of the as-rolled samples. It may be noted that additions of higher amounts of B and Ti virtually eliminate pearlite patches (Fig. 1b). With additions of Cb and Cu the microstructure becomes somewhat acicular (Fig. 1c and Fig. 1d) with a reduced tendency for pearlite formation. In the vanadium-containing composition the microstructure was polygonal ferrite with distinct patches of pearlite.

Effect of re-austenitization and cooling at different rates.

Since the cooling rate (after re-austenitization) has a tremendous influence on the hardenability effect of boron, effects of different cooling rates (250C/min – air cooling, 1000C/min – forced air cooling, 5000C/min – oil quench, 18000C/min – water-quench) on the final microstructure have been studied quite extensively. Results indicate that high dislocation density "acicular" or "bainitic" microstructures can be obtained easily in boron steels by controlling the cooling rate and the alloying additives. For example, in the (Cb+B) composition, water-quenching results in an acicular ferrite type of matrix (Fig. 1e) whereas, in (Cu+B) composition on air cooling, a massive ferrite matrix results (Fig. 1f).

Isothermal Transformation Studies

The isothermal transformation behavior of boron-steels also has been studied quite intensively to evaluate the influence of boron on austenite decomposition (including pearlite formation) in the presence of other alloying elements. In general, pearlite formation was retarded by alloying as evident from Fig. 2 which compares the microstructure of the base composition (steel I) after 1 min. of transformation with that of steel III (containing higher B and higher Ti) after 1 hour of transformation. Additions of higher amounts of B and Ti apparently have delayed pearlite formation. Microhardness measurements indicate islands of hard carbides (cementite and/or boron-carbides).

Figure 3 shows the effect of Cb on ferrite nucleation. In the base steel, nucleation is mainly along prior-austenite boundaries (Fig. 3a) whereas, in the (Cb+B) composition, nucleation occurs both along and within the austenite (Fig. 3b). Cb also retards pearlite formation. The effect of Cb depends upon the re-austenitization temperature. At higher temperatures ferrite nucleation is retarded considerably whereas at lower temperatures it is enhanced. Thus by properly selecting the re-austenitization temperature and cooling rate, desired microstructures (fine-grained ferrite, acicular ferrite, bainite, large grained ferrite) to suit the final application may be achieved.

Continuous Cooling Transformation Studies

A systematic and thorough investigation has been undertaken to evaluate the influence of boron on the transformation kinetics (on continuous cooling) of all the experimental HSLA steels because of its practical importance. Results will be discussed only briefly here – it is intended to publish a detailed paper at a later date.

Surprisingly enough, the results indicate that copper has a more powerful effect in delaying the $\gamma \rightarrow \alpha$ transformation of boron-steels than Cb; and, in fact, the effect

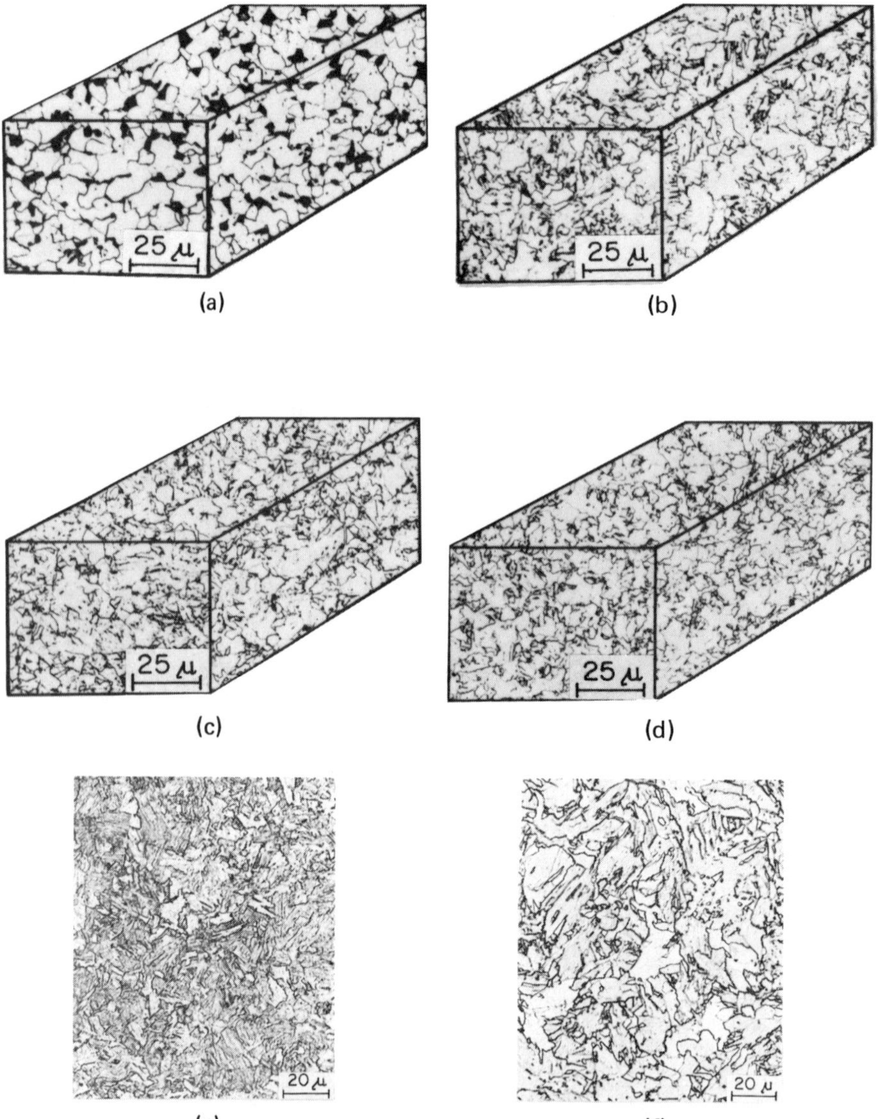

Fig. 1. Optical microstructure of boron-HSLA steels; (a) to (d) as-rolled, a) steel I (base composition), b) steel III (base + B + Ti), c) steel IV (base + Cb), d) steel V (base + Cu), e) steel IV re-austenitized and water-quenched, f) steel V, re-austenitized and air-cooled.

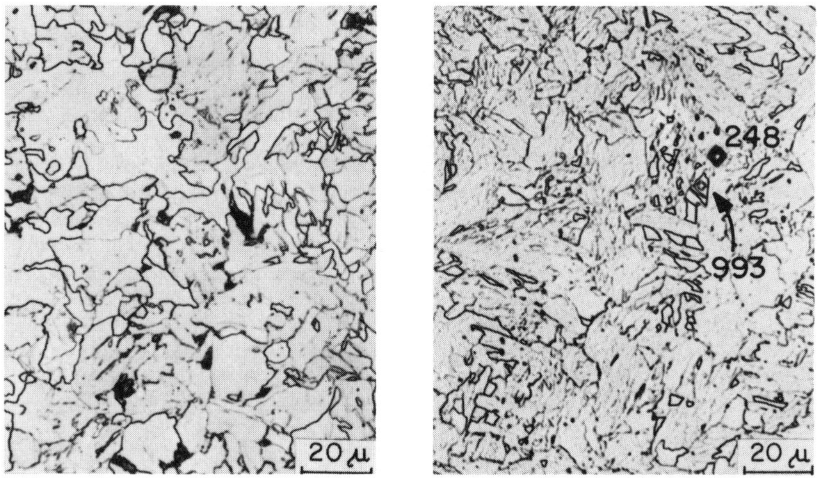

(a) (b)

Fig. 2. Influence of boron additions on isothermal transformation at 600 C (t_{aust} =950 C–5 min. → 600 C–τ → WQ) ; a) steel I (0.0010 %B, 0.037 %Ti), after 1 min. of transformation, b) steel III (0.0024 %B, 0.068 %Ti) after 1 hour of transformation, (HV_{5g}).

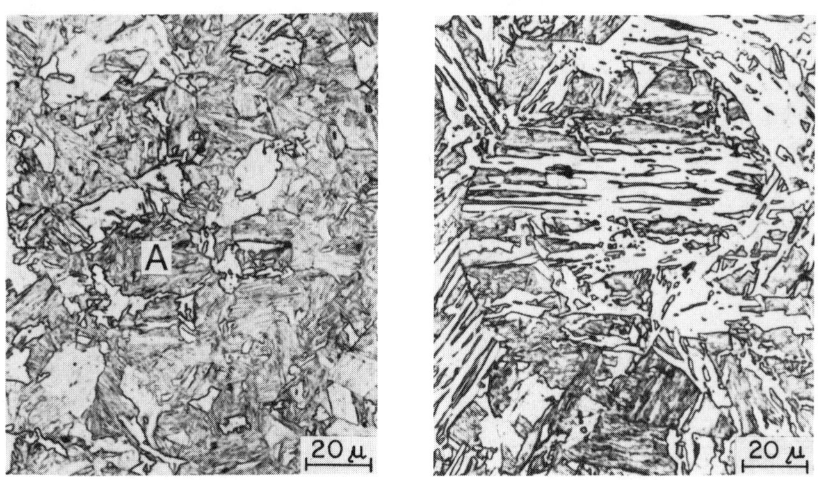

(a) (b)

Fig. 3. Influence of Cb on ferrite nucleation; a) steel I (base), (t_{aust} = 950 C–5 min. → 600 C–5sec. → WQ). b) steel IV (base + Cb), (T_{aust}=1100 C–5 min. → 600 C–10 sec. → WQ).

observed is more than that reported for Mo in boron-steels. For example, for a cooling rate of 19C/min (furnace cooling) transformation range shifted from 799–749C to 689–666C. This indicates that copper lowers the transformation temperature by

110C. This lowering as a result of combined additions of Cu and B is greater than that due to the summation of effects from individual additions. The reasons for this synergistic effect is not very clear. It has been suggested that copper segregates to prior austenite boundaries. This segregation might lower the grain-boundary energy of austenite as Cu can easily replace missing Fe-atoms in the "dis-ordered" grainboundary region since the atomic radius of copper and iron atoms are similar. This would lower the grain boundary energy locally and retard ferrite nucleation. Copper may also have an effect on the growth of segregated boron-carbides. If it delays the growth of these carbides, i.e. delays the formation of an incoherent interface between the carbide and the matrix, ferrite nucleation would also be retarded. Another way of interpreting the synergistic effect may be in terms of the influence of the elements (i.e. Cu and B) on the activation energy for ferrite nucleation. It has been suggested that boron in solution can increase the activation energy for ferrite nucleation which partly depends on the activation energy for carbon diffusion[4]. Probably, when both B and Cu co-exist, the activation energy requirement for ferrite nucleation also is synergistically increased. All these hypotheses need to be experimentally verified.

Aging Kinetics

Figure 4 shows the aging of as-rolled boron HSLA Steels. It is quite evident that curves are of the classical shape with a single peak. It is interesting to note that there is an increase in hardness although the specimens were air-cooled from a rather high finish-rolling temperature of 1020C. This is probably due to the effect of boron in delaying the $\gamma \rightarrow \alpha$ transformation which indirectly increases the solubility of alloying elements as they are more soluble in austenite than in ferrite.

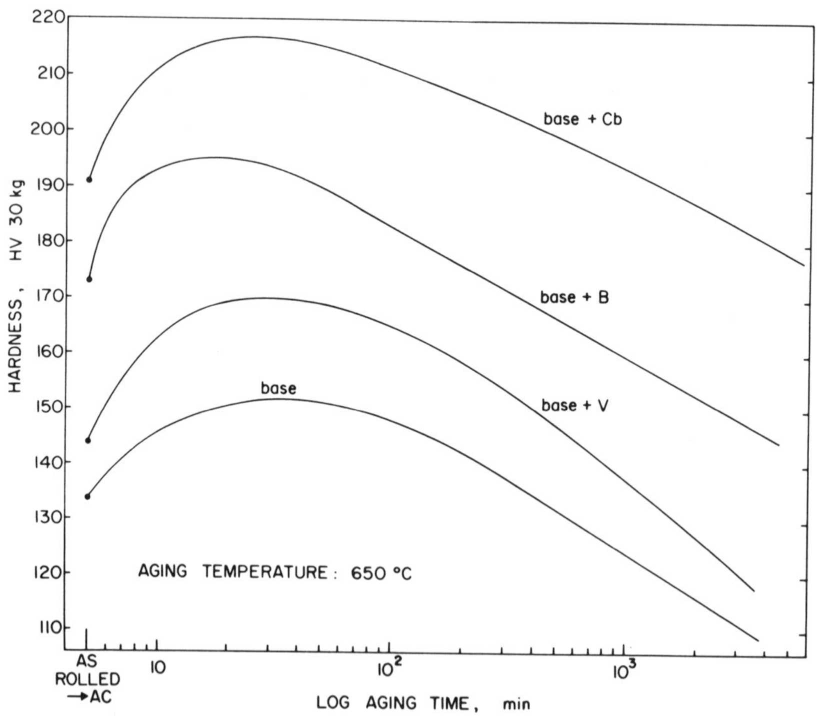

Fig. 4. Isothermal aging kinetics of boron-HSLA steels.

Mechanical Properties

Table II shows typical mechanical properties. Even in a simple composition containing 0.009 percent B and 0.037 percent Ti strength levels of over 600 MPa with transition temperatures lower than –80C can be obtained after a quench and temper treatment. With Cb additions in the as-rolled condition, a yield strength of over 400 MPa with quite a good upper shelf energy may be obtained. This is commercially attractive considering that finish-rolling was carried out above 1000C; hence, load on the rolling mill would be reduced, and a high production rate can be maintained. Another advantage would be that, if higher strengths are needed, parts can be formed in the soft, as-rolled condition and subsequently quenched and tempered to strength levels of over 650 MPa while achieving very good impact properties.

Fractographic Studies

Fracture characteristics of the broken Charpy specimens have been carried out to relate the toughness to microstructural features. For example, "fracture facets" or "unit fracture path" is finer for the quenched condition than for the as-rolled condition (Fig. 5) indicating the greater inherent resistance of the quenched specimens due to their finer "effective grain size"[5]. Figure 6 shows the section through broken Charpy specimens. It clearly illustrates the intrinsic toughness of boron-steels: cracks are transgranular and do not propagate catastrophically. Thus, our work makes it clear that boron per se is not embrittling and that good toughness can be achieved in boron steels by proper heat treatment.

Autoradiography (Boron – Distribution Studies)

Autoradiography has been carried out on selected samples with a view of delineating the distribution of boron as a function of alloying and heat treatment. Figure 7 shows typical results. It may be noted that in the water-quenched condition boron is distributed rather uniformly whereas in the tempered sample boron has started to coalesce and form distinct precipitates. It is worthwhile mentioning here that this sample has excellent toughness (T.T. < 80C). In the literature it has been reported that boron carbides may be detrimental to toughness[4]; but we believe that if boron-distribution is controlled prior to transformation, excellent toughness can be achieved in boron-steels.

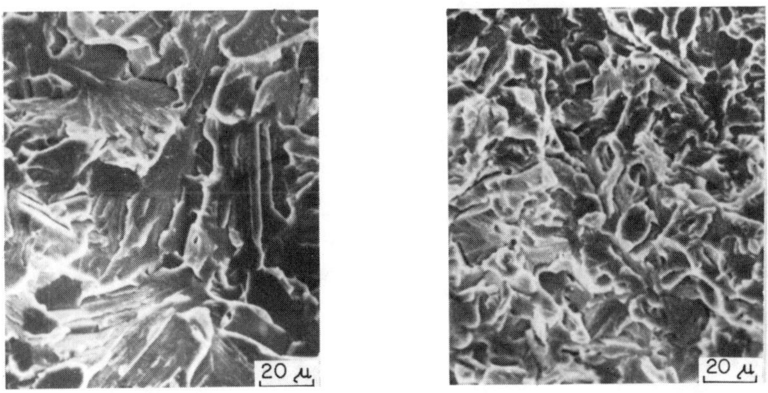

(a) (b)

Fig. 5. SEM fractographs of Charpy specimens of steel I fractured at –196°C; a) as-rolled; b) water-quenched.

Table II — Effect of Cb on the mechanical properties of Boron-Steel

		YS	UTS	RA	EL	IMPACT ENERGY				
						Joules				
		MPa		%		20	-20	-40	-60	-80
						°C				
I A [B]	AS ROLLED	296 x 318 +	431	81	46	>325	>325	16	8	3
	QUENCHED + TEMPERED	619 x 635 +	659	72	21	183	180	145	98	96
IV A [B + Cb]	AS ROLLED	418	563	77	30	141	23	11	7	4
	QUENCHED + TEMPERED	659 x 661 +	695	68	21	138	110	72	60	57

x Lower yield point

\+ Upper yield point

Note: YS = Yield strength
UTS = Ultimate tensile strength
RA = Reduction of Area
EL = Elongation at fracture

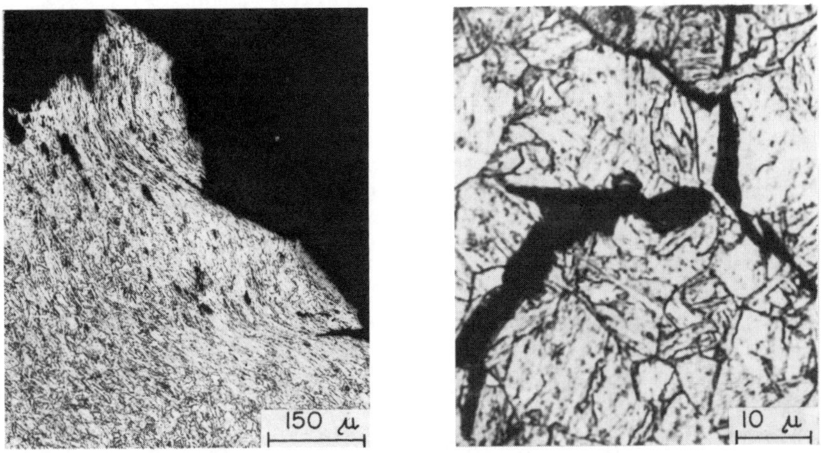

(a) (b)

Fig. 6. Crack morphology in broken Charpy specimens of steel I (quenched and tempered); a) fractured at 17°C; b) fractured at -80°C.

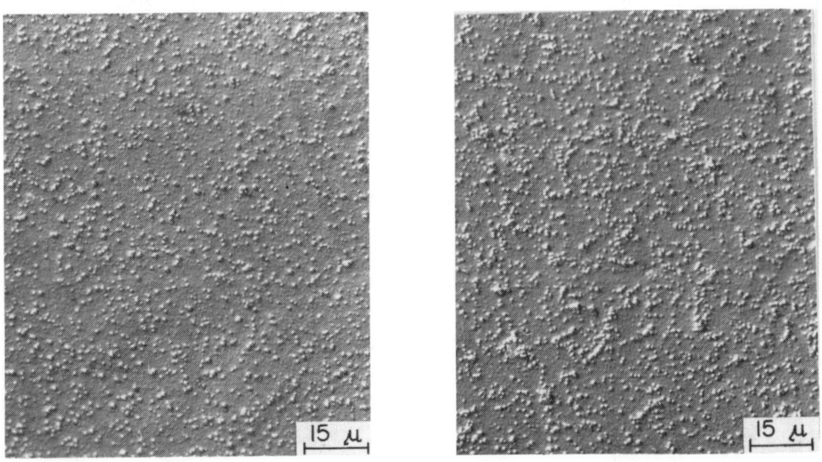

(a) (b)

Fig. 7. Boron distribution in steel I as a function of heat treatment as determined by autoradiography; a) water-quenched; b) quench and tempered (600°C – 1 hr).

SUMMARY

The overall results indicate that (i) highly dislocated transformation substructure can be obtained with boron additions provided cooling rates are adequately controlled, (ii) Cu has a synergistic effect on the transformations characteristics, (iii) age hardening response of boron-containing HSLA steels is of the classical type, and (iv) there is considerable scope for developing very economical low carbon HSLA steels with good strength levels (450 to 650 MPa) and toughness (T.T. <80C) with minute additions (0.009 percent) of boron.

ACKNOWLEDGEMENTS

The authors are grateful to the National Research Council of Canada, Quebec Government, and Foote Mineral Company for the financial support provided. The authors also are grateful to Dr. K.M. Koul for helpful discussions. The work reported here forms a part of the doctoral thesis of Mr. G.J. Sojka.

REFERENCES

1. G.J. Sojka, M.R. Krishnadev and S.K. Banerji, "Boron in HSLA Steels: Potentialities, State of the Art and Prospects" in *Boron Steels,* (to be published by AIME).
2. G.J. Sojka, "Influence of Boron on the Microstructure and Mechanical Properties of HSLA Steels", Ph.D. Thesis, Laval University, (1980).
3. E.A. Wilson, "Copper Managing Steels", *JISI,* Vol. 206, pp. 164–168 (1968).
4. Ph. Maitrepierre, J. Rafes-Vernis, D. Thivellier and R. Tricot, "Influence of $M_{23}(B,C)_6$ Precipitation on the Properties of Low-Alloy Boron-Containing Steels" in "Heat Treatment 76", Metals Society Publication No. 181, London, U.K., pp. 177-188 (1976).
5. A. Galibois, M.R. Krishnadev and A. Dube, "Control of Grain Size and Sub-Structure in Plain Carbon and High Strength Low Alloy Steels – the Problem and the Prospect", *Met. Trans.,* (Aug. 1979) (in press).

Stevens, Vander Voort, McCall, eds. Microstructural Science, Volume 8

THE SUSCEPTIBILITY OF WELDS TO EROSIVE WEAR BY SOLID PARTICLES

A.A. Sagues*, M.K. Keshavan** and G.A. Sargent**

ABSTRACT

As part of a program seeking to evaluate materials for coal conversion systems, the erosive wear of welds relative to that of the base metal has been studied in a 1050 carbon steel. Alumina particles of 240 grit (mean size: 30 μm) were carried by an air stream at a velocity of 60 m/s, impacting the surface of a test weld specimen. The weld and adjacent base areas were exposed to the same total particle load. A specially constructed surface profilometer measured the erosion profile across the specimen. Specimen exposures were carried on at room temperature and at various particle impingement angles (α). Significant differences between the erosion behavior of the welds and that of the base metal were observed. For example, at stream velocities of 60 m/s and without a post-weld-heat-treatment the ratio of erosion rate at the weld to that of the base metal was as high as ~ 1.6 for $\alpha = 90^\circ$ (perpendicular impact). That ratio decreased with α and became close to unity for $\alpha \sim 15^\circ$. This kind of behavior is correlated with weld microstructure, hardness and thermal history. The relationship between microstructure and erosive wear resistance is emphasized.

INTRODUCTION

The erosion of process components has been recognized as one of the most important materials problems standing in the way of reliable coal conversion energy technologies[1]. The phenomenon has been linked to nearly 20% of the component failures reported during the last three years by the Materials Components in Fossil Energy Applications newsletter, which is compiled and published by the U. S. Department of Energy. Examples include the deterioration of pressure letdown valves[2], cyclones[3] and transfer lines[4]. The eroding particles usually originate from sources inherent to the coal technology such as coal minerals, ashes and slags.

The case of weldments is of particular interest since they constitute a preferred joining method in erosion-prone areas of coal conversion systems. An example of unusual erosion behavior at a weld can be seen in ref. [3]. In addition, one of the authors (A.S.) has witnessed preferential erosion patterns in a welded conical component at the Hygas coal gasification plant.

In the past few years a number of investigations have addressed the fundamental problems of erosion. It has been found[5] that particle velocity, flux, time of erosion and angle of impingement all play a role in defining the rate of material removal. Recently, alloy microstructure has been recognized as another significant factor. Sargent et al[6] have reported that the dependence of erosion on the angle of particle impingement is in turn affected by the metallurgical condition of the alloy. That concept has been used in the present investigation to study the erosion of the different microstructural areas found in a weldment.

*Institute for Mining and Minerals Research, Kentucky Center for Energy Research Laboratory, Lexington, Kentucky 40583 USA.
**Department of Metallurgical Engineering and Materials Science, University of Kentucky, Lexington, Kentucky 40506 USA.

PROCEDURE

The starting material was AISI 1050 plain carbon steel, supplied as a 3.2 mm (1/8 inch) thick annealed plate. Rectangular coupons (see Fig. 1) were cut from the plate. The as-received material had a spheroidized pearlitic structure. One of the coupons was normalized for 1 h at 1000°C in order to develop a lamellar pearlitic structure. As-received and normalized coupons were then subjected to a single longitudinal pass with a TIG welding device, using a bead diameter of ~ 3 mm and a nominal speed of 25 mm/minute. The sides of the coupon were clamped to keep the edge temperature at near ambient. Argon gas was used to prevent oxidation. The simulated weld consisted of a fused zone of base metal only, with no filler used. Two specimens were also subjected to a post weld heat treatment (PWHT) by annealing them for 1 h at 600°C in a flowing He atmosphere and then cooling slowly to room temperature. Table 1 shows the four different metallurgical conditions in which specimens were exposed in the erosion apparatus.

Table I

Metallurgical Condition of Welded Coupons

Condition I Weld on spheriodized pearlite base	Condition II Condition I plus PWHT
Condition III Weld on normalized, fine lamellar pearlite base	Condition IV Condition III plus PWHT

The specimens were exposed to a jet of air carrying alumina particles of 240 grit (mean size: 30 μm), at a speed of 60 m/s. The particle flux corresponded to ~ 5 g of alumina per second. The jet came out of a nozzle 6 mm in diameter situated 25 mm away from the area of the coupon where particle impact took place. The particle incidence angle could be set anywhere between 10 and 90 degrees (perpendicular incidence). In order to obtain the planar erosion pattern suitable for profilometric measurements, the specimen was made to move parallel to the plane of the sheet during exposure at a uniform speed of ~ 0.5 mm/s, in a direction perpendicular to that of the weld line. Masked areas near the edges of the coupon remained uneroded as a reference. Figure 2 shows schematically the experimental arrangement, as well as the (exaggerated) appearance of an eroded specimen.

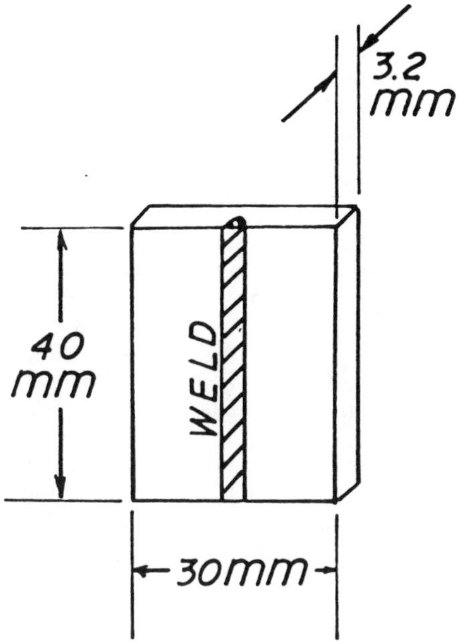

Fig. 1. Dimensions of welded coupons.

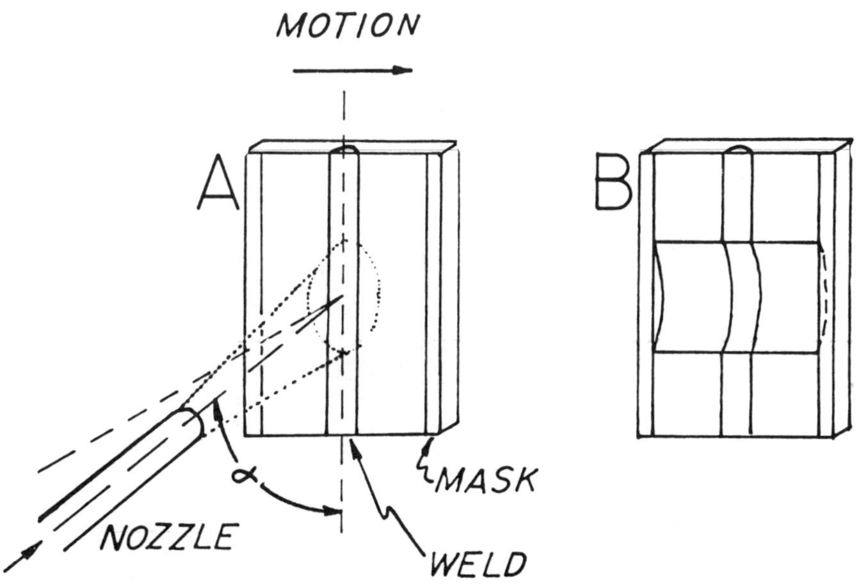

Fig. 2. A: Erosion exposure arrangement. B. Appearance (exaggerated) of eroded coupon.

The surface of the specimen to be eroded was first ground flat to a 600 grit paper finish to eliminate weld irregularities. The depth of material removed during this and subsequent procedures was negligible when compared with the weld depth. The surface profile was then measured along a direction perpendicular to that of the weld pass by means of a profilometer using a sapphire stylus with a 10 μm radius tip, capable of detecting height changes in the 100 nm (1000 Å) range. Figure 3A shows an example of this initial profile. The slight specimen curvature and protuberance of the weld (due to differences in hardness) after the 600 grit abrasion are clearly seen.

After exposure in the erosion apparatus for the time interval indicated above and the desired angle of impingement, the specimen was positioned again in the profilometer and the profile of the area of maximum erosion (corresponding to the region scanned by the center of the jet) was obtained. In order to ensure that the area of maximum erosion was being examined, a series of profiles one millimeter apart were usually obtained and the deepest profile selected. Figure 3B shows the profile measured after exposure of the same specimen as in Fig. 3A. Using the masked areas as references to the profile of the unexposed specimens, the depths of erosion for the base metal and the weld can be easily measured. It is clear that for the experimental conditions used for the specimen of Fig. 3, the weld has eroded at a rate higher than that of the case material. Figure 3G shows the corresponding values of Vickers hardness.

The weight of the specimen was also recorded before and after exposure.

The specimens were usually reground to a 600 grit finish and taken through the same procedure, either repeating the conditions or using a different angle of particle impingement. Selected specimens were examined metallographically. Vickers hardness profiles across some of the welds were also obtained.

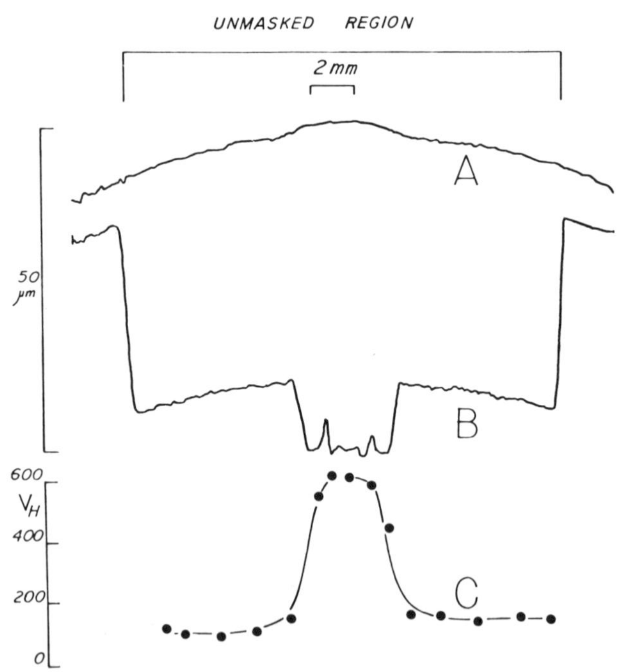

Fig. 3. A: Surface profile of ground specimen (Condition I) before erosion. B: Surface profile of the same specimen after erosion with α=90°. The profile corresponds to the area scanned by the center of the jet. C: Vickers hardness values across the weld region for the same specimen.

RESULTS AND DISCUSSION

Figures 4 and 5 show the microstructures and hardness profiles of the welded specimens used in this investigation. The center of the weld regions exhibited a martensitic structure. The metallographic appearance did not change appreciably after the PWHT; therefore only conditions I and III are represented. The hardness profiles did respond to the PWHT and are shown for all conditions.

The erosion of a material by solid particles is defined as the weight loss of material in time t divided by the weight of particles impinged in time t, with units (g/g). The weight loss in a specimen could be computed by evaluating the erosion crater volume from profilometric measurements and multiplying by the density of the material. Indeed, a preliminary fixed-beam calibration of the erosion unit used here showed an agreement to within 10% between gravimetric and profilometric weight loss determinations. However, since the present study is concerned mainly with the relative erosion of weld and base metal, only the maximum crater depth needed to be measured. The erosion is proportional to that crater depth d for normal particle impingement, or proportional to $d/\sin\ \alpha$ for oblique incidence at an angle α provided that the beam divergence is not large and that a steady erosion state exists. Given the nozzle geometry and erosion times used here, both conditions are easily satisfied to a reasonable degree[6].

Figures 6 and 7 show the magnitude $\delta = d/\sin\alpha$ measured as a function of α and metallurgical condition, for the base metal and also for the center of the weld line. All other parameters such as time of exposure and air stream velocity were kept constant at the values indicated earlier. Each experimental point represents up to three separate measurements, but the same specimen was used for all the measurements within a given metallurgical condition. The δ values show experimental scatter, especially at the lower angles. The profilometric technique is more sensitive and accurate when comparing the relative erosion of weld and base metal, as is exemplified in the lower parts of Figs. 6 and 7. The ratio of δ at the weld to δ at the base metal (δ_W/δ_B) follows well defined trends.

Observations on the results summarized in Figs. 3 to 7 can be listed as follows:

(1) The untreated welds tend to erode at a faster rate than the base metal over most of the angular range investigated. It is important to note that this is so in spite of the welds being several times harder than the base metal. The degree of pearlite spheroidization in the base metal appears to affect little its erosion behavior relative to the weld.

(2) As the impingement angle decreases, the erosion of the untreated welds approaches that of the base metal. At very low angles there is an indication that the weld may be eroding less than the base metal. It is significant to notice in this respect that the abraded ($\alpha = 0$) specimen of Fig. 3A shows actually a protruding weld.

(3) Differences in hardness by as much as a factor of six between the weld and the base metal result in differences in the respective rates of erosion by a factor of not more than 1.6.

(4) The PWHT greatly reduced the differences in erosion behavior between weld and base metal. It must be noted, however, that the PWHT used here (600°C for 1 h) is more severe than would normally be applied in industrial practice.

The present results emphasize the importance of a material's microstructure as a determining factor in its erosion performance. They are also a clear example that an increase in the hardness of a material can actually degrade its resistance to erosion.

The difference in erosion between weld and base metal could be traced to earlier observations[5,7] on the angular dependence of erosion of brittle and ductile materials. A relatively ductile material such as the pearlitic base metal will tend to show a maximum of erosion at an intermediate angle of impingement (typically 25°) and

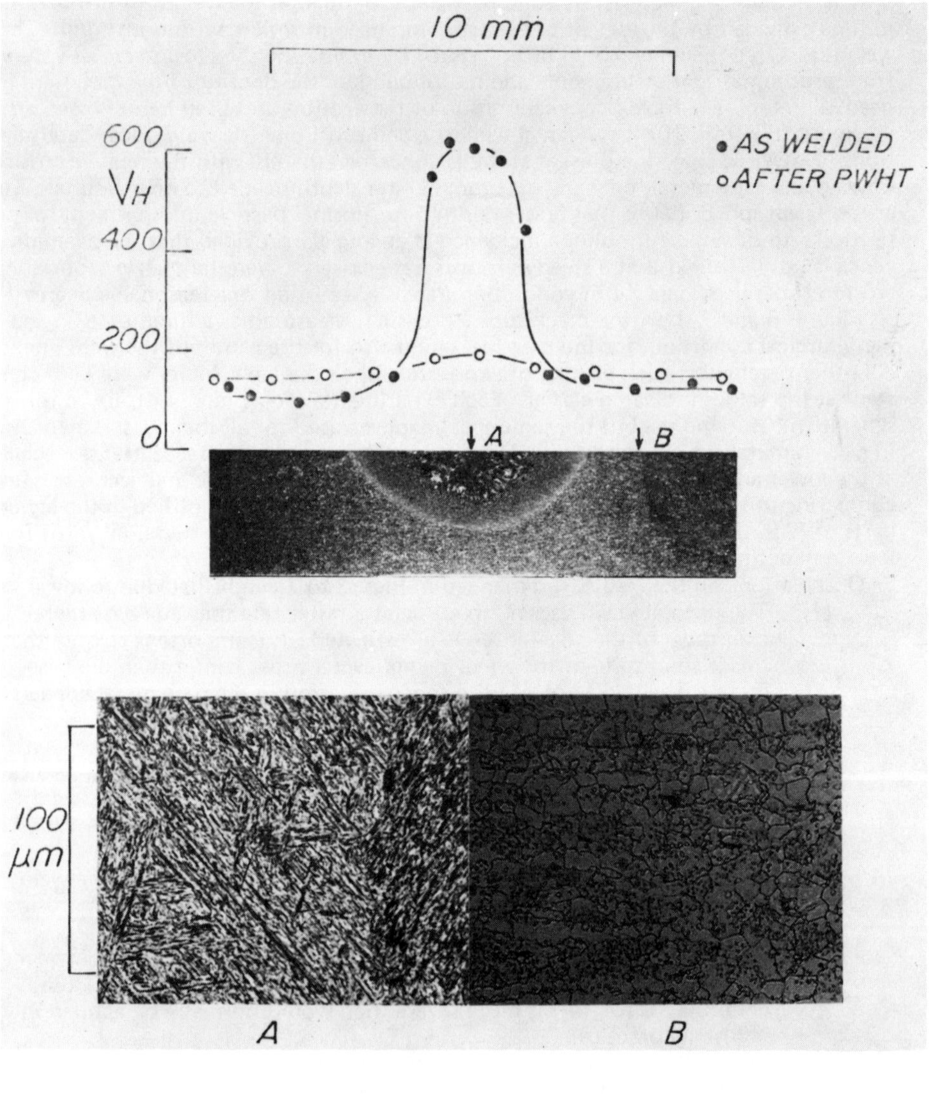

Fig. 4. Microstructure for Condition I. A shows martensite at the weld. B shows the spheroidized pearlitic, fine-grained microstructure of the base. The hardness profile of the weld is also shown. The open circle data represent the hardness profile in Condition II, after PWHT.

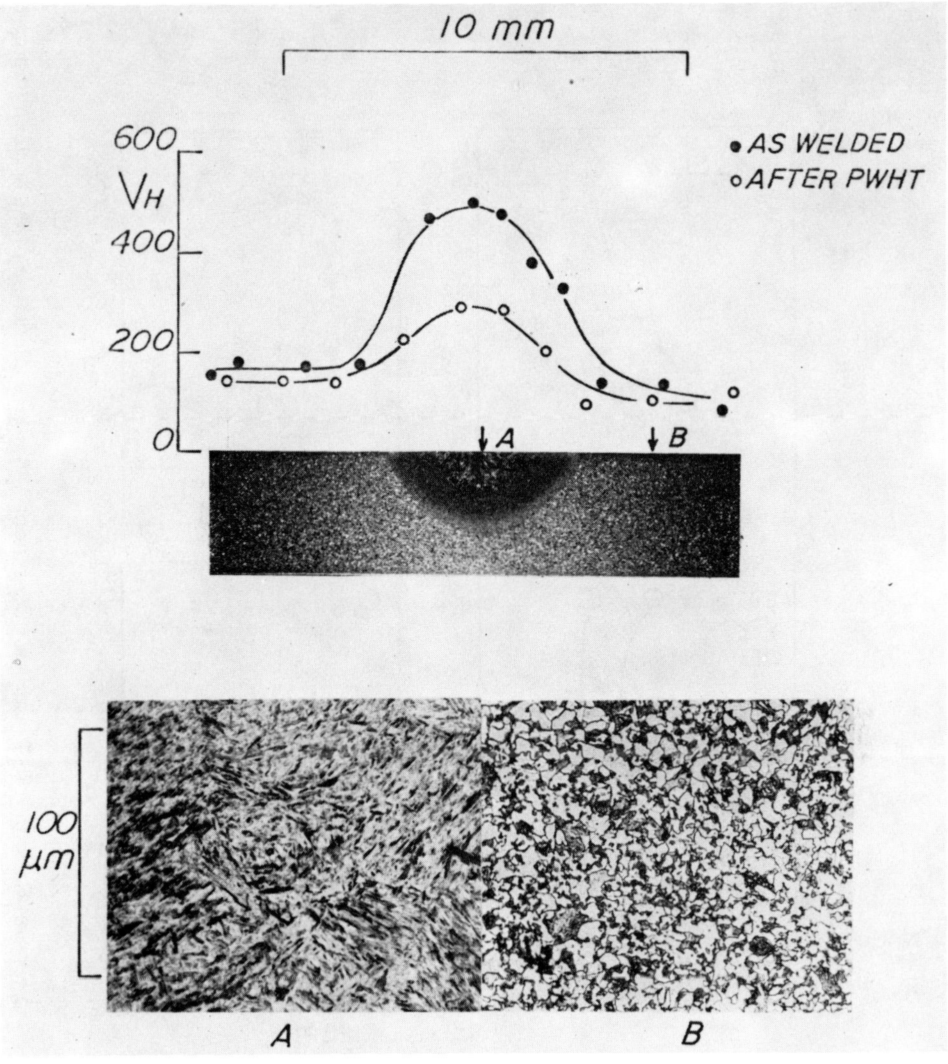

Fig. 5. Microstructure for Condition III. A shows martensite at the weld. B shows the very fine lamellar pearlite and fine-grained microstructure of the base. The hardness profile of the weld is also shown. The open circle data represent the hardness profile in Condition IV, after the PWHT.

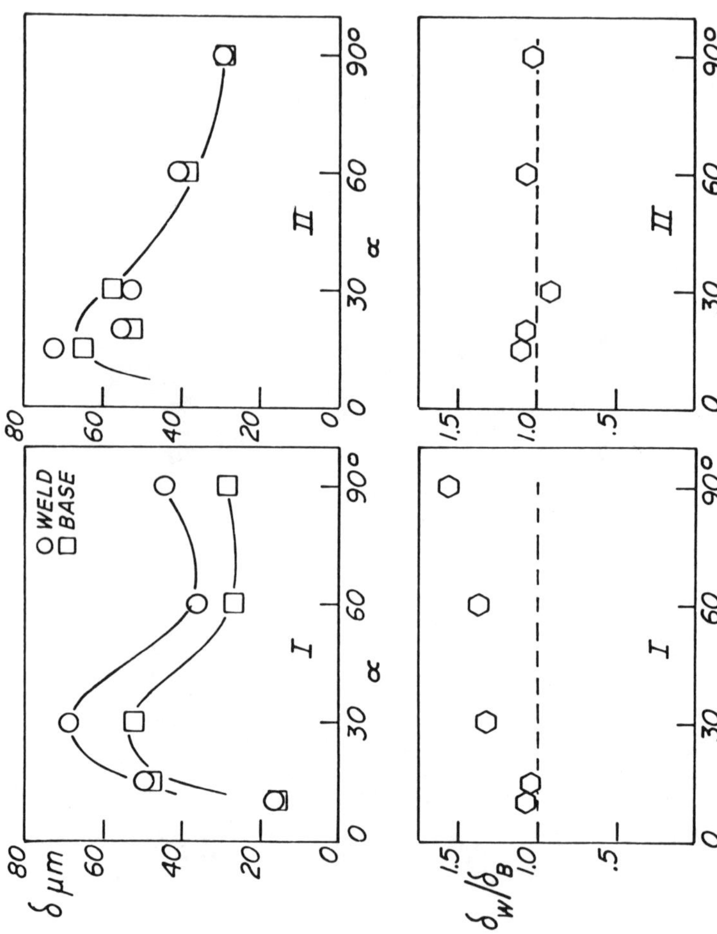

Fig. 6. Erosion (δ) and Erosion Ratio (δ_W/δ_B) results for Conditions I and II, as a function of the angle of impingement *α*.

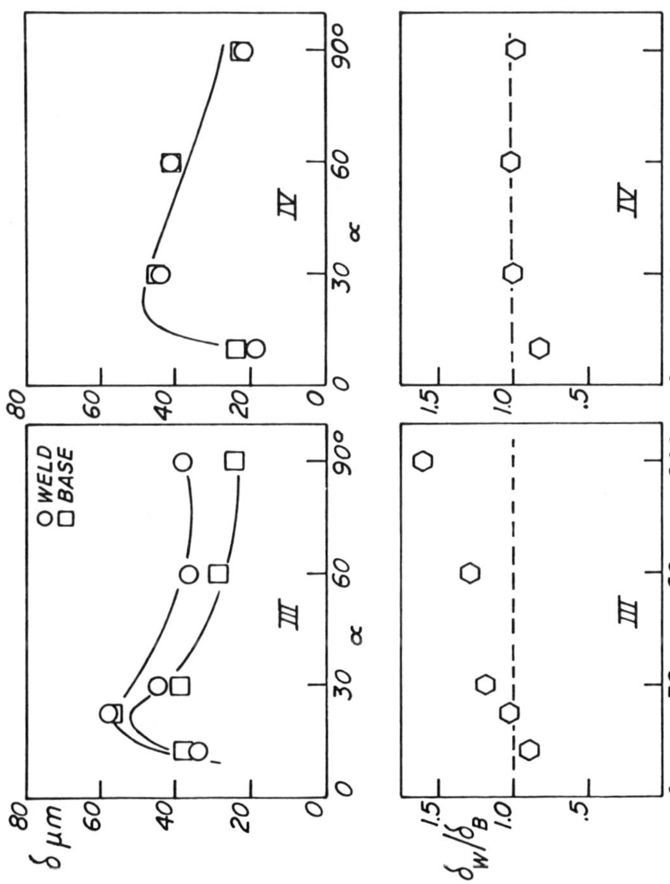

Fig. 7. Erosion (δ) and Erosion Ratio (δ_W/δ_B) results for Conditions III and IV, as a function of the angle of impingement α.

less erosion at 90° (Fig. 8A). Brittle materials such as ceramics or hard martensite have been shown to present erosion maxima at 90° [8] (Fig.8B). When both conditions are present in different parts of the same piece of material a differential behavior of the kind reported here is to be expected: higher erosion of the hard, brittle weld at high impingement angles, and a diminishing or even reverse effect at the lower angles.

The experimental erosion versus angle of impingement curves shown in Figs. 6 and 7 for the welds before PWHT do not follow the extreme behavior of a completely brittle material shown in Fig. 8B, although a moderate increase in erosion is seen near 90°. Similar results have been reported elsewhere[8]. The shape of the curves is probably due to the fact that the martensite presents a certain degree of ductility, since cooling after solidification is not instantaneous. Another contributing factor for the mixed brittle-ductile characteristics has been advanced by Keshavan et al[8]. It is based on the assumption that impinging particles of small size (such as the 240 grit alumina used in this investigation) are more likely to create greater localized stresses in the material upon collision than those resulting from the impact of larger particles. Those larger stresses promoted localized ductility and thus a more ductile erosion behavior would be observed. The fact that curves such as in Fig. 8B result more readily from experiments where larger particles were used supports this interpretation.

The hardness measurements are useful in determining the degree of annealing of the weld. However, as it has been pointed out in a previous paper[6], static measurements of that type do not necessarily afford a conclusive prediction of the erosion performance of a material. As a final observation, it should be noted that, in certain applications, machining of welds in order to obtain smooth flow patterns may not be effective if preferential erosion takes place; an especially thorough PWHT may be necessary.

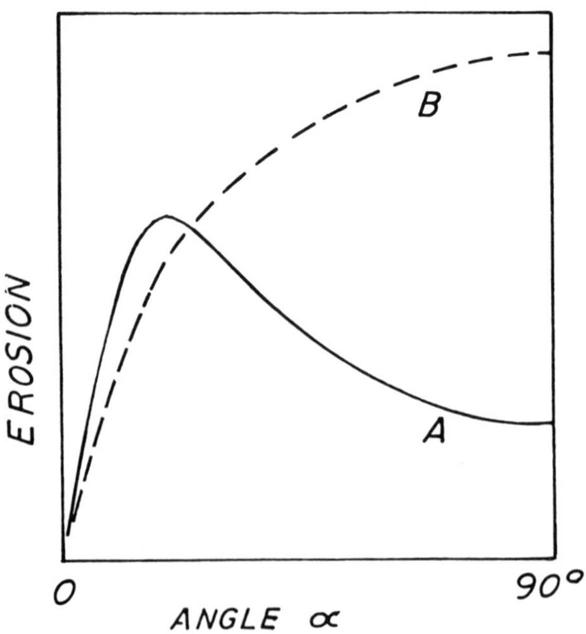

Fig. 8. Erosion versus angle of impingement α curves expected for an ideally ductile material (A) and for an ideally brittle material (B).

CONCLUSIONS

Welds can erode preferentially (as much as 1.6 times faster than the base metal), especially at high particle impingement angles. This appears to be a result of differences between the angular dependence of erosion for brittle and ductile microstructures. Preferential erosion can be reduced by an adequate post-weld heat treatment.

ACKNOWLEDGEMENT

The authors are indebted to Linda Cox for conducting the erosion tests and profilometric measurements.

REFERENCES

1. T.B. Cox and H.E. Frankel, in "Innovations in Materials and Their Applications", *Proc. of the ASM Materials Conf.,* 25–27 Oct. 1977, Chicago, IL. p. 55, (1977).
2. *Materials & Components Newsletter,* No. 22, DOE/ET–0053/22, (June 1, 1979).
3. Materials Technology for Coal-Conversion Processes, 13th Quarterly Report, January-March, 1978, Argonne National Laboratory, ANL–78–54, p. 90, (1978).
4. *Materials & Components Newsletter,* No. 20, DOE/ET–0053/20, (February 1, 1979).
5. I. Finnie, in *Source Book on Wear Control Technologies,* American Society for Metals, p. 220 (1978).
6. G.A. Sargent, M.K. Keshavan and H. Conrad, to be published in *Journal of Metals,* (June, 1979).
7. I. Finnie, *Proc. Soc. Exp. Stress Analysis,* Vol. 17 (No. 2) p. 65 (1960).
8. M.K. Keshavan, G.A. Sargent and H. Conrad, to be published.

STRUCTURE–PROPERTY–BEHAVIOR CORRELATIONS: II, FAILURE ANALYSIS

Chairpersons: C. Bagnall and R.E. Witkowski

Stevens, Vander Voort, McCall, eds. Microstructural Science, Volume 8

GULF OF MEXICO OCEAN THERMAL ENERGY CONVERSION (OTEC) BIOFOULING AND CORROSION EXPERIMENT†

Brenda Little*

ABSTRACT

A field test was undertaken in the Gulf of Mexico to correlate decreased heat transfer efficiency with the characteristics of the biofouling/corrosion film on heat exchanger surfaces of aluminum (5052), titanium (grade 2), copper-nickel (90—10), and stainless steel (AL 6X). Subsurface water was pumped through the heat exchangers at a flow rate of 6 ft/sec for approximately fourteen weeks. The biofouling/corrosion of sections from the sample loops was characterized using the following parameters: inorganic constituents, wet-film thickness, adenosine triphosphate (ATP), and total organic carbon (TOC). Scanning electron microscopy (SEM) coupled with energy-dispersive x-ray spectrometry was used to describe the topographical features and elemental composition of the film. Heat transfer efficiency across the materials was measured using a device developed at Carnegie-Mellon University.

The various metals exhibited dissimilar heat transfer responses and biofouling rates from the onset of the experiment. Early degradation of the heat-transfer efficiency was measured in the copper-nickel and aluminum heat exchangers. Both surfaces were covered with a corrosion layer and the aluminum surfaces were also heavily biofouled. The stainless steel and titanium exhibited a much slower initial loss in heat transfer efficiency, minimal corrosion, and patchy colonization by microorganisms.

INTRODUCTION

The efficiency of OTEC heat exchangers is dependent on the maintenance of a surface with a high heat-transfer coefficient. Materials exposed to natural seawater typically develop a layer of attached organisms that is referred to as biofouling. A biofouling film can contain dead cells and cellular debris, organic secretions, and inorganic precipitates, in addition to viable cells. Metal surfaces in contact with seawater may be further altered by the formation of a corrosion layer. The resulting biofouling and/or corrosion deposits have poorer thermal conductivity than the original metal surface, introducing resistance to heat transfer between seawater and the working fluid.[1] The objective of the Gulf of Mexico OTEC biofouling and corrosion experiment was the same as similar experiments in Hawaii[2], St. Croix, U.S. Virgin Islands[3], and a previous Gulf of Mexico experiment.[4] to characterize the biofouling and corrosion in heat exchangers exposed to flowing oceanic water as a function of time and to correlate these phenomena with the measured decay in heat-transfer efficiency.

A model system incorporating the basic elements of an ocean thermal energy conversion (OTEC) plant was mounted on a National Oceanic and Atmospheric Administration (NOAA) 40-foot discus buoy and deployed in the Gulf of Mexico 160 nautical miles weat of Tampa, Florida (27°33'N, 85°31'W) on 2 December 1978. The flow-through system consisted of four individual heat exchangers, one each of 5052 aluminum, grade 2 titanium, 90—10 copper-nickel, and AL 6X stainless steel, through which water from a 75-foot depth was pumped at a flow rate of 6 ft/sec.

---

†Some aspects of this paper were presented at the 6th Annual OTEC Conference, Washington, D. C., June 19-22, 1979.
*Naval Ocean Research & Development Activity,National Space Technology Laboratory Station, Mississippi 39529 USA.

A fifth heat exchanger experiment developed at The Applied Physics Laboratory (APL) John's Hopkins University was included to provide biofouling data for Alclad aluminum 3004 heat exchanger tubes over which seawater flowed. Those data will be presented in the APL report.[5]

METHODS AND MATERIALS

Flow System

Each of the four flow-through heat-exchange experiments consisted of a 90-inch sample, vertically oriented and designed to contain standing water during sample collections and flow interruption due to pump failures. All pipe samples were degreased with trichloroethylene before installation in the experiment. The 90-inch sample loop was prescored at 9-inch intervals to facilitate sample removal. The sample section was prescored at ½-inch intervals for subsampling. The *in situ* heat transfer surfaces were composed of lengths of tubing identical to those of the sample loops and connected in series to them. The pressure water system consisted of three pumps located below the deck, with their suction inlets below the level of the buoy waterline. Each pump took in water through a common all-plastic suction strainer. The synthetic rubber suction hose was 6 inches in diameter and 75 ft long. All pump components in contact with seawater were polyvinylchloride (PVC) or an equivalent inert nonmetallic material. The discharge line of each pump has a check valve to prevent backflow, a pressure indicator, and a vacuum gage for the detection of obstructions. Each pump was driven by a close-coupled, totally enclosed, fan-cooled motor. The electric motors were 115-V single-phase and developed approximately 3 horsepower each. Figures 1 and 2 provide details of the flow schematic.

The pressure water system contained a Kates flow-control valve to maintain a constant flow rate regardless of upstream or downstream pressure changes. The primary flow measurement was obtained with a (RAMAPO) strain-gage-type electronic flowmeter located downstream of the heat transfer measurement system.

The power system consisted of two diesel-powered alternators with 115-V, 60-Hz, 30-amp capacity each.

Heat Transfer Measurement

The heat transfer monitor was designed by John Fetkovich of Carnegie-Mellon University.[6] On command, an electrically heated ring clamped on the outside of the metallic tube being tested heated the tube to 1.8°C above the water temperature or applied heat for 10 time constants, and then cut off. The time series of the temperature difference between the tube wall and water was used to accurately establish a time constant.

The heating collar was positioned 36 inches from the upstream end of the test pipe. The temperature was determined some distance from the center of the pipe and compared to the inlet water temperature 12 inches upstream. Ten seconds after the heat was turned off, data collection began. As the biofouling film formed, the cooling time increased. Decreased heat transfer efficiency or increased resistance due to fouling was reported as R_f in hr ft^2 °F/BTU units. R_f was defined as the difference between 1/h where h is the measured heat transfer coefficient at any given time and the initial value of 1/h. Reference 6 discusses the corrections, based on water temperature and flow rate, used to calculate the heat transfer coefficient and fouling factor.

The Shore Collection Station at NSTL Station, Mississippi was used to control the OTEC–GOME experiment. Data were received at the NSTL and printed out immediately to permit evaluation of data quality before the experiment continued. A data tape was produced for Argonne National Laboratories from which the resistance due to fouling (R_f) was calculated.

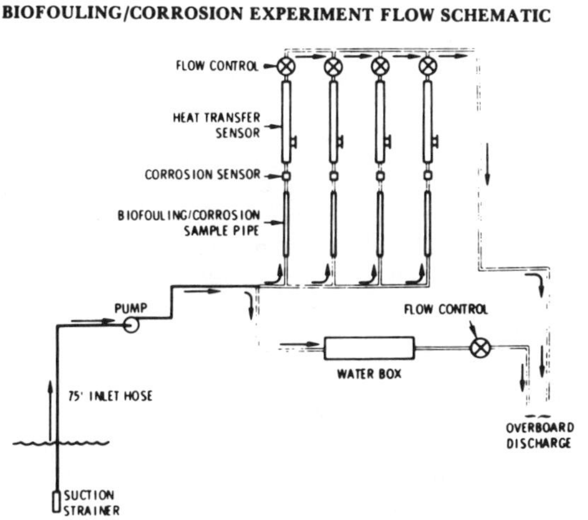

Fig. 1. Biofouling/corrosion experiment schematic.

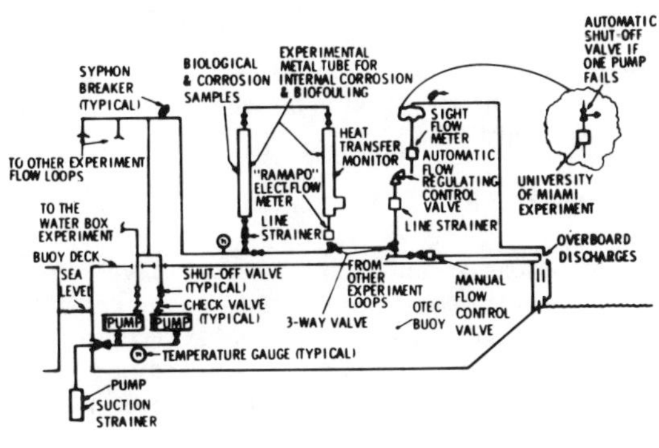

Fig. 2. Biofouling/corrosion – heat transfer experiment flow loop.

Sample Removal

To sample a pipe, water flow was diverted through a by-pass line, and the 9-in sections were removed with a modified tube cutter. Removed sections were filled with filtered (0.45 μm pore size) seawater, capped and put on ice until they could be transferred to the support ship for processing. On the support ship, sections were secured vertically in a vise (still filled with seawater) and subsections were removed for analysis with a pipe cutter. All pipes were sampled immediately after cleaning and before exposure to flowing seawater. The aluminum pipe was sampled after 15 min, 60 min and 24 hr exposure to flowing seawater. Subsequent samples for the aluminum pipe and for those of the other materials were taken at 48 days, 78 days and at 99 days.

Light Section Microscopy

Wet film thickness of 4% glutaraldehyde-fixed specimens was determined with a Model M-308 Light Section Microscope* following the method of Schmaltz[7] with a transparent film, reflections from the top and bottom of the film are observed, and their separation measured. The correction factor which must be applied to the observed separation to account for the refractive index of seawater is 1.6.

Coupons fixed in 4% gluteraldehyde in seawater were placed on filter paper circles wet with the same solution in tightly covered polystyrene dishes. Cotton thread was draped across the specimen with the ends in contact with the wet filter paper. This allowed excess fluid on the coupon to drain overnight, and equilibrium hydration to be established. The next morning thickness of the hydrated film was measured in the closed dish.

Total Organic Carbon (TOC)

One-inch frozen pipe sections were equilibrated to room temperature, measured with a vernier caliper for area determination and the surface films removed with a Teflon spatula. Extraction of the surface films was accomplished in 10% phosphoric acid with potassium persulfate (20g/1) added. The extract was transferred to a 10 ml precombusted glass ampule. The extract was purged with oxygen at 80 ml/min to remove inorganic carbon and then sealed in the ampules. Duplicate rings were extracted for all samples. After sealing ampules, conversion of organic carbon to carbon dioxide occurred in a pressure vellel maintained at 175°C ± 5°C.

An Oceanography International Corporation** non-dispersive infrared analyzer was used to determine the carbon dioxide concentration resulting from wet oxidation of organic matter in each ampule.

Adenosine Triphosphate (ATP)

The cut and outside edges of half-inch pipe section were wiped with tissue saturated with 100% ETOH. The section was then rinsed in a beaker of filtered seawater and placed in 24 ml of ice cold 0.6N H_2SO_4. A magnetic stirring bar was placed in the center to provide agitation during extraction. After about 20 min (while other sections were cut from the pipe and processed) the extracted section was removed, and 4 ml of 0.06OM EDTA in 0.05M Tris (pH7.8) added, followed by 4 ml of 3N NaOH. Final pH was adjusted to 7.8 and the extract frozen for later analysis.[8] After extraction, each section was measured with a vernier caliper to determine sample area.

Each thawed extract was reconstituted to 45 ml with 0.05M Tris (pH7.8) and assayed with a photometer, integrator and DuPont*** luciferin-luciferase enzyme-substrate system. At least three determinations were made on each extract. Variability of results were evaluated using a Q test to eliminate spurious values[9] and Youden's [10] methods for small data sets.

Inorganic Chemical Analysis

The procedure used for the determination of concentration of inorganic scale components on OTEC heat exchanger test pipes was that recommended by Morse and Craig.[11] A 1-inch pipe sample was given a quick rinse in double-distilled water to remove excess sea salt. It was then immersed in a 3% acetic acid solution for 24 hours. The initial volume of the acid solution was 40 ml. The pipe sample was removed and

*Gaertner Instr., Chicago, Illinois, USA.
**Oceanography International Corporation, P.O. Box 2980, College Station, Texas 77840, USA.
***E.I. DuPont Co., Concord Plaza, Quillen Bldg., Wilmington, Delaware 19898, USA.

a rinse of the pipe was added to the acid solution. The solution was then transferred to a 50-ml volumetric flask and the beaker rinsed into the volumetric flask until a volume of 50 ml was obtained. Selected metal concentrations in the solution were then determined by graphite furnace and flame atomic absorption spectroscopy.[12] The geometric surface area of the pipe was calculated and the metal concentration in nannomoles per cm^2 of surface area determined.

Two corrections were applied to the calculated surface concentrations. The first was a blank correction based on pipe samples which were cleaned, but not exposed to seawater. The concentration of the selected metals on the control pipe were subtracted from those on the test pipes. The second correction was applied to major sea salt components. Chloride and sodium were present on the pipe surfaces in ratios close to that found in sea salt, and were presumably left on the surface during drying procedures. At present it is not possible to determine what fraction of the seawater is present in a porous surface scale and what portion is merely seawater that was left on the scale surface in spite of rinsing with distilled water. Using seawater concentration ratios, the sea salt component of calcium and magnesium was subtracted from their surface concentrations to yield "excess" calcium and magnesium surface concentrations. These "excess" concentrations were assumed to represent calcium and magnesium concentrations in the solid scale.

Scanning Electron Microscopy

Coupons fixed in 4% glutaraldehyde in filtered (0.45 μm pore size) seawater were transferred through several filtered seawater/distilled water washes, dehydrated in acetone, critical point dried with liquid CO_2. Specimens were sputter-coated with Au-Pd before examination with an AMRAY* 1000A scanning electron microscope equipped with the Kevex** 7000 energy dispersive x-ray fluorescence spectrometer.

RESULTS

System Performance

The Gulf of Mexico biofouling/corrosion experiment was plagued with equipment failures. Biofouling/corrosion samples were collected three times during the course of the 14-week experiment. Each of the samples was collected after a period of decreased flow (< 6 ft/sec) or flow stoppage (Fig. 3). The system operated without interruption for the first 31 days. A hose failure at 32 days resulted in flow fluctuations as low as 2.6 ft/sec until the suction hose was replaced at day 54. The first biofouling/corrosion sample was collected at day 48. Flow was resumed at approximately 5.5 ft/sec for 14 days, when both of the diesel generators failed due to fuel contamination. The diesels were restarted at day 72, functioned for 4 days, failed on day 76, restarted on day 77, and functioned at a flow of 6 ft/sec until day 86, when the pump seals failed and the experiment was terminated. The second biofouling/corrosion sample was collected at day 78 and the final sample was collected on day 99 (13 days after the pump failure).

Heat Transfer Resistance

Heat transfer data are presented in Fig. 4. The 90—10 copper-nickel (Cu-Ni) surfaces exhibited immediate loss of heat transfer efficiency but reached a plateau of about $R_f = 1.3 \times 10^{-4}$ at about 34 days. Though the R_f increased slightly after that, it remained less than 2.0×10^{-4}. After the first flow stoppage at 72 days, the R_f values for the Cu-Ni pipe fluctuated slightly in comparison to much larger decreases in R_f values in the other heat exchangers.

*AMRAY, Inc., 160 Middlesex Turnpike, Bedford, Massachusetts 01730, USA.
**Kevex Corp., 1107 Chess Drive, Foster City, California 94404, USA.

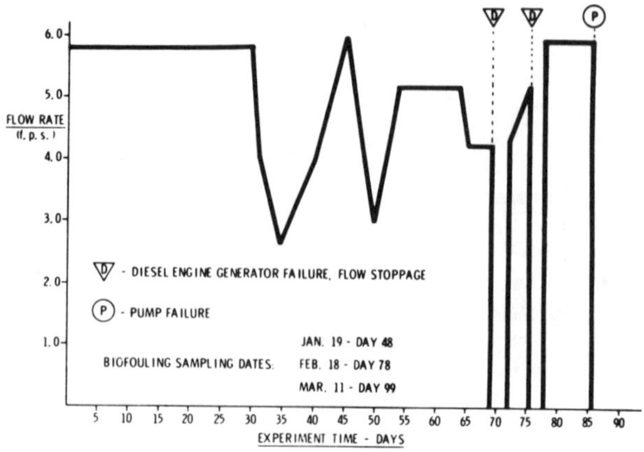

Fig. 3. OTEC – GOME – Flow rate versus time.

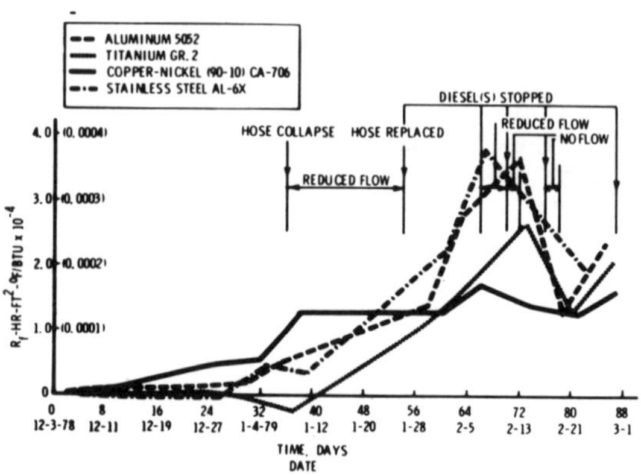

Fig. 4. Gulf of Mexico experiment (GOME).

The aluminum surfaces also experienced early decay in heat transfer efficiency and at day 72 had reached $R_f = 3.6 \times 10^{-4}$. After the flow stoppage at 72 days, the next measured R_f showed a decrease to 1.3×10^{-4}. Resumed flow resulted in an increasing R_f when the experiment concluded.

The heat transfer properties of the titanium surface remained essentially unaffected by biofouling/corrosion until day 37. The maximum observed R_f was at day 72, as in aluminum, at a value of 2.7×10^{-4}. An abrupt decline in R_f coincided with the flow stoppage at day 72 with an increase immediately after the resumption of flow.

Stainless steel had a slow initial loss of efficiency, but experienced the greatest loss of efficiency in the shortest time ($R_f = 3.7 \times 10^{-4}$ at day 66), that also declined during the stagnant period.

Wet Film Thickness

Because of the opaque nature of the corrosion film on the copper-nickel surface, the light-section microscope could not be used to measure wet film thicknesses. Film thickness measurements for the other surfaces are presented in Fig. 5. The film thickness in the aluminum flow-through system was predominately corrosion product, while that of stainless steel and titanium was due primarily to biofouling. It appears that some sloughing of the fouling film took place in the aluminum pipe after the period of stagnation of 13 days. The film thickness decreased approximately 5 μm during the period.

Total Organic Carbons (TOC)

Total organic carbon data are presented in Fig. 6. Total values generally increased from the start of the experiment to day 78. After the 13-day stagnation period drops in the total organic carbon values were observed for the copper and titanium surfaces and slight increases for aluminum and stainless steel surfaces. TOC values for all samples are included in the appendix.

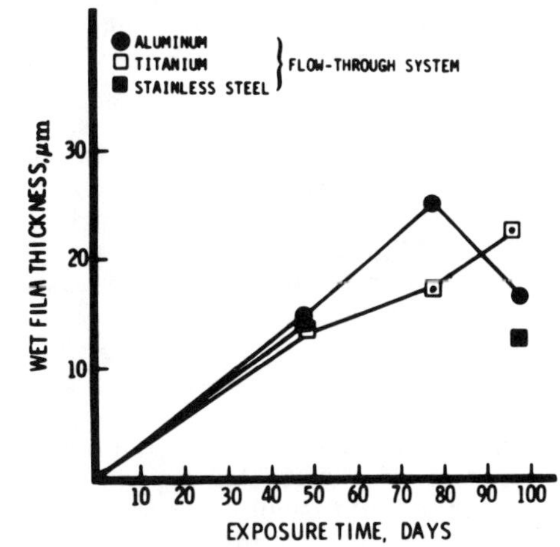

Fig. 5. Wet - film thickness measurement versus time.

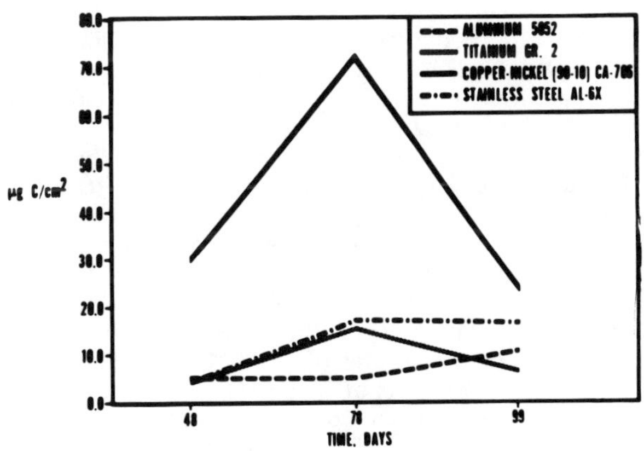

Fig. 6. Total organic carbon versus time.

Adenosine Triphosphate (ATP)

ATP, a constituent of living cells only, was below detection limits for all samples taken during the experiment.

Inorganic Constituents

The blank-corrected surface concentrations are presented in Table 1. The aluminum 5052 alloy corroded vigorously during the first few minutes of exposure to flowing seawater. The amount of aluminum present in the scale after 15 minutes exposure was 70% of the amount present at the end of the 48-day exposure. Therefore, the initial build-up of corrosion product on the aluminum surface is very rapid. Scale formation on the copper-nickel was also rapid. The scale on the copper-nickel surface was enriched in iron with respect to the original alloy. The titanium and AL6X stainless steel pipe surfaces accumulated less than 10% as much inorganic material as the copper-nickel pipe surfaces. Inorganic deposits on the surfaces of all materials decreased after the 13-day stagnation period.

Values for the magnesium and calcium concentrations, in excess of the maximum contribution from sea salt, are presented in Table 2. These values were calculated assuming either all of the sodium or potassium present was representative of sea salt. All test pipes, except the aluminum pipes, exhibited a strong enrichment with respect to Ca over Mg. This may be indicative of $CaCo_3$ formation. The aluminum pipes had an initial enrichment with Mg over Ca but on extended exposure averaged close to seawater.

Scanning Electron Microscopy

The pipes, as received from the manufacturer had macroscopically and microscopically different surface preparations. After cleaning with trichloroethylene the surfaces of the aluminum, copper-nickel and titanium appeared shiny and smooth, while that of the stainless steel was dull and rough to the touch.

Extrusion lines were evident on the 0-hr aluminum surface (Figs. 7 and 8). Within fifteen minutes after exposure to seawater the surface was smoother (Figs. 9 and 10) due to a uniform oxide formation. Drying of the samples caused cracking of the oxide layer (Fig. 11). Slime, exuded by microorganisms, covered the oxide layer of the 48-day exposure. Particulate matter and filamentous bacteria were attached to the surfaces of the 48, 78 and 99-day exposures (Figs. 12, 13, 14 and 15).

The 0-hr titanium surface was relatively smooth (Fig. 16) compared to 0-hr for aluminum (Figs. 7 and 8). Bacterial attachment and proliferation was well underway by the 48-day exposure and continued through the 78 and 99 day exposures (Figs. 17, 18, 19 and 20). In addition to the biological fouling, the titanium surface was covered with crystalline formations attached to the surface by means of a filament (Fig. 19).

The AL 6X stainless steel surface at the 0-hr exposure exhibited surface irregularities typically, 1–5 μm wide, 20 μm deep and ranging from 10 to hundreds of μm long, present as a result of the commercial preparation of the pipe (Figs. 21 and 22). Filamentous bacteria had colonized the surface by the 48-day exposure (Figs. 23 and 24). The 78 and 99 day exposure surface features were obscured by slime (Figs. 25 and 26).

The copper-nickel surface (Figs. 27 and 28) at 0-hr was relatively smooth compared to aluminum (Fig. 7) and titanium (Fig. 16). A brittle, porous inorganic scale predominated the 48, 78 and 99 day exposures (Figs. 29, 30 and 31). The oxide was composed of 65% Fe and lesser amounts of Cu and Ni (Fig. 31). The corrosion product was loosely attached to the underlying pipe surface and was easily detached (Figs. 32 and 33).

Table 1. Selected metal concentrations in inorganic scale

Sample		Na	K	Ca	Mg	Fe	Cu	Ni	Cr	Al
Aluminum	15 min.	232	9.8	18.5	913	71.3	2.59	--	--	5820
	1 hr.	51.3	6.1	34.3	670	194	0.019	--	--	5020
	24 hrs.	83.5	8.7	39.1	560	138	1.56	--	--	5490
	48 days	103	2.10	211	816	199	0.70	--	--	8120
	78 days	71.6	1.36	202	1240	292	49.6	--	--	13500
	99 days	0.244	-0.008	0.322	1.25	-0.066	0.002	--	--	7.9
Stainless Steel	48 days	105	1.95	34.3	60.9	391	34.5	2.57	48.1	--
	78 days	225	4.62	46.2	54.8	177	16.5	12.5	13.7	--
	99 days	0.186	-0.010	0.129	0.077	-0.23	1.10	0.158	-0.0222	--
Copper-Nickel	48 days	322	7.12	191	365	3680	9180	2370	--	--
	78 days	96.3	2.13	141	237	3600	8810	2070	--	--
	99 days	0.566	-0.0083	0.462	0.548	5.56	8.5	2.53	--	--
Titanium	48 days	83.6	1.78	12.2	36.5	118	44	--	--	--
	78 days	70.7	1.55	18.5	38.2	83.3	7.6	--	--	--
	99 days	0.309	0.0001	0.041	0.049	-1.18	0.058	--	--	--

All concentrations in nmoles/cm^2. All Ti values <0.003.

Table 2. Calcium and magnesium ratio in inorganic scale

Pipe Sample	Exposure Time	C_{Mg}	C_{Ca}	C_{Mg}/C_{Ca}	R*	C_{Mg}	C_{Ca}	C_{Mg}/C_{Ca}	R*	C_{Mg}/C_{Mg}
Aluminum	15 min.	886	13.4	66.1	12.9	862	8.63	99.9	19.44	1.00
	1 hr.	664	33.2	20.0	3.90	638	28.2	22.7	4.41	1.04
	24 hrs.	550	37.3	14.8	2.87	514	30.3	17.0	3.30	1.07
	48 days	804	209	3.85	0.75	805	208	3.85	0.75	1.00
	78 days	1231	200	6.15	1.20	1232	201	6.15	1.19	1.00
	99 days	1.22	0.32	3.86	0.75	1.29	0.33	3.91	0.76	0.95
Stainless Steel	48 days	49.1	32.0	1.53	0.30	50.8	32.3	1.57	0.31	0.97
	78 days	29.5	41.3	0.71	0.14	30.9	41.5	0.74	0.14	0.95
	99 days	0.06	0.12	0.44	0.09	0.13	0.14	0.92	0.18	0.46
Copper-Nickel	48 days	329	184	1.79	0.35	328	184	1.79	0.35	1.00
	78 days	226	139	1.63	0.32	226	139	1.63	0.32	1.00
	99 days	0.48	0.45	1.08	0.21	0.59	0.47	1.26	0.24	0.81
Titanium	48 days	27.1	10.4	2.61	0.51	27.3	10.4	2.62	0.51	0.99
	78 days	30.2	17.0	1.78	0.35	30.2	16.9	1.78	0.34	0.10
	99 days	0.01	0.03	0.42	0.08	0.05	0.04	1.19	0.23	0.20

Values corrected for seasalt using Na. Concentration (C) is in $nmoles/cm^2$.

$$* \; R = \frac{(C_{Mg}/C_{Ca}) \text{ Measured}}{(C_{Mg}/C_{Ca}) \text{ Seawater}}$$

If R is greater than 1 the pipe surface is enriched with Mg; if R is less than 1 the pipe surface is enriched in Ca.

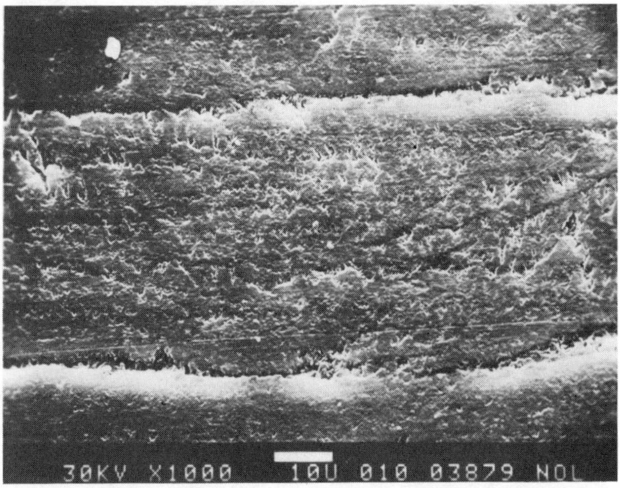

Fig. 7. Aluminum 0-hr exposure. Surface with extrusion lines.

Fig. 8. Aluminum 0-hr exposure. Surface with extrusion lines. Magnification uncertain.

DISCUSSION

There are major problems in interpreting the results of the Gulf of Mexico experiment. Each of the biofouling/corrosion samples was collected during or after a catastrophic equipment failure.

The composition of a biofilm is strongly dependent on environmental factors - flow conditions, oxygen and nutrient concentrations. The titanium, stainless steel, and aluminum surfaces had been colonized by filamentous bacteria before the first sample was collected at 48 days. Even small perturbations of short duration can alter the constituents of the film. The absence of measurable quantities of ATP, a component found in living cells, in the samples for this experiment is an indication that the flow interruptions adversely affected the microorganisms present in the film. Microorganisms colonizing solid surfaces grow until these surfaces are completely covered. Subsequent growth is from the surface outward with aerobes in areas where the oxygen supply is

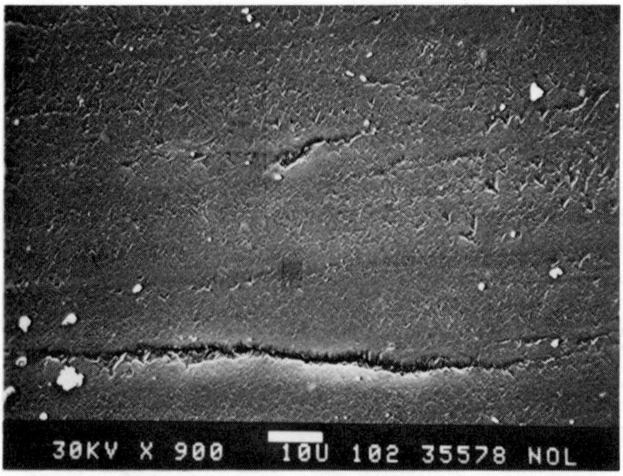

Fig. 9. Aluminum 15-min exposure. Surface covered with oxide layer.

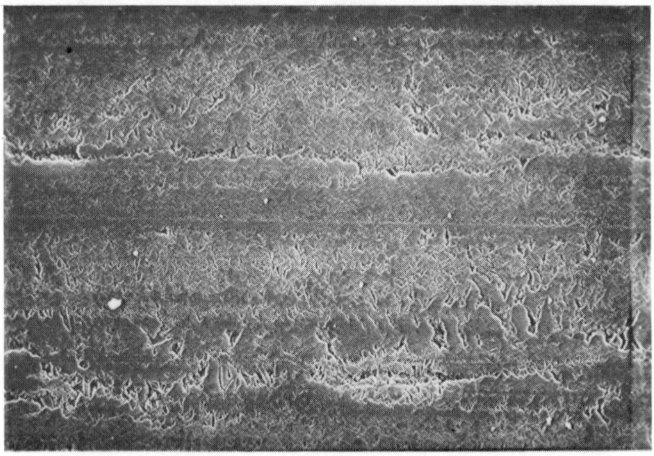

Fig. 10. Aluminum 60-min exposure. Surface covered with oxide layer. Magnification uncertain.

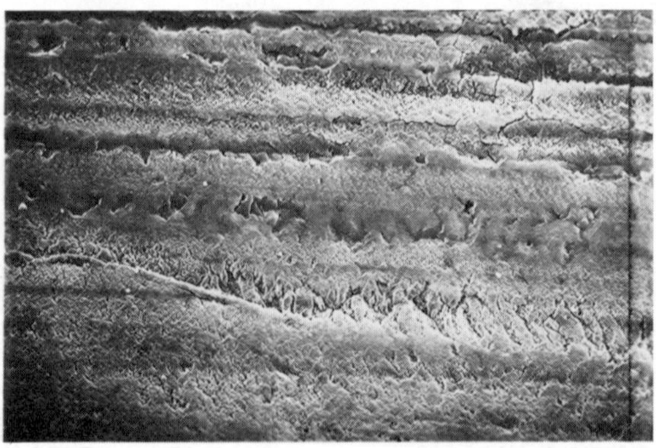

Fig. 11. Aluminum 24-hr exposure. Cracks formed in oxide layer as a result of the drying process. Magnification uncertain.

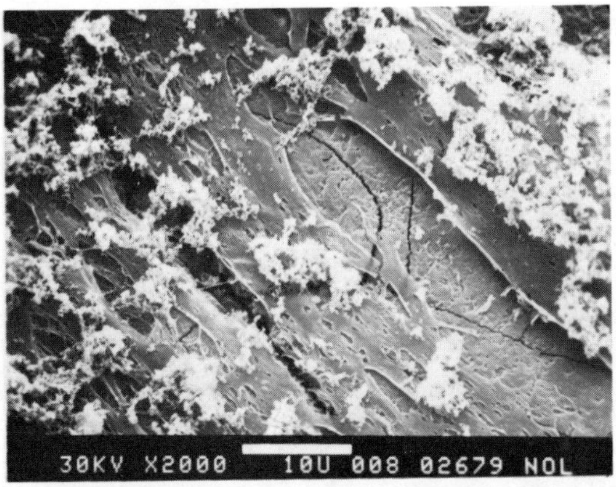

Fig. 12. Aluminum 48-day exposure. Oxide layer covered by the slime and particulate matter.

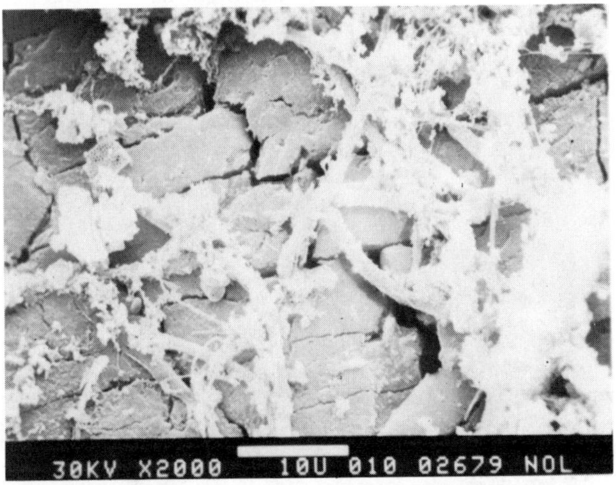

Fig. 13. Aluminum 48-day exposure. Filamentous bacteria and particulate matter on surface.

sufficient and anaerobes in the depths of the film into which oxygen cannot diffuse. When nutrients are no longer available for those basal microorganisms, they die or lyse and thus destroy the attachment of the film to the solid surface so that sloughing occurs.[13] The flow stoppages between days 65 and 78 resulted in substantial drops in R_f for those surfaces that were colonized (aluminum, stainless steel and titanium); a much slighter decrease was observed in the copper-nickel pipe where fouling was predominately due to corrosion.

Film thickness decreased on the aluminum and stainless steel surfaces after the 13-day stagnation period and the inorganic chemical analyses of the corrosion film indicated sloughing of the corrosion film from all surfaces.

The high total organic carbon values (a measure of adsorbed, dissolved organic matter; living and dead biomass; and associated secretions) in the films from copper-nickel surfaces indicated that microorganisms had been present, although no intact individuals were found. Microorganisms attached to copper-base surfaces secrete copious quantities of polymer[14], and a comparatively high carbon-to-nitrogen ratio is

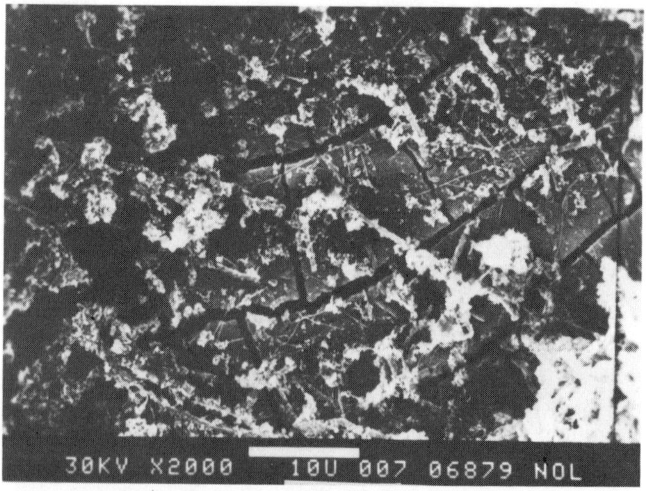

Fig. 14. Aluminum 78-day exposure. Filamentous bacteria and particulate matter on surface.

Fig. 15. Aluminum 99-day exposure. Filamentous bacteria and particulate matter on surface.

characteristic of the fouling on these materials.[15] Decreases in total organic carbon for the Cu-Ni and Ti samples following the stagnation period indicate sloughing or removal of organics from those surfaces.

Stainless steel, which exhibited the greatest apparent loss of heat-transfer efficiency for the period of investigation, is inert in seawater. It forms very little corrosion product, nor does it resist fouling.[16] The large R_f increase can be attributed to the insulating properties of the biofilm layer, the development of which may have been enhanced by the highly irregular surface of the particular pipe specimens that were used. The aluminum surface underwent an early degradation of heat-transfer efficiency and reached an R_f slightly less than that of stainless steel at about 79 days. The 5052 aluminum alloy corroded vigorously during the first few minutes of exposure — approxi-

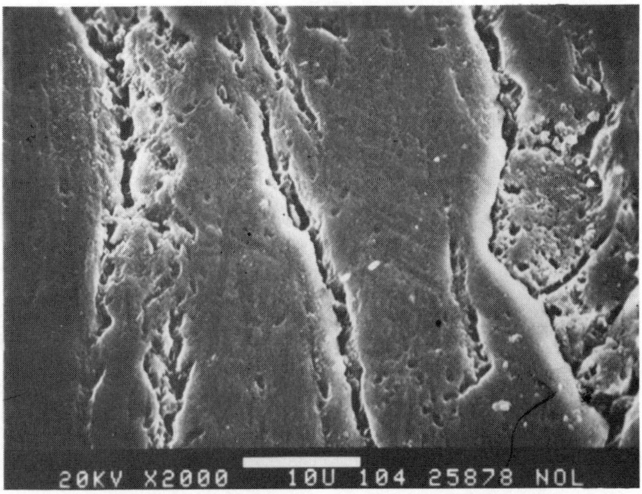

Fig. 16. Titanium 0-hr exposure. Surface relatively smooth as compared to aluminum and stainless steel.

Fig. 17. Titanium 48-day exposure. Filamentous bacteria on surface.

Fig. 18. Titanium 78-day exposure. Microfouling of surface.

Fig. 19. Titanium 78-day exposure. Crystalline structures occurring singly and in clusters.

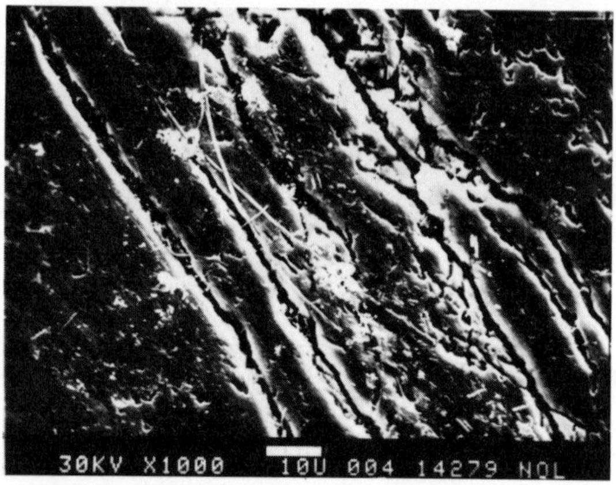

Fig. 20. Titanium 99-day exposure. Biofouling on surface.

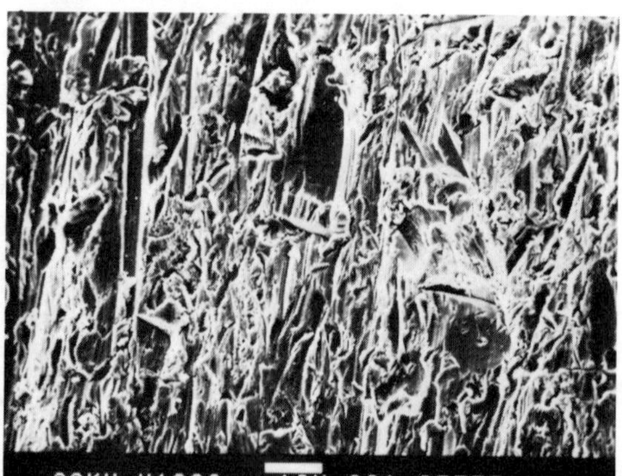

Fig. 21. Stainless-steel 0-hr exposure. Surface irregularities due to manufacturing process.

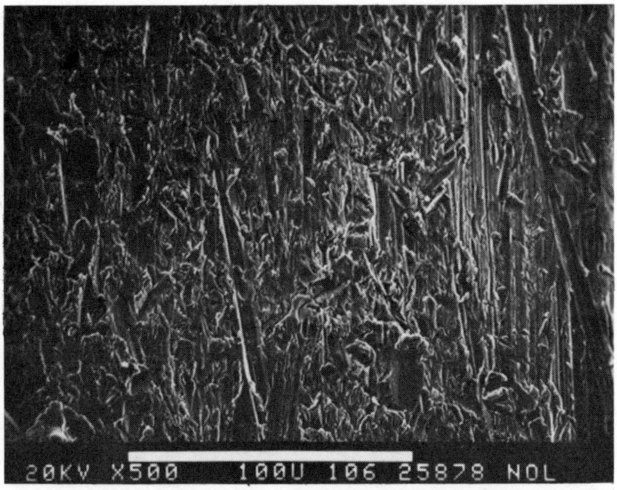

Fig. 22. Stainless-steel 0-hr exposure. Surface irregularities due to manufacturing process.

Fig. 23. Stainless-steel 48-day exposure. Filamentous bacteria colonizing surface.

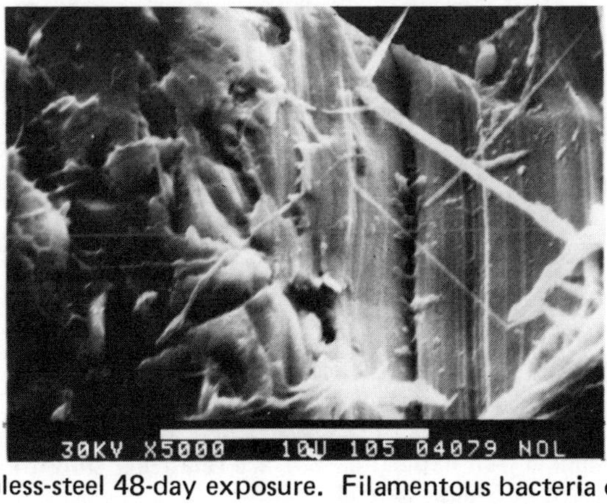

Fig. 24. Stainless-steel 48-day exposure. Filamentous bacteria on surface.

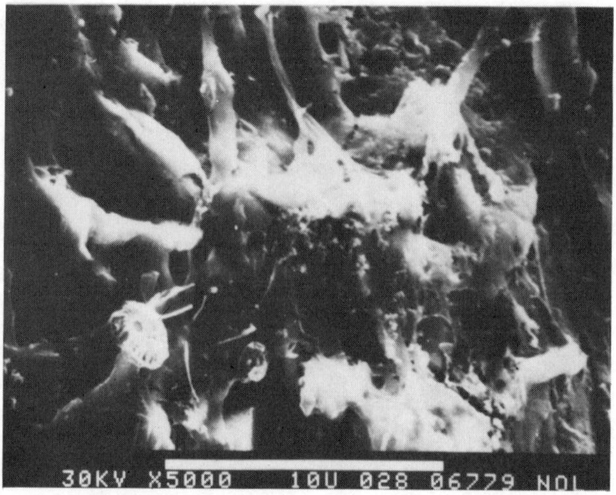

Fig. 25. Stainless-steel 78-day exposure. Slime with microorganisms entrapped obscuring surface irregularities.

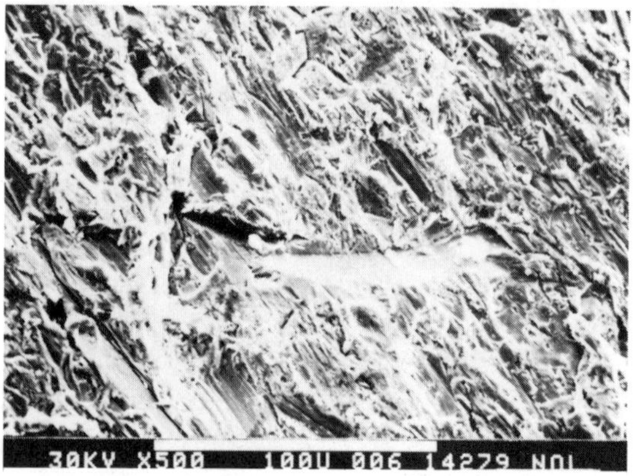

Fig. 26. Stainless-steel 99-day exposure. Surface irregularities evident.

Fig. 27. Copper-nickel 0-hr exposure. Surface relatively smooth as compared to Aluminum, Titanium or Stainless-steel.

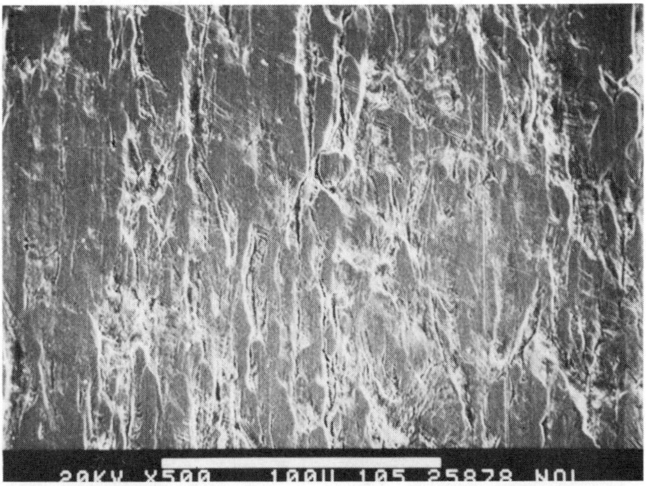

Fig. 28. Copper-nickel 0-hr exposure. Surface relatively smooth as compared to Aluminum, Titanium or Stainless-steel.

Fig. 29. Copper-nickel 48-day exposure. Corrosion film with particulate material.

Fig. 30. Copper-nickel 78-day exposure. Particulate matter on surface.

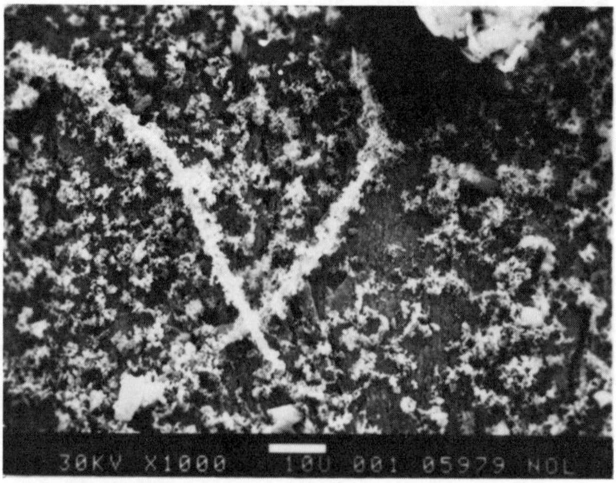

Fig. 31. Copper-nickel 78-day exposure. Brittle, porous corrosion layer with particulate matter on surface.

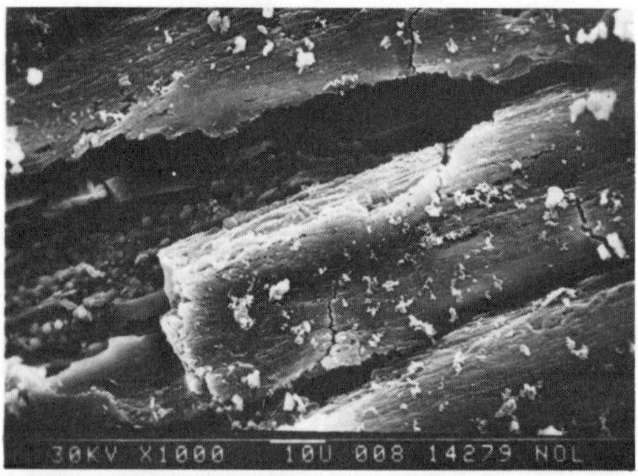

Fig. 32. Copper-nickel 99-day exposure. Cracks in corrosion layer.

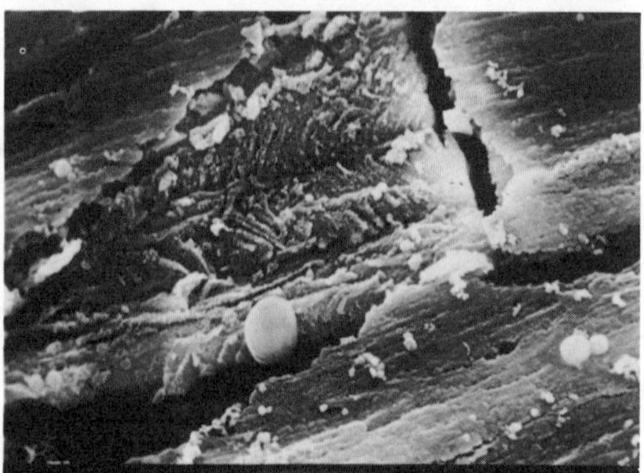

Fig. 33. Copper-nickel 99-day exposure. Brittle, porous corrosion product partially detached from surface.

mately 70% of the aluminum scale component found at 48 days was present after only 15 minutes. The corrosion film is hydrated and may be responsible for the initial minimal losses in efficiency. Greater losses appear to be due to the presence and constituents of the biofilm.

The copper-nickel pipes exhibited the greatest initial loss of efficiency. During the first 32 days, R_f reached a value of 1.5×10^{-4}. All losses were attributable to the rapidly forming corrosion film. The corrosion film was enhanced in iron with respect to the original copper-nickel pipe, as indicated by the EDAX fluorescence spectrum and atomic absorption analysis of the scale (Table 1). A similar observation was reported by Wilson using Auger electron spectrometric analysis.[17]

The titanium surface experienced slow initial losses and a maximum R_f value of 2.7 $\times 10^{-4}$. Titanium, like stainless steel, is resistant to corrosion and is nontoxic to marine organisms.[18] The loss of efficiency was due to biofouling.

The fouling resistance data from the Gulf of Mexico Experiment (prior to the first shutdown) and those from the KeaHole Point, Hawaii[15], experiment have been compared in Figs. 34, 35, 36 and 37. Except for the longer lag period for all surfaces in the Gulf of Mexico experiment, the data are remarkably similar. The slopes of the curves after the lag phase are identical. The second and third curves for the Hawaii data represent the R_f data after cleaning.

CONCLUSIONS

1. Titanium and AL 6X stainless steel pipe surfaces accumulated less than 10% as much inorganic material as the aluminum and copper-nickel pipe surfaces.
2. The formation of inorganic deposits on the aluminum pipe surfaces was initially very rapid, with approximately 70% of the material found after 48 days exposure being deposited during the first 15 minutes.
3. Reduced heat-transfer efficiency across the stainless steel and titanium surfaces was due to biofouling.
4. The increase in R_f of the copper-nickel surface was due to corrosion.
5. Corrosion products on the aluminum surface may impede heat transfer in the early stages of exposure, but biofouling and its related processes appear to control the observed increase in R_f.

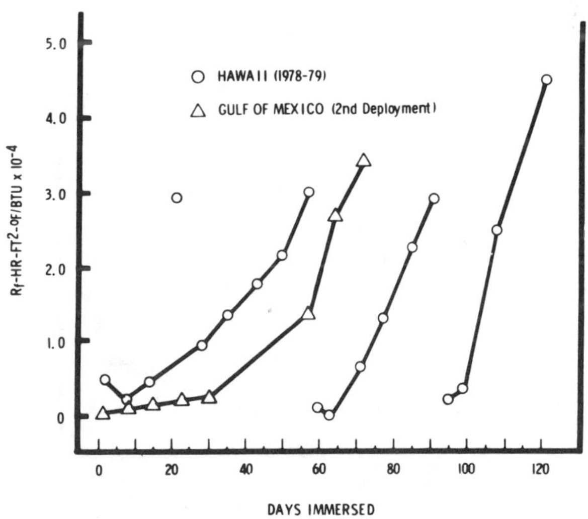

Fig. 34. Fouling resistance versus time — aluminum.

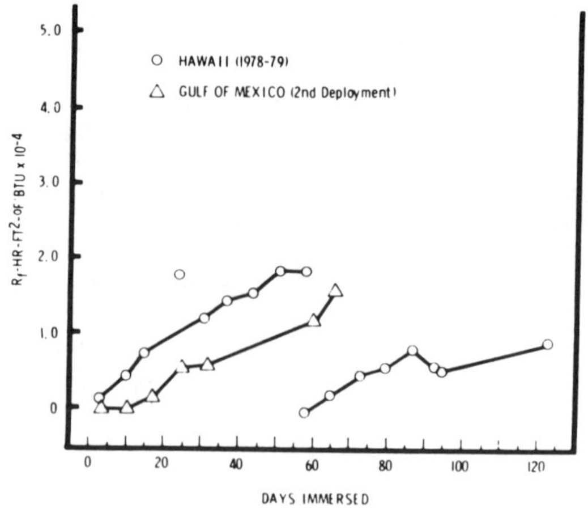

Fig. 35. Fouling resistance versus time — copper-nickel.

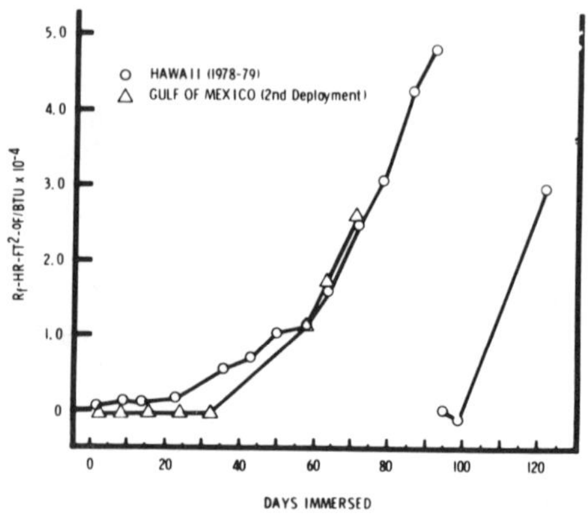

Fig. 36. Fouling resistance versus time — titanium.

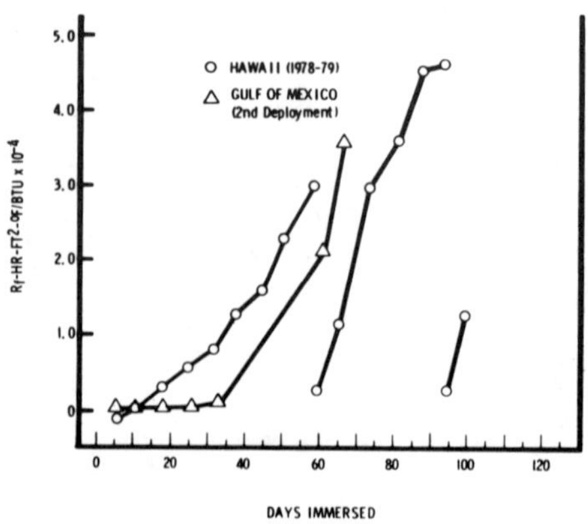

Fig. 37. Fouling resistance versus time — stainless steel.

6. After a 13-day stagnation period inorganic and organic deposits are lost from fouled surfaces. A short stagnation period of four days resulted in a decrease in R_f for surfaces colonized by microorganisms. All decreases may be due to the sloughing of the biofouling/corrosion film.
7. Surface irregularities may enhance biofouling.
8. The rates of heat transfer degradation from the Gulf of Mexico and Hawaii heat transfer data are virtually the same although the initial lag periods differed.

REFERENCES

1. Bell, K.J., "The effect of fouling on heat exchanger design, construction and operation," Proceedings of the Ocean Thermal Energy Conversion (OTEC) Biofouling and Corrosion Symposium, Seattle, Washington (October 10–12, 1977).
2. Berger, L.R., W.F. McCoy and J.A. Berger, "Biofouling Assays for OTEC Pipes, Proceedings of the OTEC Biofouling," Corrosion and Materials Workshop, Rosslyn, Virginia (January 8–10, 1979).
3. Aftring, R.P., D.G. Capone, L. Duguay, J.W. Fell, I.M. Master and B.F. Taylor, "Biofouling and site characterization studies in Ocean Thermal Conversion (OTEC) experiment at St. Croix, U.S. Virgin Islands." Report to Department of Energy. (1978).
4. Little, B.J. and D.M. Lavoie, "Gulf of Mexico Ocean Thermal Energy Conversion (OTEC) Biofouling and Corrosion Experiment", Proceedings of the OTEC Biofouling, Corrosion, and Materials Workshop, Rosslyn, Virginia (January 8–10, 1979).
5. Pandolfini, P.P., W.H. Avery, and F.K. Hill, "Experiments on ultrasonic cleaning of a shell-less, folded-aluminum-tube, OTEC heat exchanger", Paper 8A–1 at 6th OTEC Conference, Washington, D.C., (June 19–22, 1979).
6. Fetkovich, J.G., G.N. Granneman, L.M. Mahalingham and D.L. Meier. Degradation of heat transfer rates due to biofouling and corrosion at KeaHole Point, Hawaii. Proceedings of the OTEC Biofouling and Corrosion Symposium, Seattle, Washington (October 10–12, 1977).
7. Schmaltz, G. *Techn. Ober Flachenkunde.* (1936).
8. Karl, D.M. and P.A. LaRock. Adenosine triphosphate measurements in soil and marine sediments. *Fish. Res. Board,* Canada 32: 599–607, (1975).
9. Dean, R.B. and W.J. Dixon, "Simplified statistics for small numbers of observations," *Anal. Chem.* 23: 636–638, (1969).
10. Youden, W.J. *Statistical Methods for Chemists,* J. Wiley & Sons, New York, 126 pp., (1951).
11. Morse, John W. and H. Lee Craig, Jr., "Test procedures to define the ranges of balance between dissolved CO_2 and carbonates in seawater that could lead to deposition of calcareous deposits on heat transfer surfaces within the temperature range of OTEC heat exchangers", Final Report to Department of Energy, Contract No. EY–76–C–1830; Pacific Northwest Laboratories, Battelle Memorial Institute, Special Agreement No. B–7455–A–E, 35 pp., (1978).
12. Segar, D.A. and A.Y. Cantillo, "Direct determination of trace metals in seawater by flameless atomic absorption spectrophotometry", In: *Analytic Methods in Oceanography,* T.R.P. Gibb, Jr. (ed.), American Chemical Society, Washington, D. C., 56–81, (1975).
13. Hohen, C.H. and A.D. Ray, "Effects of thickness on bacterial film", *Journal WPCF* 45:11, (November, 1973).
14. Gerchakov, S.M., F.J. Roth, B. Sallman and L.B. Udey and D.S. Marszalek, "Observations on Microfouling Applicable to OTEC systems," Proceedings of the Ocean Thermal Energy Conversion (OTEC) Biofouling and Corrosion Symposium, Seattle, Washington (October 10–12, 1977).

15. Liebert, B.E., L.R. Berger, H.J. White, J.M. Moore, W.M. McCoy, J.A. Berger and J. Larsen-Basse,"The effects of biofouling and corrosion on Heat Transfer Measurements," Proceedings of the Ocean Thermal Energy Conference, Washington, D.C., (June 19–22, 1979).
16. LaQue, F.L., "Qualifications of stainless steel for OTEC heat exchanger tubing," ANL/OTEC Prepared for the U.S. Department of Energy Division of Central Solar Technology under contract W–31–109. (January 1979).
17. Wilson, J.W., "Composition of seawater corrosion product films on copper base alloys: Auger electron spectrometric analyses," *Corrosion,* (March, 1979).
18. Franson, I.A. and L.C. Covington, "Titanium and the OTEC environment," Proceedings of the Ocean Thermal Energy Conversion (OTEC) Biofouling and Corrosion Symposium, Seattle, Washington (October 10–12, 1977).

Total organic carbon data for OTEC pipe samples

Sample #	μg C/Ring	Ring Area (cm^2)	μg C/cm^2	$\bar{x}$ μg C/cm^2
AL 48 day	61.5	10.71	5.7	5.6
	65.5	10.85	6.0	
	54.5	10.58	5.2	
AL 78 day	28.2	10.74	2.6	4.0
	35.9	10.61	3.4	
	64.8	10.87	6.0	
AL 78 day	78.5	10.78	7.3	6.6
	86.0	10.74	8.0	
	48.0	10.71	4.5	
AL 99 day	117.0	10.76	10.9	10.8
	120.5	10.70	11.3	
	112.0	10.89	10.3	
Ti 48 day	41.5	10.67	3.9	4.6
	54.0	10.71	5.0	
	53.5	10.78	5.0	
Ti 78 day	197.4	10.85	18.2	15.7
	145.6	10.99	13.2	
Ti 99 day	200.8	31.28	6.6	6.6

Sample #	μg C/Ring	Ring Area (cm^2)	μg C/cm^2	$\bar{x}$ μg C/cm^2
Al 6X 48 day	74.0	10.90	6.8	4.7
	30.6	10.18	3.0	
	42.0	9.77	4.3	
AL 6X 78 day	174.0	10.60	16.4	17.5
	195.0	10.55	18.5	
	188.4	10.74	17.5	
AL 6X 99 day	273.0	18.41	14.8	16.9
	192.8	10.13	19.0	
Cu-Ni 48 day	412.9	11.11	37.2	30.7
	295.0	11.17	26.4	
	316.4	11.12	28.4	
Cu-Ni 78 day	1050.0	11.05	95.0	72.4
	885.0	11.05	80.1	
	460.0	10.91	42.2	
Cu-Ni 99 day	335.0	21.01	15.9	23.9
	267.5	10.12	26.4	
	325.0	10.96	29.6	

AL - Aluminum 5052

Ti - Grade 2 titanium

AL 6X - Allegheny Ludlum stainless steel

Cu-Ni - 90-10 Copper-nickel

Stevens, Vander Voort, McCall, eds. Microstructural Science, Volume 8

METALLOGRAPHIC STUDIES OF HYDROGEN INDUCED CRACK GROWTH IN A–106 STEEL*

T.A. Adler**, M.R. Louthan, Jr.**, and R.P. McNitt**

ABSTRACT

Crack propagation tests with compact tension samples of A–106 steel pipe in gaseous oxygen and hydrogen show that hydrogen increases the ease of crack growth and causes a change in the mode of fracture. In oxygen the crack propagates by shear bands connecting the large voids associated with inclusions, producing fracture by microvoid coalescence. In hydrogen the crack propagates transgranulary in a direction perpendicular to the maximum normal stress, producing a cleavage-like fracture. A significant amount of plastic deformation occurs prior to and during crack growth in hydrogen, but the amount of plastic deformation is much less than in oxygen. This difference in plastic strain is apparent in the stretch zones separating the overload failure from the fatigue starter cracks and in surface roughness associated with crack propagation.

INTRODUCTION

Contemporary energy storage and transport systems frequently utilize mild steel structural components which may be required to be in contact with a hydrogen environment in future applications. In particular, ASTM A–106 steel pipe has been suggested for use in hydrogen environments. Also, recent proposals to either replace or supplement natural gas and petroleum with hydrogen gas in present pipelines have raised questions. These questions are compounded by complications which may result in extrapolation of laboratory data to engineering situations where stress state, loading rates and hydrogen pressures may differ from the laboratory environment. Although mild steels, including A–106, show little or no evidence of catastrophic delayed failure due to hydrogen, studies[1–4] have shown that exposure to and testing in hydrogen environments causes losses in both the fracture stress and ductility of ferritic-pearlitic steels. Accompanying such mechanical degradation are alterations of the topological features of the fracture surfaces and changes in the mode of failure.[2–4] Fracture mechanics type tests have shown that exposure to hydrogen environments increase fatigue crack growth rates,[5,6] although no hydrogen induced subcritical crack growth was observed in statically loaded samples[6]. These results indicate that additional work is needed to insure that utilization of hydrogen in existing energy systems does not lower engineering parameters and compromise system integrity. This paper describes metallographic studies of crack nucleation and growth in A–106 steel in both hydrogen and oxygen environments.

---

*Work supported by DOE Contract No. E(40–1)5255.

**Virginia Polytechnic Institute, Blacksburg, Virginia 24061 USA.

MATERIALS AND PROCEDURE

Compact tension samples were machined from hot worked ASTM A–106 grade B steel pipe with a composition in the range shown in the table below.

Composition of A–106 steel, %

C	Mn	Si	Fe
0.25 max	0.27–0.93	0.10 min	balance

The pipe stock was heavily banded (Fig. 1) and inclusions were flattened in the axial-circumferential plane with the longest dimension in the axial direction. The 1.14 cm thick samples were cut so that the crack plane was either the radial-circumferential plane or the radial-axial plane (Fig. 2). The specimen geometry conforms to the ASTM E399 standards for fracture toughness testing[7] and the specimens were fatigue pre-cracked in room air. Precracked samples were monotonically loaded while exposed to either 13.8 MPa oxygen or hydrogen. The crosshead speed was 0.0127 cm/min (0.005 in/min). The crack opening displacement during loading was determined from compliance calibrations and measurements of crosshead displacements. The samples were loaded to predetermined crack openings which were selected to provide varying degrees of crack growth in the environment. After the environmentally affected growth was completed, samples were cooled to liquid nitrogen temperatures and cracked to failure. This test procedure provided three regions of crack growth; fatigue, environmentally influenced and brittle overload.

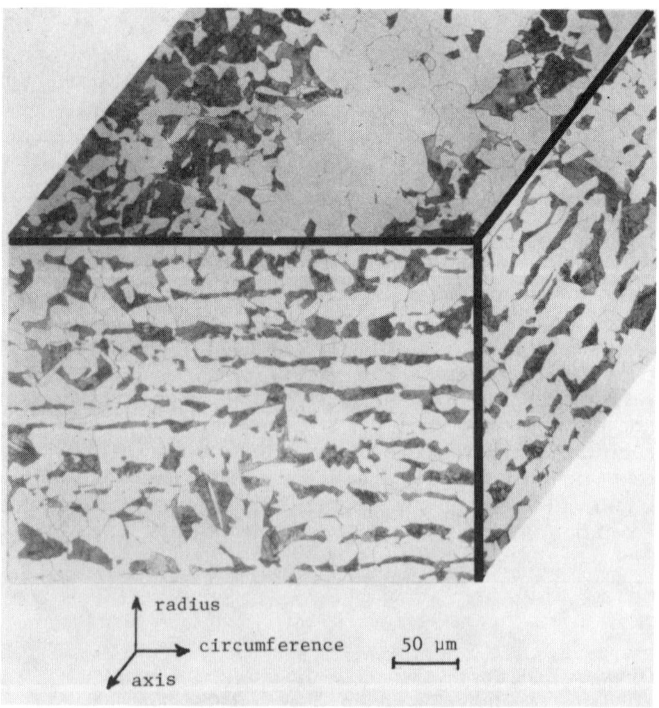

Fig. 1. Microstructure of ASTM A–106 steel pipe. 2% nital etch.

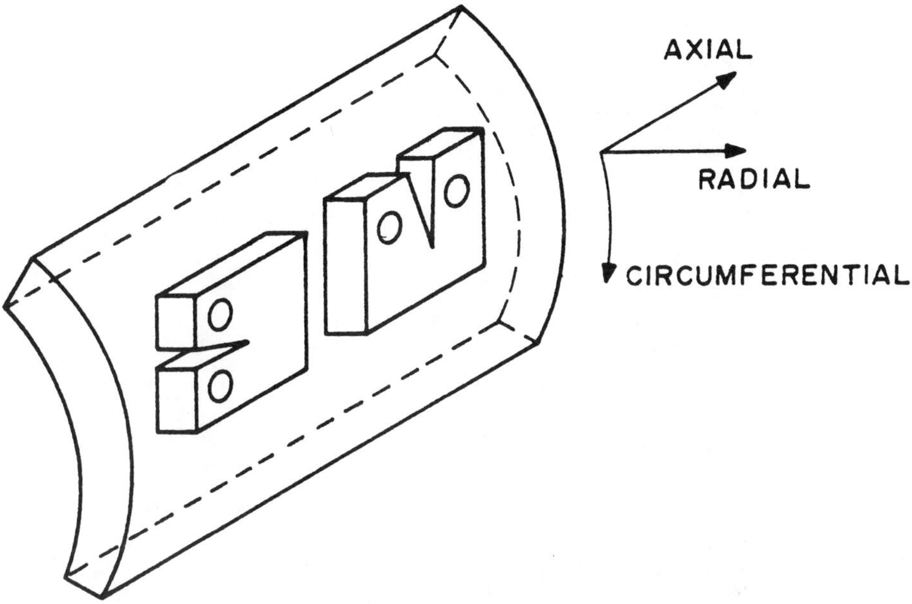

Fig. 2. Schematic illustration of the sample orientations relative to the pipeline stock.

Fracture surface topography was examined utilizing both optical and scanning electron microscopy. The amount of crack growth in the environment, the shape of the crack front, Lüders band sizes, and stretch zone sizes were determined by various metallographic techniques.

CRACK GROWTH IN ENVIRONMENT

The amount of crack growth which accompanied the application of a given crack opening displacement was dependent on the test environment (Fig. 3). A straight line extrapolation of test data to zero crack growth showed that the displacement required to initiate cracking was dependent on test environment and crack growth was much easier in the presence of hydrogen. Furthermore, as evidenced from the slope of the growth curves, crack growth rates were generally faster in hydrogen than in oxygen and were influenced by sample orientation. Growth rates were generally faster when the longest dimension of the inclusions was parallel to the direction of crack growth regardless whether the test was in oxygen or hydrogen environment. However, the orientation of the inclusions apparently had no effect on the displacement required for initiation of crack growth in hydrogen even though a strong dependence was seen in the oxygen tests. In oxygen, the initiation of crack growth was favored when the longest dimension of the inclusions was parallel to the direction of crack growth. These results show that the environment has a major influence on both the initiation and continuation of cracking in A–106 steel. Additionally, the lack of orientation effects in the initiation of cracking in samples tested in hydrogen suggests that inclusions and inclusion-matrix interfaces do not play an important role in the initiation of hydrogen-induced cracking.

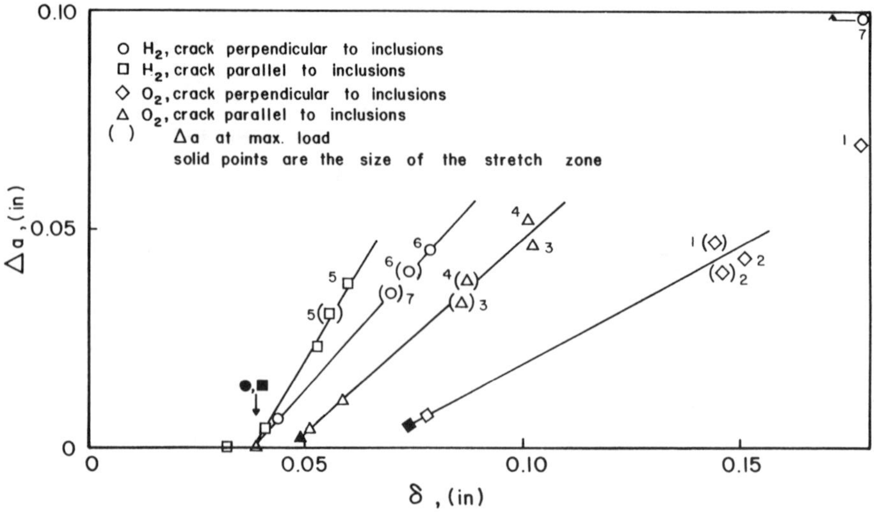

Fig. 3. Crack growth, Δa, as a function of crosshead displacement, δ. The samples were loaded in 13.8 MPa (2000 psi) H_2 or O_2. The crack growth at maximum load was estimated by drawing strength lines through the data.

Fracture Mode

The fracture features of the test samples shown in Figs. 4–9 are from sets of three orthogonal sections. They demonstrate the role of environment and inclusions in the fracture process. The orientation of these sections is illustrated in Fig. 10. Fracture of samples tested in oxygen was almost entirely by microvoid coalescence (Figs. 4a, 4b, 5a) while samples tested in hydrogen showed mostly a cleavage-like fracture process although a few, large microvoids were observed (Figs. 4c, 4d, 5b). Metallographic sections near the crack tip show that crack tip blunting precedes crack growth in samples tested in oxygen (Figs. 6a, 7a). This blunting causes a stretch zone to develop at the crack tip while microvoids nucleate and grow near inclusions ahead of the crack. After a sufficient amount of plastic strain the crack begins to grow by coalescence of the voids and intersection of the voids with the stretch zone. The smaller microvoids are in line with the pearlite bands in the microstructure and probably originate at the cementite platelets (Fig. 8a). This type of failure is common in mild steels.[8,9]

Prior to crack growth in hydrogen, plastic deformation occurs near the crack tip and a few voids associated with inclusions grow ahead of the crack. During crack propagation, the voids which had initiated are connected by matrix failure along paths which macroscopically are perpendicular to the maximum normal stress. Secondary cracks emanate from the voids, and, on a macroscopic level, these cracks may be on planes nearly perpendicular to the maximum normal stress (Fig. 8b). Crack propagation on planes perpendicular to the maximum normal stress is typical of results from other tests of mild steels in hydrogen environments.[2,4]

Microscopically, the path of an advancing crack in a sample tested in hydrogen is transgranular and the propagation appears to disregard grain boundaries and pearlite colonies (Figs. 8b, 9) and has no preference or adversion for any particular microstructural feature. The crack tips remain sharp and little evidence of large scale crack blunting is observed.

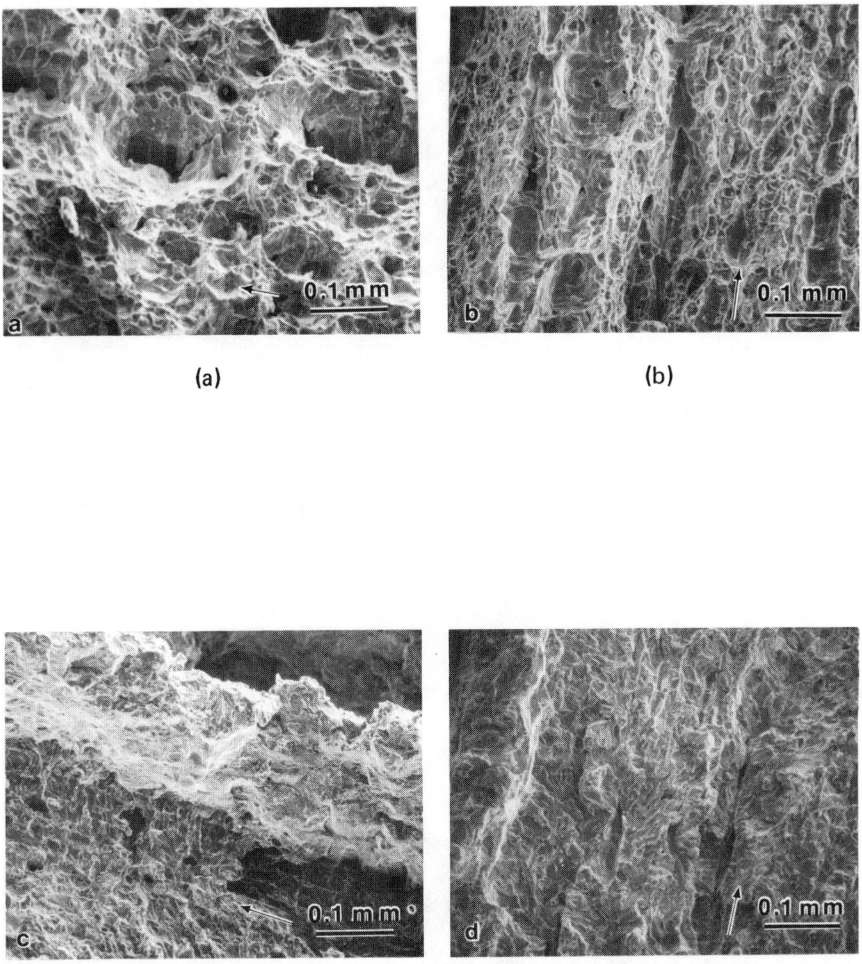

(a) (b)

(c) (d)

Fig. 4. SEM of the fracture surfaces of compact tension samples. (a) O_2 environment and longest dimension of the inclusions perpendicular to crack growth direction; (b) O_2 environment and longest dimension of the inclusions parallel to the crack growth direction; (c) H_2 environment and longest dimension of the inclusions perpendicular to the crack growth direction; (d) H_2 environment and longest dimension of inclusions parallel to the crack growth direction. The crack growth direction is indicated by the arrows.

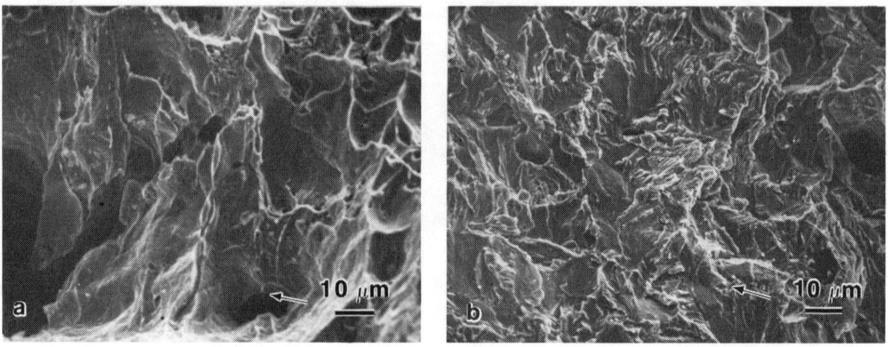

(a) (b)

Fig. 5. SEM of the fracture surface of compact tension samples with the longest dimension of the inclusions perpendicular to the crack growth direction: (a) O_2 environment; (b) H_2 environment. The arrow indicates the crack growth direction.

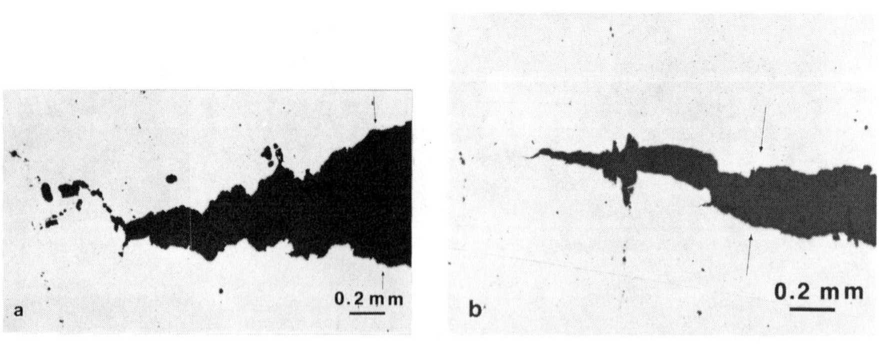

(a) (b)

Fig. 6. Side view of compact tension samples with the longest dimension of the inclusions perpendicular to the crack growth direction. (a) O_2 environment; (b) H_2 environment. The arrows mark the crack tip prior to loading in H_2. Optical microscope; no etch.

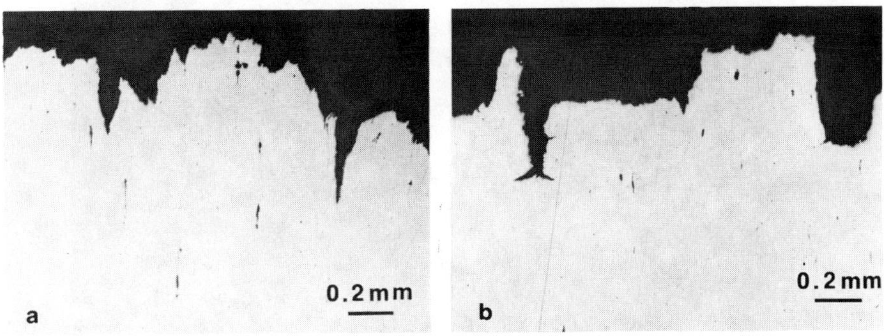

(a) (b)

Fig. 7. Front view sections of compact tension samples with the longest dimension of the inclusions perpendicular to the crack growth direction. (a) oxygen environment; (b) hydrogen environment. Optical microscope; no etch.

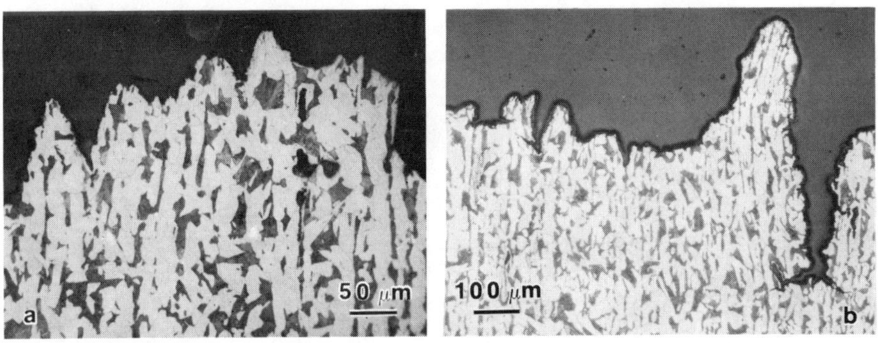

(a) (b)

Fig. 8. Front view sections of compact tension samples with the longest dimension of the inclusion perpendicular to the crack growth direction. (a) oxygen environment; (b) hydrogen environment. Optical microscope; 2% nital etch.

Fig. 9. Side view section of a compact tension sample load in hydrogen with the longest dimension of the inclusions perpendicular to the crack growth direction. The arrow shows the crack growth direction. Optical microscope; 2% nital etch.

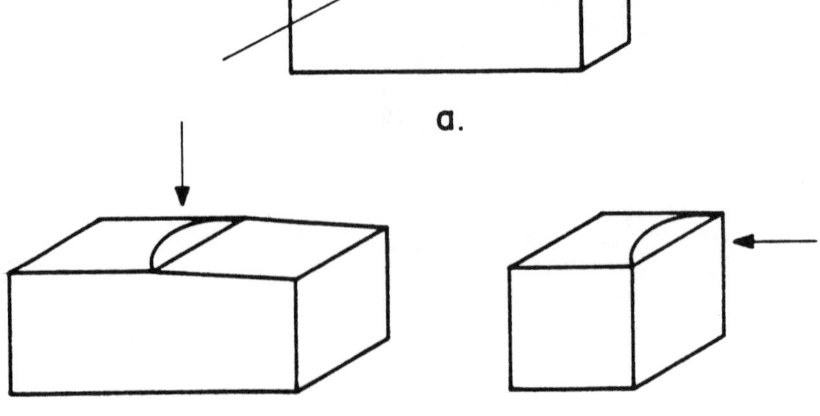

Fig. 10. Schematic illustration showing the orientation of the sections through the crack.

Crack Front Shape

Crack fronts developed in samples tested in hydrogen are macroscopically quite different in shape than those typically developed in samples tested in oxygen (Fig. 11). Traces of the crack front for a series of samples tested to different crack opening displacements are shown in Fig. 12. Inclusion orientation caused no noticable change in the crack front shape regardless of the test environment. In samples tested in hydrogen, crack growth initiates near the center of the fatigue precrack and the growth front quickly extends to the edges of the sample. The crack propagates slightly faster in the center of the sample than at the edges; however, extensive edge cracking is observed. The macroscopic fracture surface is flat (no shear lips) and the microscopic features are the same at the center and edges of the sample (Fig. 13).

Crack growth in oxygen is quite different. The crack begins to grow in several regions along the fatigue precrack. These individual growth fronts coalesce and the crack front extends extensively in the central region of the sample with little or no accompanying growth at the sample edges until large scale crack tunneling is developed. Crack growth at the specimen edges, when it finally occurs, is on macroscopic shear planes and large shear lips are developed.

The edge region of a compact tension sample is in a biaxial state of stress compared to the triaxial state of stress which is induced at the specimen center because of geometrical constraints. The observation of crack front development therefore is consistent with a rationalization which assumes that triaxial stresses play a more important role in failure in oxygen than in failure in hydrogen. This surprising result suggests that the often quoted role of triaxial stresses in hydrogen embrittlement processes may not be of general importance to the process.

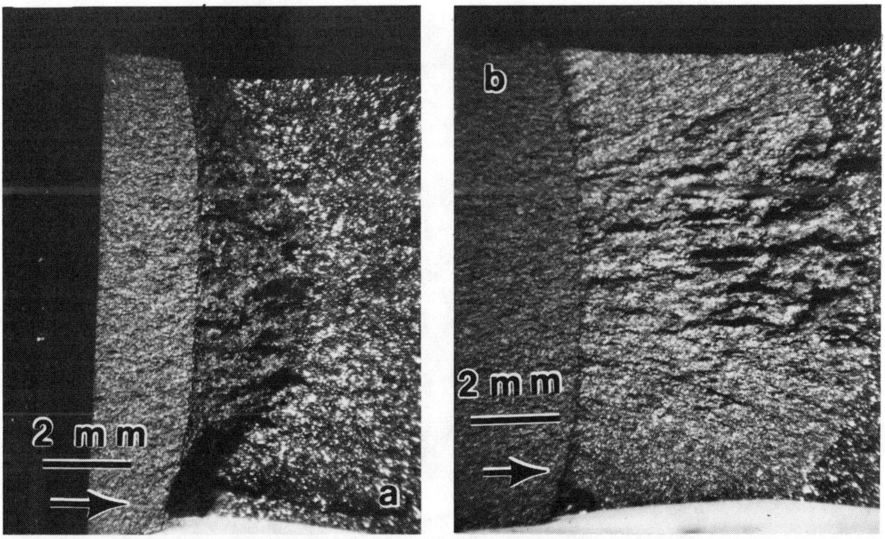

(a) (b)

Fig. 11. Optical fractographs of the fracture surfaces of compact tension samples with the longest dimension of the inclusions perpendicular to the direction of crack growth. (a) O_2 environment; (b) H_2 environment. The arrows indicate the direction of crack growth.

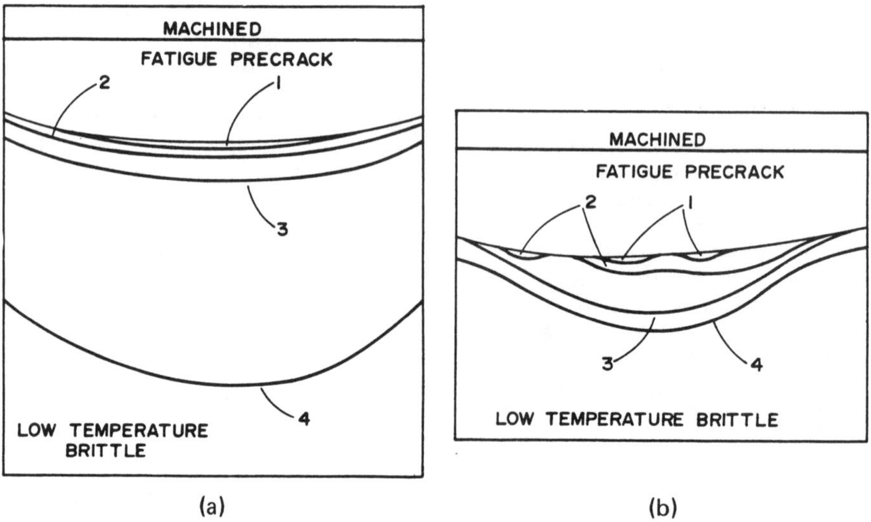

Fig. 12. Traces of the crack front shape for a series of samples loaded to different crack opening displacements. (a) O_2 environment; (b) H_2 environment.

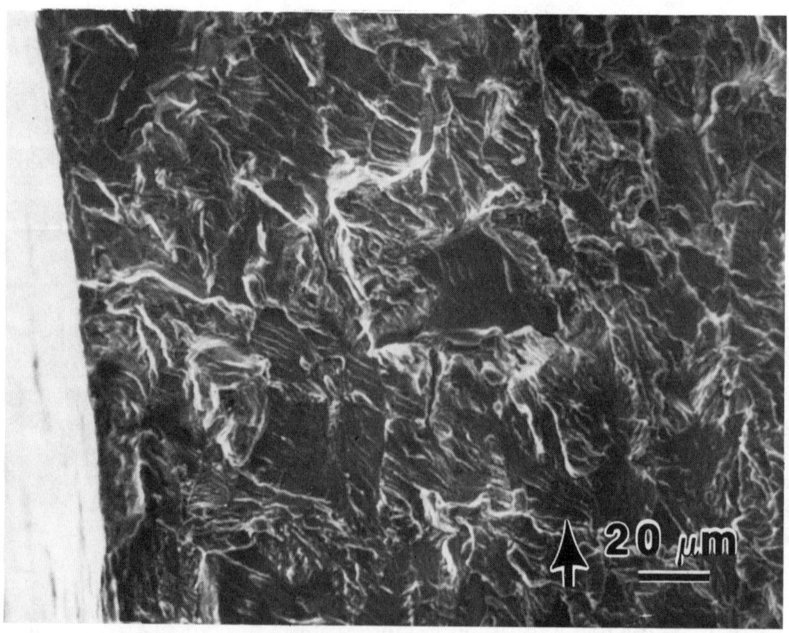

Fig. 13. SEM of the fracture surface near the edge of a compact tension sample load in hydrogen. The arrow indicates the crack growth direction.

Extent of Plastic Deformation

The extent of plastic deformation at (or near) the crack tip prior to crack growth was estimated through measurements of the Lüders bands (Fig. 14). The macroscopic appearance of the bands was the same for specimens pulled in either environments and the band width for a given crack opening displacement was insensitive to environment and inclusion orientation (Fig. 15). This result suggests that hydrogen has little effect on the macroscopic plastic flow in mild steels, even though hydrogen effects on microscopic flow are reasonably well known.

Growth of the Lüders bands occurred prior to the initiation of crack growth, indicating that extensive plastic deformation is a precursor to crack extension. The crack opening displacements at which crack growth initiated are marked on the curves (Fig. 12). In the hydrogen environment there was less plastic deformation preceding crack growth because the load required to cause crack growth was less than in specimens tested in oxygen.

Significant plastic deformation also accompanies crack growth. The extent of this deformation is apparent by the reduction in sample thickness (Fig. 11b) and the roughening of the sample surface that occurred during fracture (Fig. 16). The deformation is also microscopically apparent because of the continued growth of voids around inclusions as the crack extends.

Stretch Zones Preceding Crack Growth

The areas of initial crack growth were examined with the SEM in the backscattered electron mode and with stereo pairs made with the SEM in the secondary electron mode. No stretch zones were seen in any sample tested in hydrogen although extensive zones were apparent on the oxygen tested samples (Fig. 17). Because higher loads were required for failure in oxygen a sample was tested in oxygen to the same displacement (load) that caused crack growth initiation in hydrogen and again no stretch zone was seen. Although the stretch zone did not form preceding crack growth in hydrogen, it should be emphasized that plastic deformation had occurred.

DISCUSSION

The precise means by which hydrogen influences the various deformation processes and final failure of metals in general are imperfectly understood. Indeed seemingly contradictory hypotheses have been advanced. Some theories are based on extremizing the stress deviators (shear stresses). Others require that the dilatation (triaxial tension) should be maximized. Some believe hydrogen enhances dislocation motion while others suggest hydrogen blocks it. Some suggest that hydrogen primarily influences elastic load strengths while others imply that the main contribution is in modifying the yield and post yield effects. The present study shows that any proposed mechanism of hydrogen influence on A–106 steel must be consistent with the following observations: (1) the crack path is macroscopically perpendicular to the maximum normal stress, (2) the crack path microscopically is primarily transgranular and not planar within each ferrite grain, (3) significant local plastic deformation accompanies the crack growth, (4) no large scale crack blunting or stretch zone develops before crack growth occurs and, in fact, crack tips may be sharpened, and (5) no strong dependence on triaxial stresses is required. These observations are not consistent with existing hydrogen embrittlement theories. Thus, when these data are coupled with an earlier tabulation[10] of selected observations which any universal hydrogen embrittlement must explain, additional support is obtained for the conlcusion that the current hydrogen embrittlement mechanisms are not universal and that either no universal mechanism exists or a new theory must be developed.

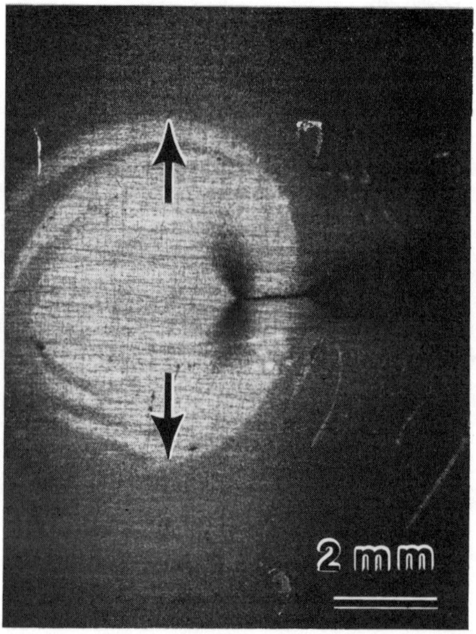

Fig. 14. Lüders band on the surface of a compact tension sample loaded in O_2. The width of the surface area covered by Luders bands is marked by arrows. Optical microscope; no etch.

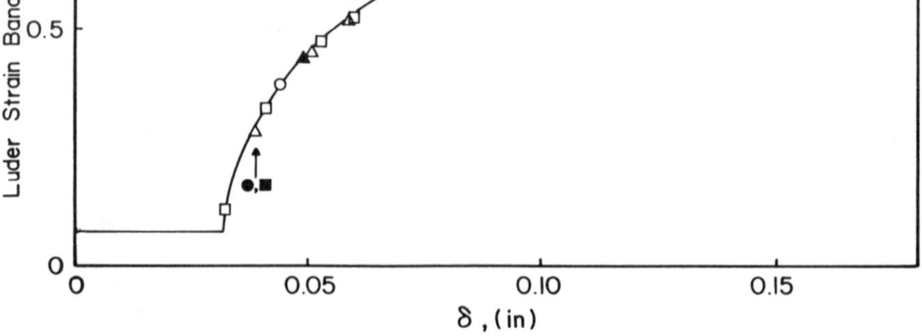

Fig. 15. Width of the area covered by Luders bands as a function of crosshead displacement, δ.

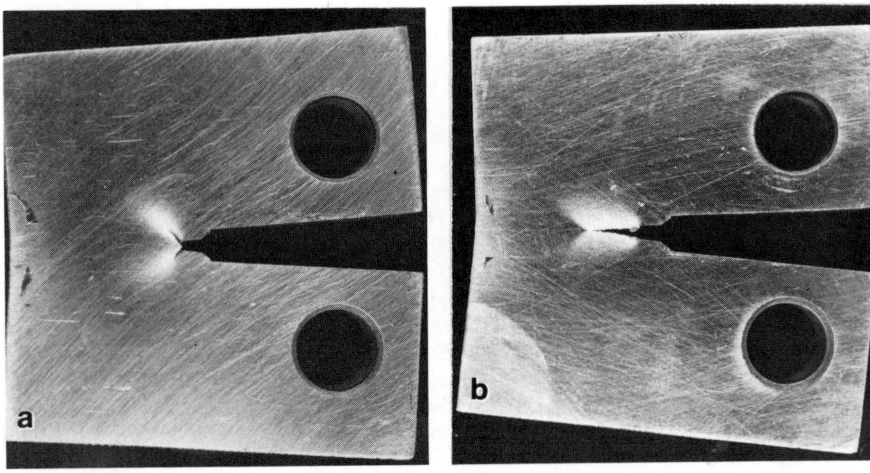

(a) (b)

Fig. 16. Surfaces of the compact tension samples loaded in (a) O_2 and (b) H_2. The bright areas are the result of surface roughness due to severe plastic deformation.

(a) (b)

Fig. 17. SEM of the initial crack growth areas of compact tension samples tested in (a) oxygen and (b) hydrogen. These fractographs were taken in the backscattered electron mode. The arrows show the crack growth direction. In (b) the lines mark the boundaries between the fatigue precrack, hydrogen assisted fracture and low temperature brittle fracture.

REFERENCES

1. Rogers, H.D., "Influence of Hydrogen on the Yield Point of Iron", *Acta Met.*, Vol. 4, p. 114 (1956).
2. Sudarshan, T.S., M.R. Louthan, Jr., and R.P. McNitt, "Hydrogen Induced Suppression of Yield Point in A–106 Steel", *Scripta Metallurgica,* Vol. 12, pp. 799–803, (1978).
3. Robinson, S.L., "Transmission of Gaseous Hydrogen – A Preliminary Technology Evaluation", *Hydrogen for Energy Distribution,* Institute of Gas Technology, Chicago, pp. 453–483 (1979).
4. Walter, R.J. and W.T. Chandler, *Effects of High-Pressure Hydrogen on Metals at Ambient Temperature,* Report NASA R–7780–1,2,3, Rocketdyne Division, Rockwell International, Canoga Park, CA, (Oct. 1973).
5. Walter, R.J. and W.T. Chandler, "Cyclic-Load Crack Growth in ASME SA–105 Grade II Steel in High-Pressure Hydrogen at Ambient Temperatures", *Effects of Hydrogen on Behavior of Materials,* A.W. Thompson and I. M. Bernstein eds., AIME, New York, pp. 273–298 (1976).
6. Nelson, H.G., "Hydrogen-Induced Slow Crack Growth of a Plain Carbon Pipeline Steel Under Conditions of Cyclic Loading", *Effects of Hydrogen on Behavior of Materials,* A.W. Thompson and I.M. Bernstein, eds., AIME, New York, pp. 602–611 (1976).
7. "Standard Test Method for Plane-Strain Fracture Toughness of Metallic Materials, E399", *ASTM Standards,* American Society for Testing and Materials, pp. 505–521 (1974).
8. Knott, J.F., "Micro-Mechanics of Fracture and the Fracture Toughness of Engineering Alloys", *Fracture 1977,* Vol. 1, D.M.R. Taplin, ed. Pergamon Press, New York, pp. 61–92 (1978).
9. Sailors, R.H., "Fracture Feature Anisotropy in a Martensitic Steel Plate", *Fractography – Microscopic Cracking Processes, ASTM STP 600,* C.D. Beachem and W.R. Warke, eds., American Society for Testing and Materials, pp. 172–189 (1976).
10. Louthan, M.R. Jr., and R.P. McNitt, "The Role of Test Technique in Evaluating Hydrogen Embrittlement Mechanisms", *Effects of Hydrogen on Behavior of Materials,* A.W. Thompson and I.M. Bernstein, eds., AIME, New York, pp. 496–506 (1976).

Stevens, Vander Voort, McCall, eds. Microstructural Science, Volume 8

GRAPHITIZATION STUDIES AT FOSSIL FUEL POWER PLANTS

William R. Pavlichko* and Alexander Solomon*

ABSTRACT

The purpose of this study was to determine the extent of graphitization in the main steam piping systems of several fossil fuel power plants. Graphitization embrittlement of welded joints in main steam piping made of carbon and carbon-molybdenum alloys has been known for more than 35 years. Generally main steam lines of high temperature plants built before 1948–1949 were made of carbon steels and many of these plants are still operating.

More than 70 welds were examined in 1978 and the results are presented including graphite morphologies, ratings and their relative lack of ductility. Procedures for repair or replacement of weld joints are discussed. Occasionally carbon steel piping is incorrectly used in areas other than main steam lines and several examples are presented. The susceptibility of friction welds to graphitization is discussed. The graphitization study has contributed to the safe operation of fossil fuel power plants.

INTRODUCTION

In January of 1943 a welded joint in a high-pressure steam line at the Springdale Station of the West Penn Power Company failed while in service. Subsequent metallurgical examination revealed that the carbon molybdenum steel piping had failed at a weld heat affected zone due to graphitization (the precipitation of graphite flakes from the steel matrix). Two possible mechanisms for graphitization are:

(1) Graphitization results as carbon atoms, migrating at high temperatures, collect over a period of time along a plane representing the AC_1 temperature of the steel. After manufacture, the AC_1 temperature (about 1333°F for low carbon steel) is reestablished in power plant piping by the welding process. Within a half-inch of either side of the weld, graphite nuclei are formed. During subsequent high temperature service, carbon atoms in random movement adhere to these nuclei to form semi-continuous layers of low strength graphite. In time, the strength, and ductility of this area is reduced to essentially zero.[1]

(2) Breakdown of carbide to graphite during service.[2]

As studies of this phenomenon continued, it was found that susceptibility to graphitization was generally related to the amount of aluminum used in deoxidizing carbon-moly steel (more aluminum – more graphitization).[2,3,4] The amount of silicon used for deoxidizing in aluminum-free steels also was believed to have an effect.[4]

The low carbon-moly piping weld HAZ were susceptible to graphitization because of the increase of operating temperatures to above 800°F. Research into the problem[2] resulted in the Metallurgy and Piping Task Force of the EEI Prime Movers Committee recommending the use of piping containing at least one-half percent chrome (1945).

---

*Consumers Power Company, Jackson, Michigan 49201 USA.

Many plants built in the 1940's with carbon-moly piping are still operating and require occasional sampling and weld rehabilitation. Thus, the problem is still present as evidenced by a recent graphitization failure in 1977 at a Penn Electric Power Plant which resulted in six deaths. Investigators of the failure concluded that severe graphitization was responsible for the fracture.[5]

Graphitization embrittles a weld section but stresses are still required for failure. [6,7]

We have been sampling main steam line piping in units believed to be susceptible, since the 1950's in order to prevent catastrophic failures.

WELD SAMPLING PROCEDURE

Boat-shaped longitudinal weld cross-sections were removed with the "weld prober" saw[6] from various weld joints in the main steam piping lines (Fig. 1). Hack-saws were used to remove sections where the saw could not fit. Over 70 weld cross-sections were removed for examination in 1978 (seven projects). The boat-section across the weld must be at least ½ inch deep from the OD (closer to 1 inch preferred) in order to obtain a sample with adequate cross-section for metallographic and bend tests. Samples from both upstream and downstream HAZ of each weld were obtained.

METALLOGRAPHIC EXAMINATION

Upstream and downstream sections of each joint were metallographically prepared and etched with Nital solution. Where the greatest amount of graphitization in the HAZ was observed, photomicrographs were taken at 100 and 500 diameters. The extent of graphitization was based upon a comparison of each of these photographs with standards illustrated in an article by Thielsch et al.[6]

The extent of graphitization is determined by distribution and morphology observed at different magnifications. Examination at 100X establishes the presence and amount while at 500X the degree and type are determined.

BEND TEST EXAMINATION

To further evaluate the extent of graphitization in the joints, slow-bend tests of the heat-affected zones were made using the guided bend test jig shown in Fig. 2. The specimens for the bend test were prepared by milling 1/8 inch thick slices from what remained of the boat-shaped segments after the metallographic specimens had been prepared. These slices were machined to 3/8 inch wide bend specimens. Specimens to be tested were bent by placing the heat-affected zone parallel to the axis of the plunger in the test jig. The specimens were bent to a point where either a 1/16 inch long crack appeared on the specimen or the specimen could not be bent further with more insertion of the plunger into the yoke (180°). The bend angle (the supplement of the angle formed by the two legs of the bent specimen) was measured for each specimen. Figure 3 illustrates a typical bend test specimen.

TEST RESULTS EVALUATIONS

If the metallographic rating was heavy or worse then weld sections were recommended for rehabilitation. If the bend test results indicated a section was brittle (1/16 inch crack or greater at less than 180°) replacement was recommended. Only one of the two criteria was necessary for weld replacement.

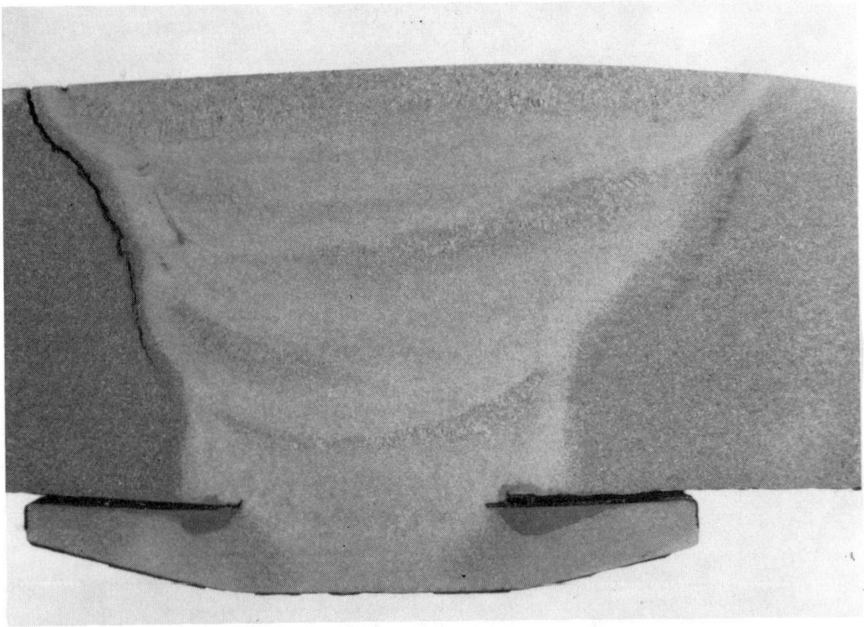

Fig. 1. Longitudinal weld cross-section taken after 230,000 hours of service at 900° F. An OD initiated crack which progressed through three-fourths of the wall may be seen in one HAZ (following the graphite). Graphite is also present in the other HAZ. 1X

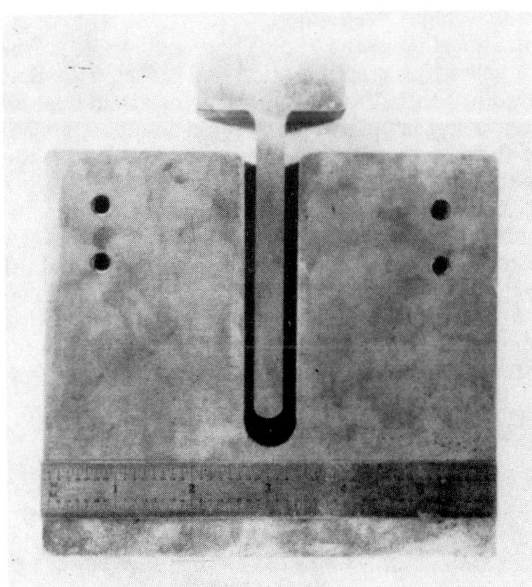

Fig. 2. Guided bend test jig.

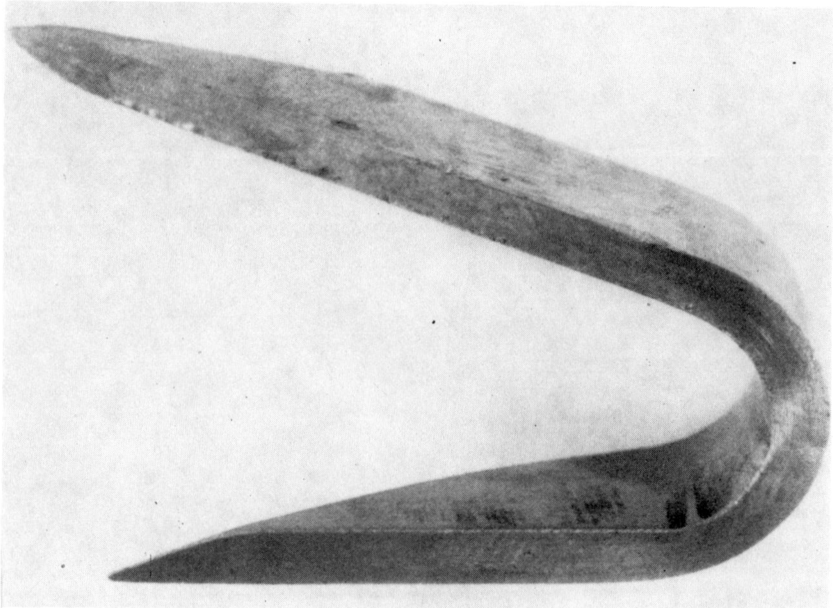

Fig. 3. Typical weld bend test specimen.

1700°F SOLUTION HEAT TREATMENT

Several sources[6,8] have recommended a solution anneal at 1700°F for two hours to put the graphite back into solution. This technique is somewhat difficult to accomplish in the plants because of the high temperature required and the sheer mass of large pipe sections which may sag if overheated.

We tried solution annealing in the 1950's and early 1960's; however, sampling in subsequent years revealed that graphitization had reoccurred. Because of this it was felt that the susceptibility to failure of any particular weld heat-affected zone could be best estimated by removing samples from the joint in question and subjecting the weld heat-affected zones to metallographic examination and bend testing followed by weld removal and rehabilitation if necessary.

Rehabilitation consisted of grooving out the heat-affected zone and rewelding followed by a postweld heat treatment:

> 1325 ± 25°F for a period of four hours per inch of metal thickness being welded. Slow heating up to temperature and controlled cooling at 400°F/hour down to 800°F and allowed to cool under insulation. Heating to be strip chart recorded with a minimum width of a circumferential heat band centered on the weld to be the width of the weld plus two-inches.

STUDY 1

The unit examined in Study 1 operated 37 years (1941 start-up) or nearly 200,000 hours at 900°F. The main steam piping was carbon—½ molybdenum, but the line was originally not considered susceptible to graphitization because the piping had a low aluminum content less than 0.05. Early researchers[2,3] showed that increasing aluminum increased graphitization rates so it was felt low-aluminum carbon-moly steel could be used.

Thus, this unit was not sampled until 1966 and at that time no heavy graphitization was found. Because of the Penn Electric 1977 failure, some old units were reexamined (1978) and two samples from this unit showed extremely severe graphitization. Eventually 21 welds were examined and 14 required immediate rehabilitation (heavy or severe rating). Most of the other welds were moderate.

Because of the extent and severity of graphitization in the main stream and auxiliary lines and to avoid future sampling programs which would be expensive in terms of outages, a total of 35 welds were finally replaced. It is expected that additional plant lifetime will be approximately ten years and no more graphitization work will be done at this unit. Various examples of graphitization observed are shown in Figs. 4 and 5. Some extremely severe sections were as serious as sections illustrated in the Penn Electric failure report.

STUDY 2

Four welds in the unit adjacent to the Study 1 unit were also examined in 1978. This unit had run 30 years (1949 start-up) and temperature was 875°F. Plant blueprints indicated the main steam line was ½ Cr—½ Mo and *no* graphitization was found after 30 years of service (Fig. 6). This was a good example of the longtime effect of chrome on retarding graphitization formation, also the effectiveness of the 1945 EEI Task Force recommendation.

STUDY 3

In late 1978 the No 2 HP heater 10 inch extraction line pipe of a fossil unit experienced a steam line leak upstream of and adjacent to the isolation steam extraction valve. Piping specifications indicated that the composition might be carbon steel. Operating conditions were 818°F and 400 psi. Service life was 23 years. A pressure gauge plug and a drain line were near the failure.

Visual Examination

The extraction line is shown in Fig. 7 "as received". "A" indicates the position of the pressure gauge plug and the drain line hole position and a pipe bulge may be seen. The position of the OD side of the throughwall crack adjacent to the valve weld may be seen on the left side of the photograph (arrow Fig. 7). The throughwall crack initiated at the ID and is shown in Figs. 8 and 9. A family of cracks may be observed with the throughwall crack (Fig. 8). The "crazed" cracking pattern is generally associated with thermal fatigue cracking. Cross-sections taken through the ID cracking showed up to 10 ID initiated cracks including the leak crack. An etched (nital) longitudinal cross-section taken through the cracks, indicates the relative positions of the family of cracks and the valve weld HAZ (Fig. 9). The throughwall crack initiated at the ID base metal, not at the weld HAZ-base metal interface. A graphite-related crack would initiate at the OD. A separate family of ID cracks were observed about 10—12 inches upstream from the throughwall crack. These crack positions are shown in Fig. 7 (B). ID cracks were also observed adjacent to both the 1 inch drain line hole and the ½ inch plug hole.

Metallography

Modern graphitization was observed in the weld HAZ of a section cut through the throughwall crack and adjacent weld. The graphite was nodular and followed HAZ contours. No rods or linkage were observed in the HAZ. Figure 10 shows the HAZ graphite at 100X.

Fatigue striations were observed on the fracture surface of the throughwall crack and are shown in Fig. 11 (SEM).

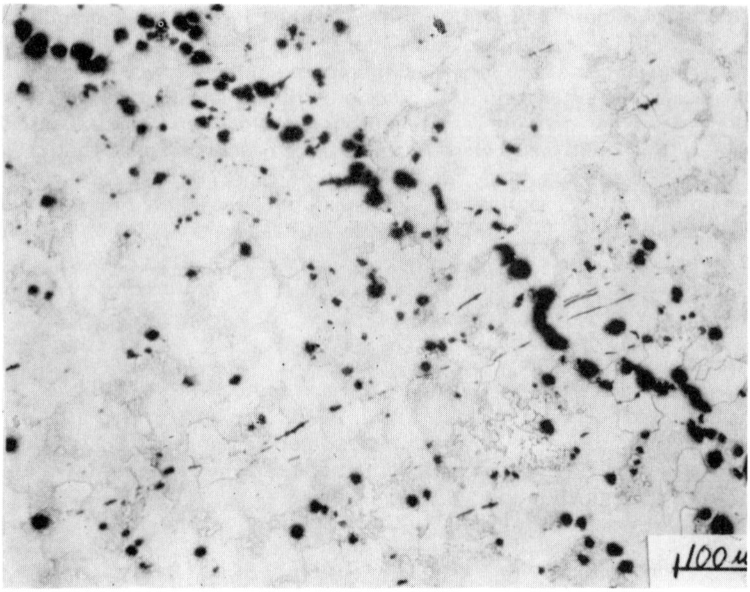

(a)

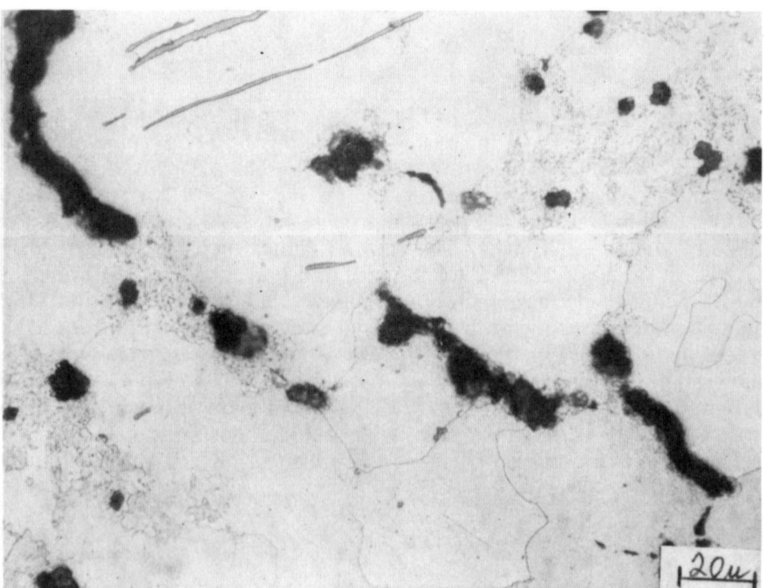

(b)

Fig. 4. Severe graphitization in the heat-affected zone in the upstream (pipe) side of FW–13, Unit 3, sample position 9:30. (a), 100X. (b), 500X.

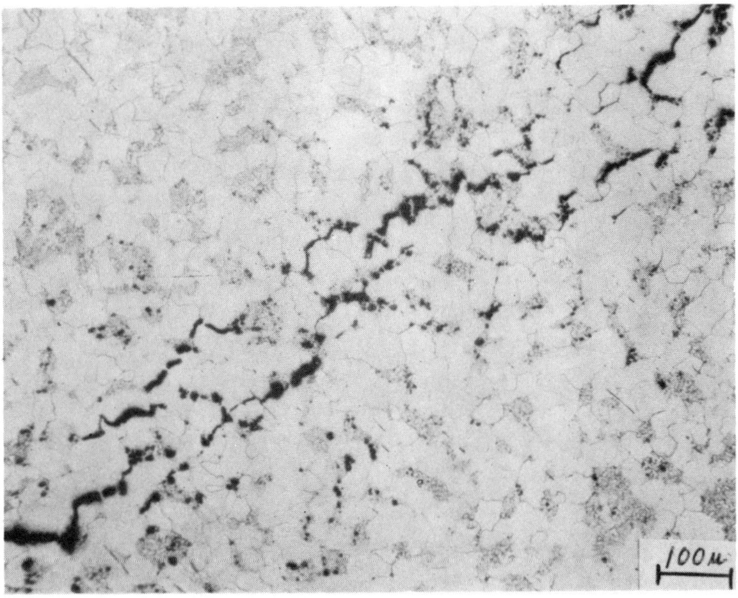

(a)

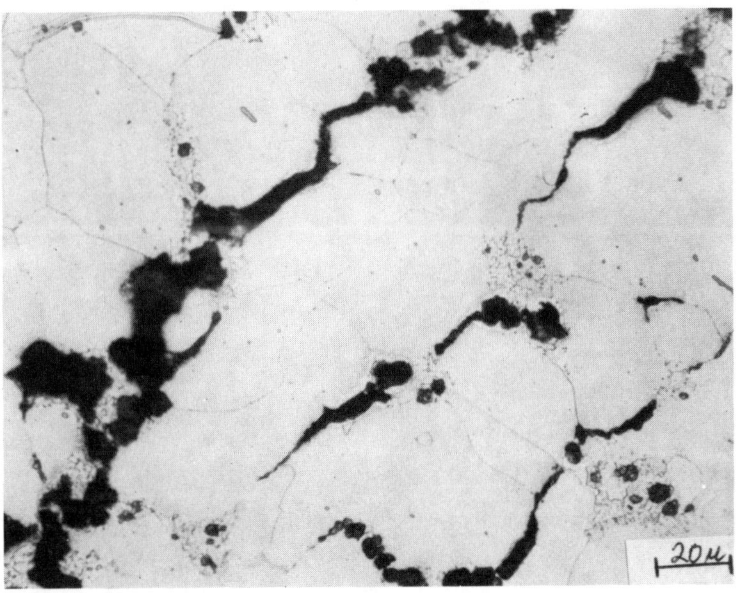

(b)

Fig. 5. Extremely severe graphitization in the heat-affected zone in the downstream (pipe) side of FW–13, Unit 3, sample position 10:00. (a), 100X. (b), 500X.

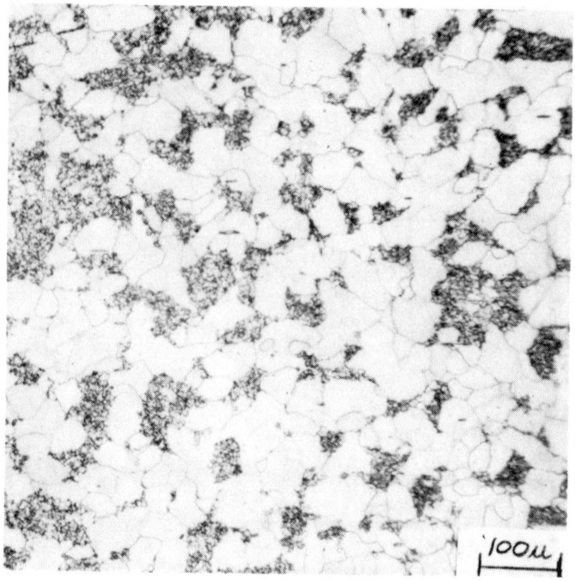

(a)

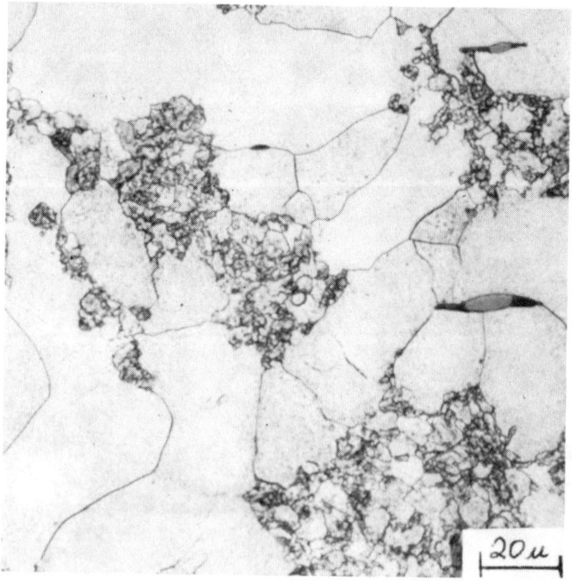

(b)

Fig. 6. No graphitization in the heat-affected zone in the upstream (pipe) side of FW–8, Unit 4, sample position 12:00. (a), 100X. (b), 500X.

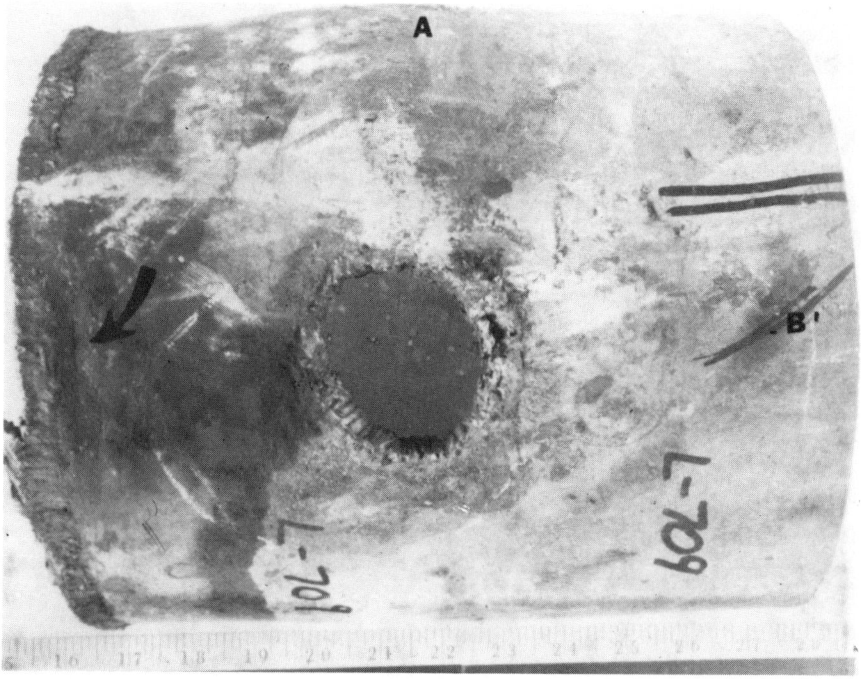

Fig. 7. A 10 inch extraction line pipe "as received". The throughwall failure OD position is indicated by an arrow. The flame cut hole is where the 1 inch drain line would be. Note bulge (A) between 1 inch line and ½ inch pressure gauge plug at top of pipe. Base metal ID cracking was observed at B. One-half inch plug is at top.

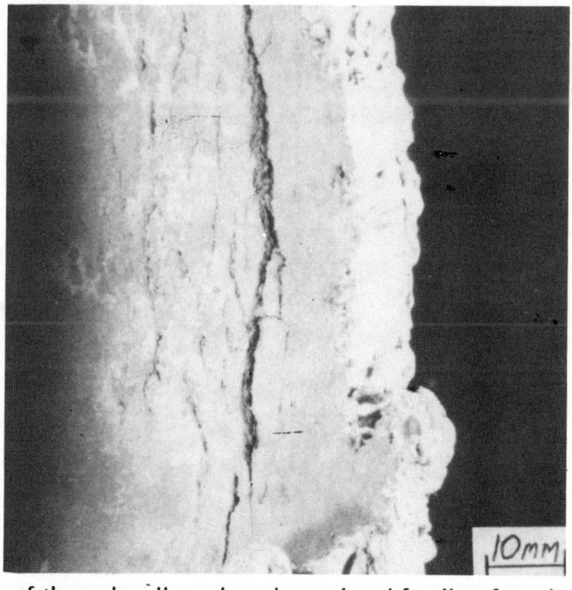

Fig. 8. ID view of throughwall crack and associated family of cracks. Right edge is flame cut through valve weld. 1X.

Fig. 9. Etched longitudinal cross-section taken through adjacent weld and failure crack. Position of crack relative to weld and HAZ zones (arrows) may be observed. The ID crack initiated in base metal not in HAZ regions.

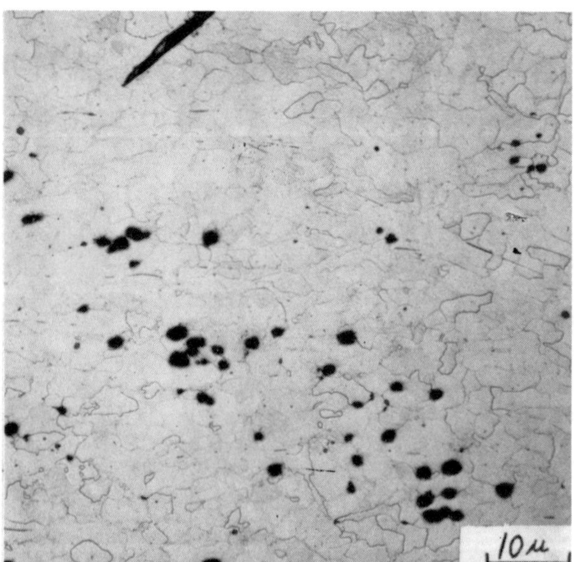

Fig. 10. Moderate graphitization in valve weld HAZ. Nodular graphite. 100X.

Fig. 11. Fatigue striations may be observed, with secondary cracking occurring at each step. High stress levels would produce secondary cracking. 230X.

Discussion

The ID initiated throughwall cracking adjacent to the valve appeared to be the result of thermal fatigue, caused either by repeated condensate carryover or plant temperature cycling or both. The fatigue cracking was not associated with nodular graphite observed in the adjacent valve weld HAZ, however, nodular graphite in base metal may have assisted crack growth. The carbon steel extraction line was operating at 818°F, somewhat above carbon steel material properties temperature limits (the bulge is indicative of some yielding occurring) and some pearlitic decomposition should occur over 23 years. The two small lines (drain and plug) may be accelerating the thermal fatigue by their close positions to the valve.[9]

Recommendations were to X-ray all welds in the line that may be candidates for ID cracking and to obtain graphitization boats from field welds exposed to 818°F temperatures to determine extent and severity of graphitization.

STUDY 4

Two graphitization samples were removed from field welds on the HP heater steam extraction line piping (Study 3) for metallographic examination to determine the extent of graphitization. The results were as follows: (Figures 12–14) The HAZ of side 1 of sample 1 had a 50% throughwall crack; the other side HAZ had extremely severe graphitization. The HAZ of side 1 of sample 2 had severe graphitization while the HAZ of side 2 had heavy to severe graphitization. Recommendation was to replace the line as soon as possible.

STUDY 5

A section of high temperature superheater tubing from near the outlet header of the superheater assembly of a coal fired boiler was sent to the Metallurgical Laboratory for

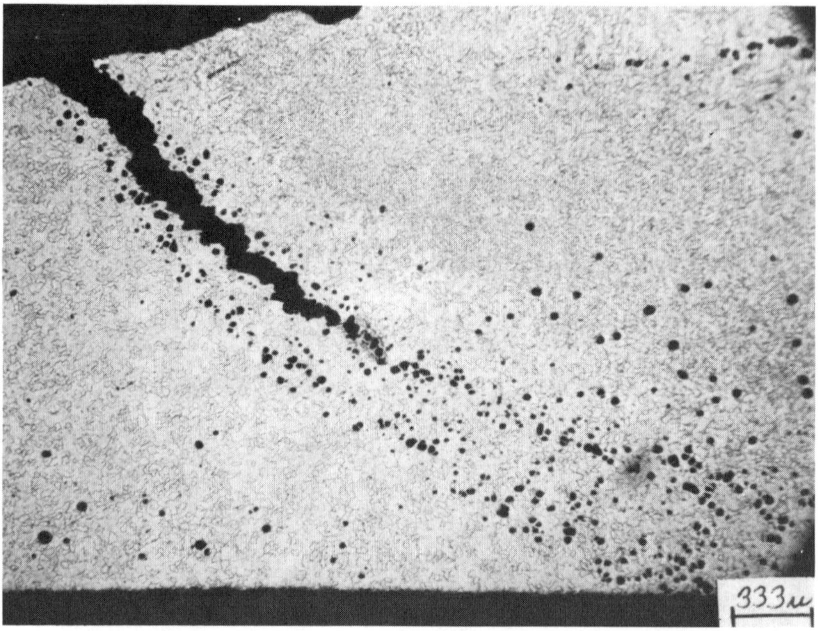

Fig. 12. The HAZ of side 1 of sample 1 boat is shown with a 50% throughwall crack (OD initiated) following the HAZ contour. No graphitization rating assigned. 30X.

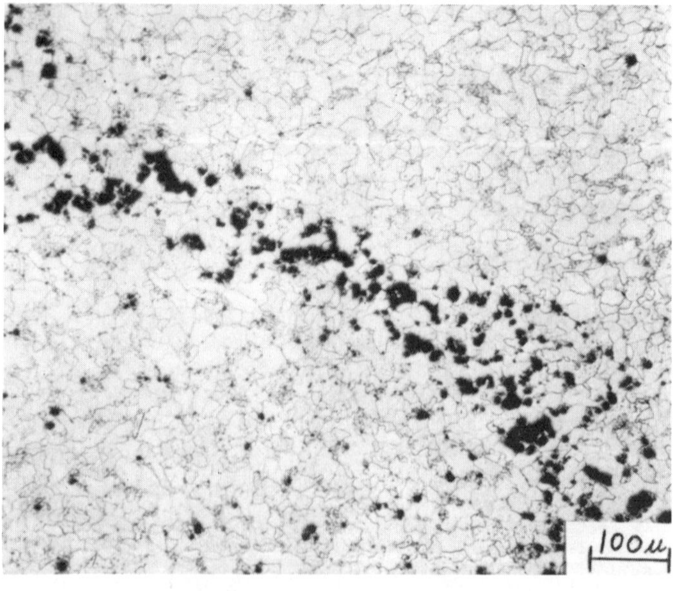

Fig. 13. Extremely severe graphitization in HAZ of side 2 of sample 1 boat. 100X.

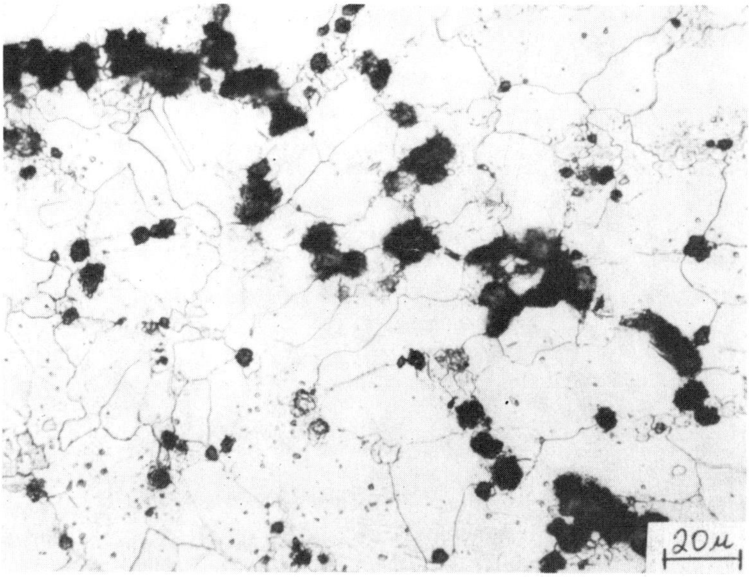

Fig. 14. Magnification of Fig. 13. 500X.

examination to determine the extent of graphitization at a friction weld and to obtain chemical analyses of the tubing on both sides of the weld. The tubing on the outlet header side was believed to be a carbon-moly grade, while on the other side of the weld, specifications indicated that the tubing was low-Cr alloy. The friction weld was in service 29 years.

The outlet header side of the friction weld was analyzed to be carbon—½ molybdenum alloy while the other side was a 5 Cr alloy.

Figure 15 shows the friction weld "as received". The friction weld, HAZ and base metal of both sides are shown in Fig. 16.

No graphitization was found on the carbon-molybdenum side (none on the 5 Cr side either).

No graphite problem was found at the friction weld examined, however, the welds made with multiple-passes (at the header) might be more susceptible to graphitization than friction welds because of the longer time at temperature and recurring reheating as every bead pass is made. The formation of initial graphite nucleii during the welding process (which acts as sites for future graphitization growth) is believed to be a diffusion process involving migration of carbon atoms to preferential sites in the HAZ-base metal regions. Diffusion movement of carbon is controlled by time and temperature of the welding operations; thus, multiple-pass welds with longer heating times could be more likely candidates for graphitization than friction welds.

Sampling of multiple-pass welds on the header is being done, also several friction welds from other unit superheaters have been examined and no graphite observed.

CONCLUSIONS

(1) Graphitization sampling programs in older steam plants are necessary for plant worker safety and equipment protection.

(2) Over a 40 year plant lifetime serious graphitization forms in carbon-moly main steam piping whether the steel has high/low aluminum or high/low silicon.

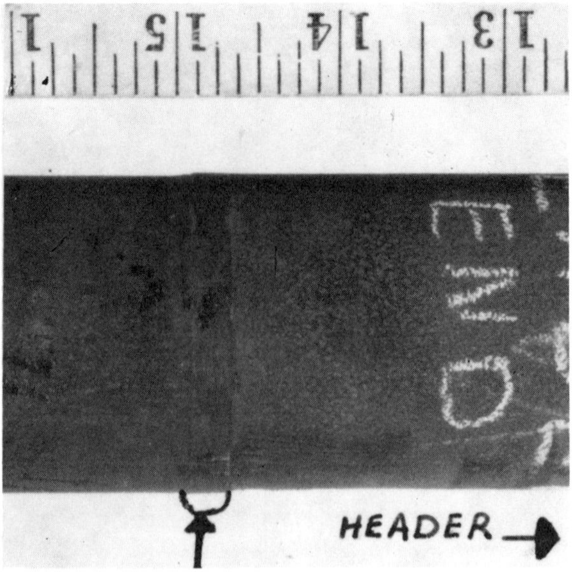

Fig. 15. Superheater tube "as received." Arrow indicates friction weld.

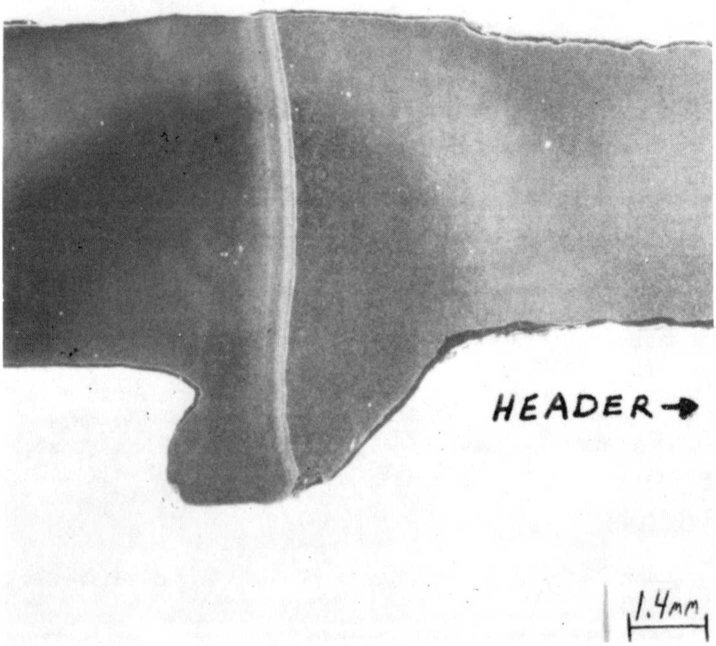

Fig. 16. Longitudinal cross-section of friction weld as etched. 7X.

(3) ½ Cr–½ Mo piping welds with 30 years of main steam line service at 875°F showed no evidence of graphitization.
(4) Friction or resistance welds seem to be much less susceptible to graphitization than multiple-pass welds.
(5) Weld normalization was abandoned in favor of weld removal and replacement.

ACKNOWLEDGEMENTS

The authors wish to thank Messrs. Arnold Keil, Ray Lasich, John Knight and Claude Sterett for their work on specimen preparation and examination. Discussions with Mr. W.D. Kurtz were helpful.

REFERENCES

1. Drakopel, B.E. and Kusterer, J.N., "Graphitization Control at Baltimore Gas and Electric." BG&E Report. (1978).
2. Conant, W.G. and Reich, W.A., "Graphitization studies of Materials for High-Temperature Service in Steam Plants," presented at the ASME Annual Meeting, New York, NY, (Dec. 2–6, 1946).
3. Rohrig, I.A. and McCutchan, A., "Influence of Postweld Heat-treatment on Graphitization," presented at the ASME Annual Meeting, New York, NY, (Dec. 2–6, 1946).
4. Weaver, S.H., "General Discussion of Graphitization Papers," *Trans. ASME,* Vol. 68, p. 631 (1946).
5. Rohrig, I.A., Williamsburg Station Accident Investigation, Investigation Task Force, (Dec. 14, 1977).
6. Thielsch, H., Phillips, E.M. and Jerome, E.R., Jr., "Considerations in the Evaluation of Graphitization in Piping Systems," presented at the AWS National Fall Meeting, Chicago, IL, (Nov. 1–5, 1954).
7. Abele, J.B. and White, A.E., "Report on Graphitization Studies on High-Temperature Welded Piping of the Philadelphia Electric Company," ASME, (1948).
8. Hopping, E.L. and White, A.E., "Report on Graphitization Studies on High-Temperature Welded Piping of the Philadelphia Electric Company," presented at the Annual Meeting, New York, NY, (Dec. 2–6, 1946).
9. Coffin, L.F., "Symposium on Effect of Cyclic Heating and Stressing on Metals at Elevated Temperatures" (ASTM Special Tech Pub. No. 165), p. 31, Philadelphia, PA. (Amer. Soc. Test. Mat.) (1954).

Stevens, Vander Voort, McCall, eds. Microstructural Science, Volume 8

FATIGUE FAILURE INVESTIGATIONS IN TURBINE WHEELS

Nathan A. Tiner*

ABSTRACT

The radial and axial inflow turbine wheels have been extensively used in cryogenic and aircraft industries and steam generating plants. The design problem in these wheels is fundamentally one of fatigue loading and the effect of corrosive environment conditions. This paper describes various fatigue failures encountered in the turbine wheels, and illustrates the use of electron microscopic observations in identifying different failure modes. A technique is also discussed to evaluate the effect of corrosion on the fatigue strength of a turbine material and how to incorporate the data into design for fatigue life.

INTRODUCTION

In general there are two types of gas turbines; axial inflow and radial inflow turbines. The axial inflow turbines are used nearly exclusively in the aircraft industry and steam generating plants. The radial inflow turbines or turboexpanders have been introduced recently in the process of modern air separation, cryogenic petrochemical gas recovery and LNG peak shaving plants to provide low temperature refrigeration. In contrast to the axial inflow turbines, the gas in the radial inflow turbines enters the stator or nozzle radially and leaves the rotor or wheel axially at a smaller radius and relatively low velocities. The result is somewhat higher efficiency and less vibration problems at variable speeds for radial inflow turbines.[1]

The radial inflow turbines used in air separation can be of two types; low power (up to 100 horse power) or high power (100 to 2000 horse power). In the low power turboexpander the energy is not recovered but absorbed by an oil brake. In the high power turboexpander the power is absorbed by a booster compressor. The turboexpanders used in the LNG peak shaving and petrochemical gas processing plants are of high power type (100 to 8000 horse power range) and usually are designed with a process compressor directly driven by the expander.[2]

Most turbine wheels are subjected to mean stresses resulting from centrifugal plus offset forces, and to alternating stresses due to different gas flow variables, like velocity and flow direction, pressure, temperature, density, and to the stress concentration factor. For a wheel the design problem is fundamentally one of fatigue loading. The design is further complicated if the normal operation of the turbine wheel involves corrosive environment conditions, such as gases containing corrosive liquids, sulfur and the like.

A complete failure analysis of the turbine wheel often is a very useful method of improving the design. The practicing design engineers frequently rely on examining a good failure for successful service life evaluation. This paper describes examples of principal types of fatigue failures investigated by the author in turbine

*Microprocess Laboratory, Laguna Beach, California 92651 USA.

wheels by the use of electron microscopy. A testing technique is also discussed to evaluate the effect of a corrosive service environment on the fatigue strength of a turbine material, and how the data can be incorporated into the design for fatigue life.

PRINCIPAL FATIGUE FAILURE MECHANISMS

Fatigue failures in most turbine wheels involve low cycle fatigue, high cycle fatigue, or corrosion fatigue. The identification of the specific characteristics of these different fatigue failures by electron microscopy are of great engineering importance. Some examples described below will illustrate the point.

Low Cycle Fatigue Failure

An example of a fatigue failure in a radial inflow turbine at low operating cycles is shown in Fig. 1. The turbine wheel was failed at the start up operation in an LNG peak shaving plant. The turbine wheel was made from 7075–T7452 aluminum alloy forging according to MIL–A–22771. One of the blades was broken near the disc. The tips of the other blades were damaged by the broken metal piece. The fracture surface of the blade exhibited coarse fatigue striation markings with striation spacing of 0.005–0.008 inch (0.12–0.20 mm) near the tip. The remaining fracture area showed predominantly dimple markings resulting from tension overloading (Fig. 2).

Each fatigue striation in general indicates one load cycle and the spacing between striations represents the progress of the crack front during one load cycle.[3] 120–200 micron striation spacing observed indicated a very high service load, near or above the yield strength of the alloy, was applied on the blade surface. Further detail analysis of the blade surface near the fracture initiation site by electron microprobe using energy dispersive X-ray spectroscopy revealed that the blade surface was impacted by solid iron particles in the gas flow causing overstressing of the turbine wheel beyond the design load.

High Cycle Fatigue Failure

High cycle fatigue failures often occur when turbine wheels operate close to the natural or critical frequency of the blades, and the wheels develop large vibrations. This problem usually is handled by protective design, that is by adjusting the mass distribution of the blades to eliminate operation near the natural frequencies and to offset flutter.[4]

An example of a high frequency fatigue failure is shown in Fig. 3. This turboexpander wheel was made of a 7075–T7352 aluminum alloy forging as in the previous example. The wheel was designed to operate at a normal speed of 33,000 rpm in a cryogenic LNG process plant. The fracture was clearly initiated at the fillet root of the blade and propagated nearly to the entire cross section by fatigue. The fracture area near the disc showed dimple rupture by overloading (see Fig. 4). The fatigue strations had 10–15 microinch (0.25–0.38 micron) spacing. This generally is indicative of high cycle fatigue crack propagation in aluminum alloys.

Metallurgical evaluation showed the wheel material was within the strength requirement for 7075–T7352 aluminum alloy, and the field records indicated the turbine was operating within the design limits. Additional vibration tests were made to evaluate the natural frequency of the wheel blades. The resonant wave trains were determined at the blade tip for different frequencies by means of a high frequency transducer, and the quality of the resonant waves ($Q=\Delta Hz/Hz$) were measured. The frequency-speed diagram (Campbell diagram) for different nodes were

Fig. 1. Turboexpander wheel made of 7075–T7352 alloy forging failed in an LNG peak shaving plant at the start up.

Fig. 2. Scanning electron micrograph showing the fracture surface of a failed blade with coarse striations. Magnification 90X.

Fig. 3. A turboexpander wheel made of 7075–T7352 alloy and a broken blade failed in service in an LNG plant.

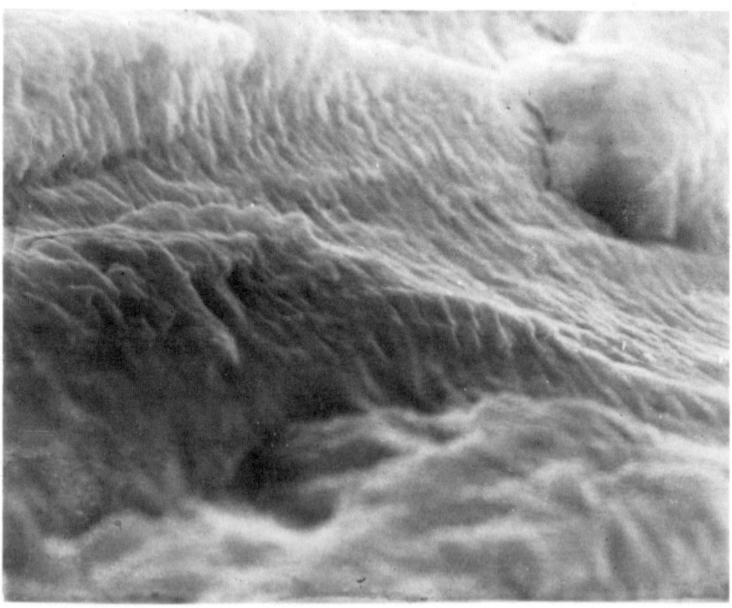

Fig. 4. Scanning electron micrograph showing the fracture surface in the fatigue region with fine striations. Magnification 3,500X.

plotted[5] (see Fig. 5). It was noted that the wheel was operating close to the dangerous critical speed at 33,000 RPM. The design of the blade geometry and mass distribution had to be modified to reduce vibration stresses in the fillet area where the fatigue crack occurred.

Another example of a high frequency fatigue failure is shown in Fig. 6. This is a low duty gas compressor wheel that operated in an LNG Tanker. The wheel was made of 304L stainless steel, and machined by an electrical discharge method with a fillet radius of 0.010–0.015 inch (0.25–0.38 mm). The fatigue cracks were initiated at the base of the blade near the tip at both the inlet and exhaust sides. The fatigue striations were distinct and the striation spacings varied from 4 to 7 microinch (0.10–0.18 micron) at the initiation site near the sharp edges and terminated near the mid section of the blades (see Fig. 7).

The compressor turbine wheel was operating at a speed of 15,000 rpm and apparently within the design stress levels in service. Three similar failures reported in the turbines prompted a critical analysis of the failed blade material. Microscopic examination of the cross section of the blade roots revealed a distinct hard surface layer about 0.001 inch (0.025 mm) deep at the exhaust side of both the leading and trailing edges. Apparently electrical discharge machining produced an extremely hard brittle layer on the blade surface, and subsequent vapor honing to remove the brittle layer from the fillet roots was not effective. It is generally known that electrical discharge machining produces a thermally induced phase transformation in metals and tends to reduce considerably their fatigue life[6].

Corrosion Fatigue Failure

Fatigue fractures when accelerated by corrosive environment is referred to as corrosion fatigue. The fracture surface resulting from corrosion fatigue generally exhibits a mixture of fatigue striations and stress corrosion features[3].

A classical corrosion fatigue failure that occurred in an axial inflow steam turbine wheel is shown in Fig. 8. The turbine wheel had 199 blades and was made of A286 austenitic nickel chromium alloy steel forging per AMS 5737. It operated in superheated steam at 700° F. One of the blades was found completely broke, and a number of blades showed evidence of root cracking on the steam inlet and exhaust tips on both concave and convex sides.

The cracked surface of the tip of a blade as observed by scanning electron microscopy is depicted in Fig. 9. The micrograph clearly shows the striation markings resulting from crack propagation under vibratory stresses and cleavage lines resulting from transgranular stress corrosion cracking. The steam was found to contain chlorides, and the fatigue cracking was clearly accelerated by the corrosive attack on the A286 alloy.

Forged aluminum alloy 7075–T7352 wheels are found susceptible to corrosion fatigue when in service in sour natural gas containing sulfur. Figure 10 illustrates a typical fracture surface of a blade failed in less than 100 hours service in an ethane extraction plant. The fracture showed fatigue striation markings and also regions of intergranular fracture indicating possible failure by corrosion fatigue.

SIMULATED CORROSION FATIGUE TESTS

Whether or not a turbine material is susceptible to corrosion fatigue in a certain service environment is best determined by comparing the fatigue limit or the rate of fatigue crack propagation in neutral atmosphere with that in the simulated service environment. Such a comparison test was made for 7075–T7352 alloy used in an LNG process plant.

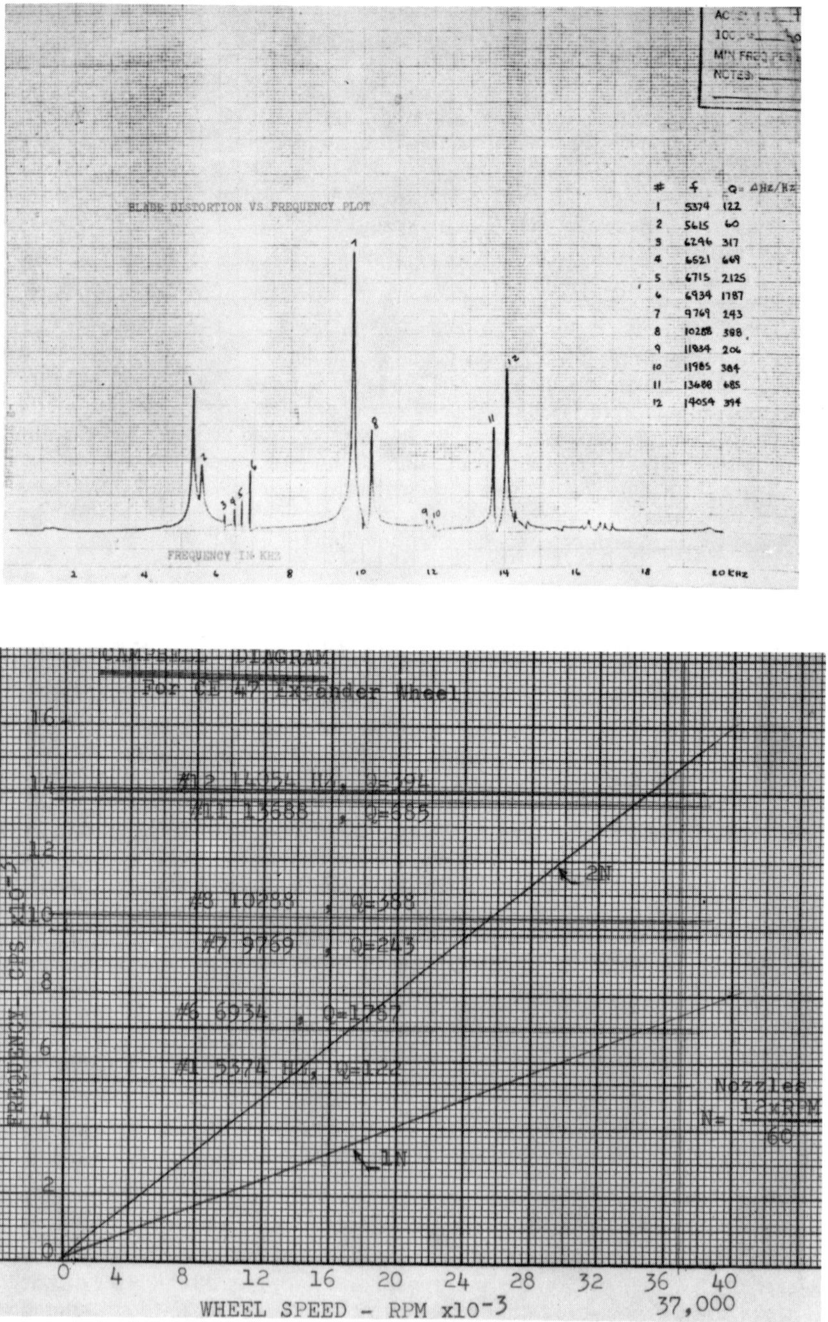

Fig. 5. The turbine blade resonance curves and the Campbell diagram to estimate dangerous critical speeds.

Fig. 6. Scanning electron micrograph showing the sharp edge of a compressor turbine blade and a fatigue crack initiation site near the exhaust side. Magnification 50X.

Fig. 7. Transmission electron micrograph of the fracture from the fatigue initiation site showing fine striations. Magnification 6,500X.

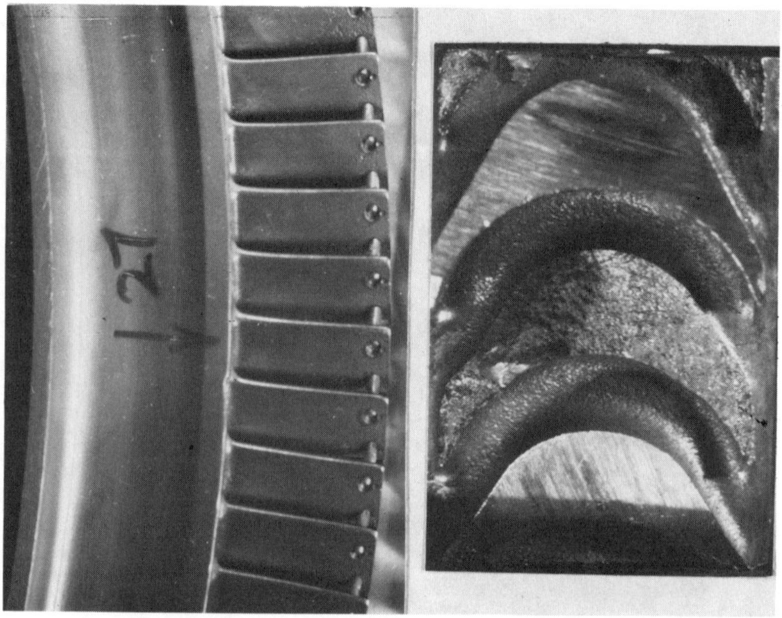

Fig. 8. A steam turbine wheel section and a blade failed in service at the inlet and exhaust tips by corrosion fatigue.

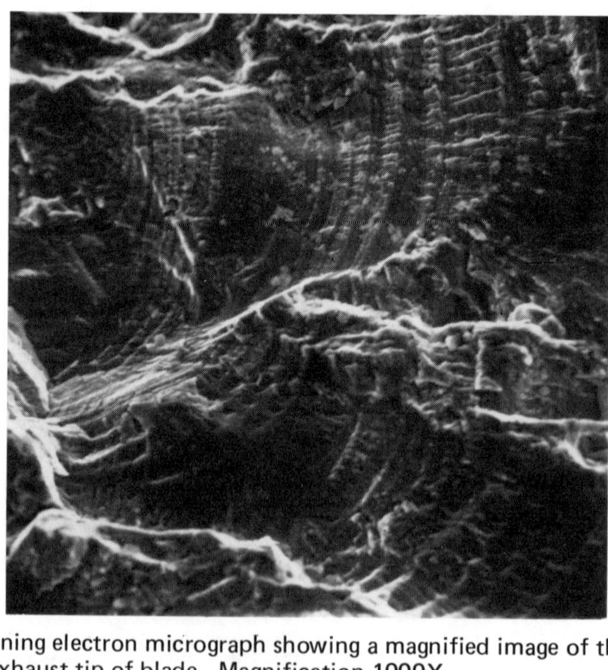

Fig. 9. Scanning electron micrograph showing a magnified image of the fracture area at the exhaust tip of blade. Magnification 1000X.

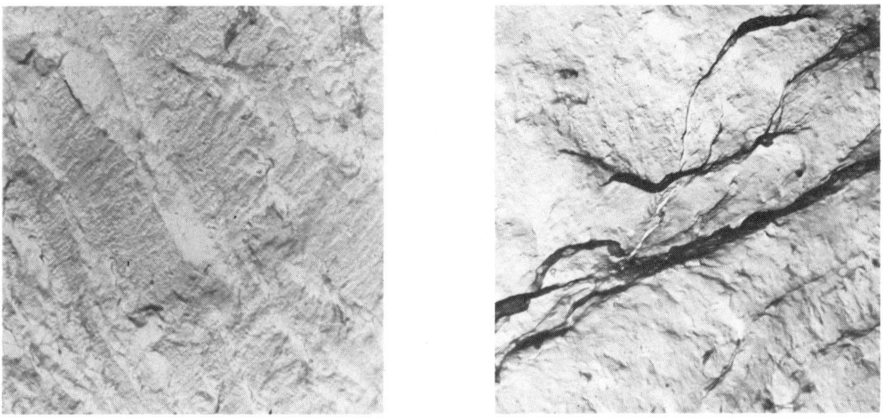

Fig. 10. Transmission electron micrographs of the fracture surface of a 7075–T7352 alloy expander wheel blade failed in service in an ethane extraction plant presumably by corrosion fatigue. Magnifications 5,000X.

Notched fatigue test coupons were prepared from the 7075–T7352 alloy forging with a stress concentration factor K_t=3.4. The tensile strength was 156 ksi and the yield strength was 146 ksi. The simulated service gas was 95% methane, 5% ethane and 15 ppm H_2S with –76° F(–60° C) dew point. The neutral atmosphere used was commercially pure 99.995% N_2 with –70° F(–56.6° C) dew point.

The test equipment consisted essentially of a thick walled tube retort fitted to a testing machine. The test coupon was attached to the fixture arm in the retort. In testing the gas was introduced into the system from pressurized steel bottle through a regulator and delivered to the retort. A gas flow rate of 10–20 ft^3/hr (283–566 liter/hr) was maintained throughout the test. The gas pressure inside the retort was 325 psig(22.8 kg/cm^2), and temperature was 120F(49C)(see Fig. 11).

Load was applied to the test coupon with a MTS Closed Loop Hydraulic Test System. The test machine had three conditioners for closed loop control, and readout of strain, load and stroke. The output signals from the signal conditioners were put into the AE1–20 Analog Computer at 0.001 to 15 Hz frequency range so that the data, such as plastic strain, total strain, hysteresis loop area, crack initiation, crack propagation, etc. can be directly plotted.

The method used in fatigue testing was a multiple step program developed at the University of Illinois with the MTS Cyclic Load Test Machine[7]. It is based on the fact that the stress-strain hysteresis loop changes when a crack is formed (see Fig. 12). In most cases, as in the present case, the hysteresis loop is so narrow that a plotter is not sensitive enough to detect crack initiation. In that case an analog computer circuit is used to calculate the plastic strain component of each hysteresis loop. The total strain is detected with a strain gage. The computer output is constant as long as there is no change in the material. When a crack is formed,

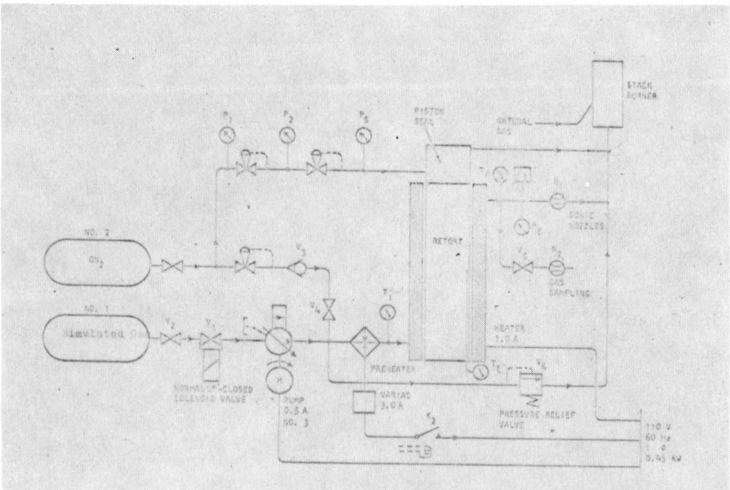

Fig. 11. Schematic of the test set up.

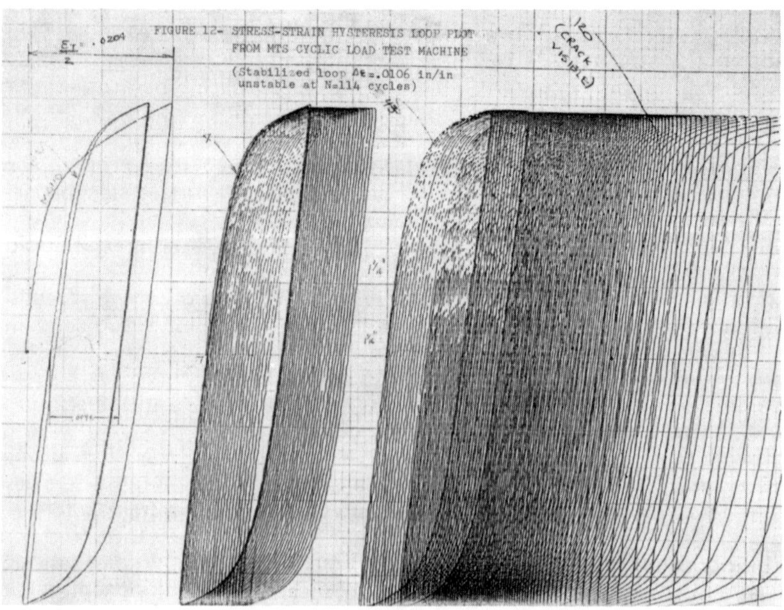

Fig. 12. Stress-strain hysteresis loop plot from MTS cyclic load test machine.

the plastic strain component becomes large and the computer output increases. The method is able to detect cracks as small as 0.002 in. (0.05 mm) in depth.

Fatigue tests were conducted using a tension-tension cyclic load with a stress ratio R=0.1. The peak loads used were 20,000 psi (1.4 kg/cm^2) to obtain a failure between the range of 10,000–40,000 cycles in the inert atmosphere. Three fatigue tests were made in an inert atmosphere and three fatigue tests in a simulated service gas environment. The computer output signals for hysteresis loop area at various time periods (or number of cycles) for each test are plotted. The number of cycles at which the plastic strain begins rapidly to increase (the crack initiation cycle), as well as the total number of cycles to fracture are determined. The plots are shown in Fig. 13.

It was noted that the number of cycles for fatigue initiation in the inert gas and in the service gas were not significantly different. However, the number of cycles to fatigue fracture (fatigue life) was reduced from 41,500 to 37,300 (about 10%) in the service gas compared to the inert atmosphere. The scatter, as judged by standard deviation, was less than 5%. The corrosive effect on fatigue noted above can be incorporated into the fatigue design curve for the 7075–T73 alloy at room temperature, shown in Fig. 14[8], by shifting the cycles to failure to the left by about 10%. This would reduce the fatigue strength of the alloy by about 8.5 percent.

CONCLUSIONS

Investigations by electron microscopy of the fracture surfaces of turbine wheels failed in service by different mechanisms, of fatigue are shown to be a useful technique in correcting errors in the design and/or fabrication of turbine wheels.

When a radial and/or axial inflow turbine is used in service in a corrosive environment, the fatigue life of the blades is often reduced. The amount of reduction can be evaluated by measuring the fatigue life in a simulated service environment and in a neutral atmosphere. This corrosion effect on fatigue can be incorporated into the design for fatigue life of the turbine wheel material.

ACKNOWLEDGEMENTS

The author gratefully acknowledges the support of Airco Cryogenics, Division of Airco Inc., for illustrations of the different fatigue failure modes in turbine wheels, and Gulf Oil Canada Limited for corrosion fatigue tests which were conducted with the help of Dr. C.B. Gilpin at the Engineering Research Center, Long Beach State University.

REFERENCES

1. H.D. Linhardt, "The Performance and Design of Radial Inflow Turbines," Fluid Dynamics Institute Thayer School of Engineering, Hanover, New Hampshire, (August 23–27, 1976).
2. H.D. Linhardt, "Cryogenic Turboexpanders," *LNG/Cryogenics,* (February/March 1971).
3. N.A. Tiner, *Failure Analysis with the Electron Microscope,* Fox-Mathis Publications, Los Angeles, California (1973).
4. D. Japikse, "Life Evaluation of Advanced Turbomachinery," Fluid Dynamics Institute Thayer School of Engineering, Hanover, New Hampshire, (Aug. 23–7, 76).
5. W. Campbell, "The Protection of Steam-Turbine Disk Wheels from Axial Vibration," Spring Meeting, Cleveland, Ohio, The Am.So.Mech.Eng. (May 1924).

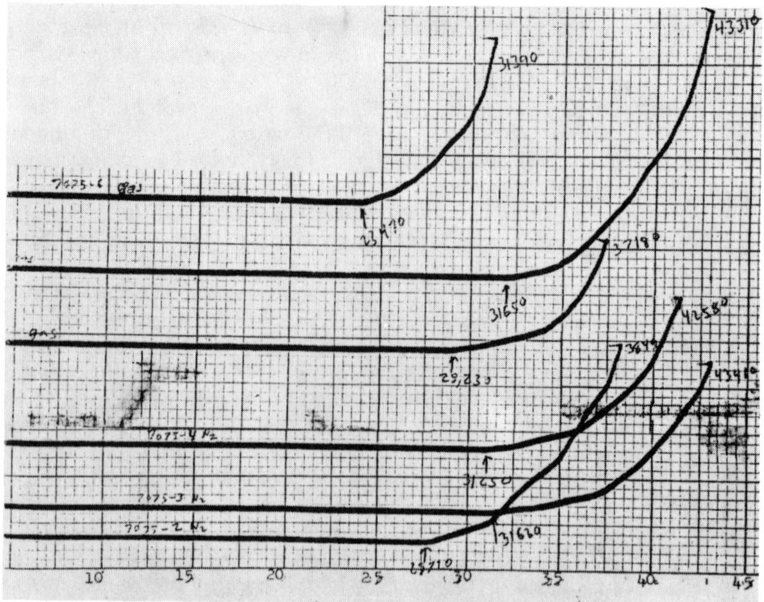

Fig. 13. Number of cycles vs. plastic strain plotted by computer output signals.

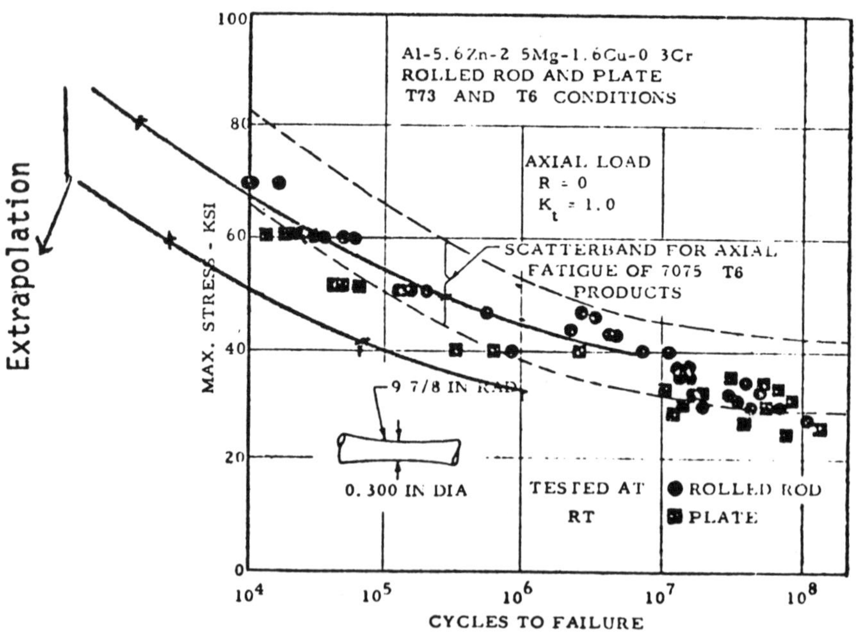

Fig. 14. Axial fatigue curves at room temperature for 7075 aluminum alloy in T73 and T6 conditions.

6. *Metals Handbook, Vol. 3 Machining,* p. 233, Am. Soc. Metals, Ohio (1967).
7. R.E. Peterson, *Stress Concentration Design Factors,* John Wiley and Sons, New York (1953).
8. *Aerospace Structural Metals Handbook,* Air Force Materials Laboratory TR–68–115, (1974).

Stevens, Vander Voort, McCall, eds. Microstructural Science, Volume 8

INVESTIGATION OF A HIGH–TEMPERATURE HERMETIC SEAL FAILURE*

Howard V. Rhude**

ABSTRACT

A high-temperature hermetic seal, incorporating two platinum lead wires brazed into an alumina insulator with a gold-copper brazing alloy, failed electrically at 1150°F (621°C), although it was designed to operate at temperatures up to 1300°F (704°C). The seal was used in a pressure transducer and the platinum leads are part of the electrical circuit of the transducer. Electrical checks indicated that the pressure transducer was shorted. Optical microscopy after disassembly of the pressure transducer showed that a bridge having a copper-red color had formed between the platinum leads.

Metallographic examination of a section through the hermetic seal revealed that the surface of the braze joint (62.5 wt% Au-37.5 wt% Cu) had degraded and material from the braze joint had formed a bridge between the two platinum leads. SEM examination (with energy dispersive X-ray analysis) showed the bridging material to be mostly copper. Ion microprobe mass analysis (IMMA) showed that lead and niobium were also deposited on the wall of the chamber enclosing the hermetic seal. Although the mechanism that caused the failure is not fully understood at this time, it is believed that residual gases in the pressure transducer, introduced during the manufacturing process, reacted with the gold-copper braze alloy at elevated temperatures causing an exothermic reaction which drove the surface temperature of the braze alloy even higher and caused copper to be evaporated from the alloy, along with impurities of lead and niobium.

As part of Argonne National Laboratory's Reactor Safety Program, it was necessary to develop a very small pressure transducer that could be attached to a nuclear fuel test assembly and be operated at temperatures up to 1300°F (704°C). Details of the pressure sensing element in the pressure transducer are proprietary information and will not be discussed. The sensor, however, is enclosed in a Kovar case to which a high-temperature hermetic seal has been welded.

Some of the details of the hermetic seal are shown in Fig. 1. The outer shell is Kovar to which has been brazed an alumina insulator. Passing through the alumina insulator are two small platinum tubes, which are also brazed to the alumina insulator. In order to braze to the alumina, a manganese-molybdenum mixture is applied to the alumina and sintered in a hydrogen atmosphere to metallize the surface of the alumina. The brazing alloy used to braze the Kovar and the platinum to the metallized alumina is 62.5 wt% Au-37.5 wt% Cu. Platinum wires are passed through the tubes and welded to the tubes to complete the hermetic seal.

As a final fabrication procedure, the pressure transducers are heated to 1300°F (704°C) for 50 h for creep stabilization and final calibration. Of the first ten pressure transducers fabricated, three failed within the 1h of this heating procedure when the temperature reached 1150°F (621°C). The failure was detected as an apparent short in the pressure transducer circuitry.

---

*Work performed under the auspices of the U.S. Department of Energy.

**Reactor Analysis and Safety Division, Argonne National Laboratory, Argonne, Illinois 60439 USA.

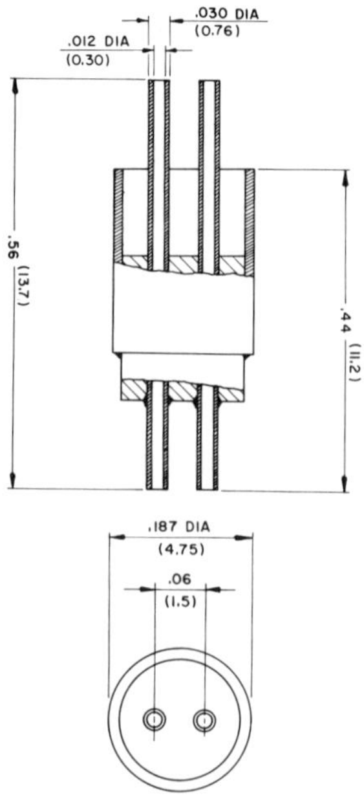

Fig. 1. Schematic of the Hermetic Seal. Open values are inches, parenthesized are mm.

Disassembly of the failed pressure transducers revealed a reddish deposit on the end of the hermetic seal within the pressure transducer case, as seen in Fig. 2. This deposit did not appear elsewhere in the pressure transducer. There was also, however, a gold-colored deposit on the wall of the pressure transducer case adjacent to the end of the hermetic seal having the reddish deposit.

The reddish deposit was identified as mostly copper by examination in a scanning electron microscope (SEM) equipped with an energy dispersive X-ray analyzer. The gold-colored deposit on the wall of the pressure transducer chamber next to the hermetic seal was identified as mostly niobium and lead by ion microprobe mass analysis (IMMA). It was thought at first that some of the Au-Cu braze alloy had vaporized and for some reason deposited out copper and gold, but this hardly seemed likely as the hermetic seals are outgassed without event at 1400°F (760°C) in a vacuum prior to assembly into a pressure transducer. For comparison, a new, unused hermetic seal that had been cycled six times at the 1400°F (760°C) temperature was also examined by SEM. It appeared on the SEM to also have a deposit, although this was not visible optically. Another pressure transducer that had not failed, but had gone through the same treatment as the failed specimen, was also examined on the SEM and it, too, appeared to have a deposit on the surface. Figures 3–8 display the above samples.

The resistances of the films between the platinum leads were measured and were as follows:

Failed pressure transducer specimen	274	ohms
Unfailed pressure transducer specimen	$\sim 10^{13}$	ohms
New hermetic seal, thermally cycled only	$\sim 10^{6}$	ohms

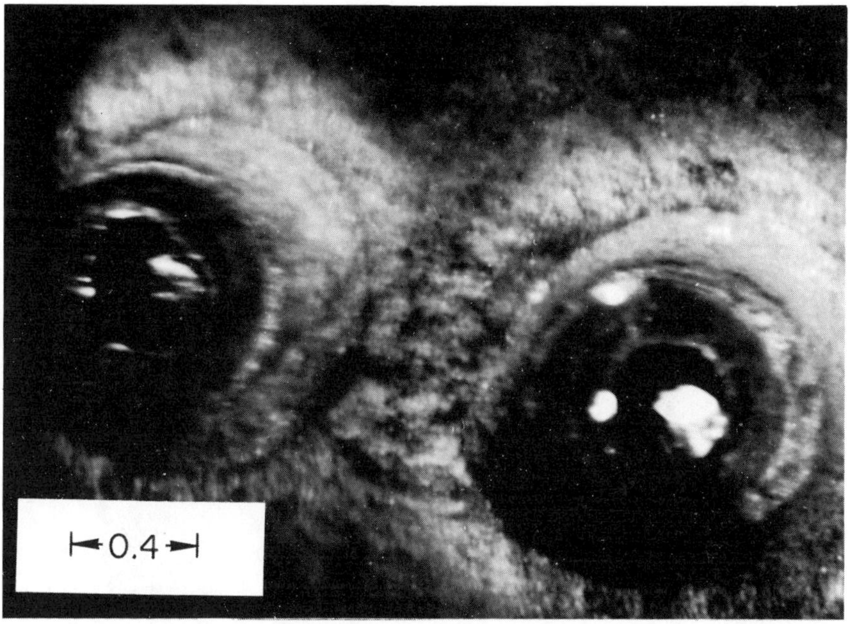

Fig. 2. Optical photograph of reddish-colored deposit on the end of the failed hermetic seal. Dimension is in mm.

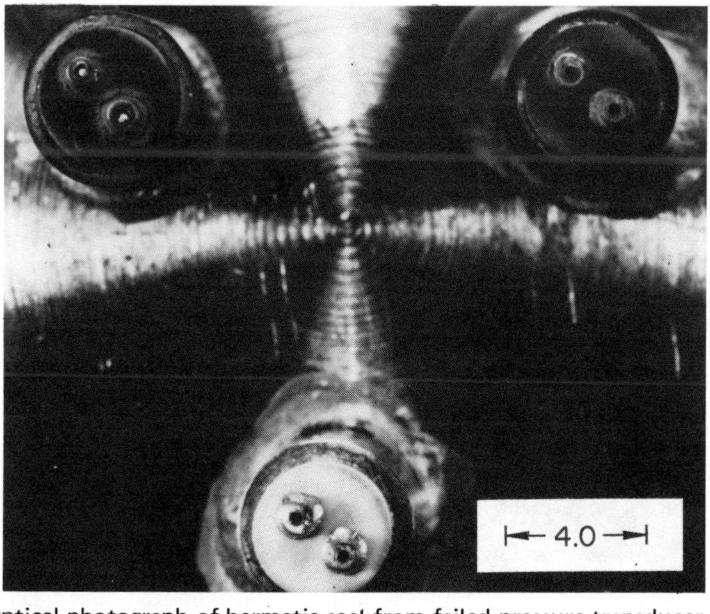

Fig. 3. Optical photograph of hermetic seal from failed pressure transducer (upper left), unfailed pressure transducer (upper right), and new, thermally-cycled hermetic seal. Dimension is in mm.

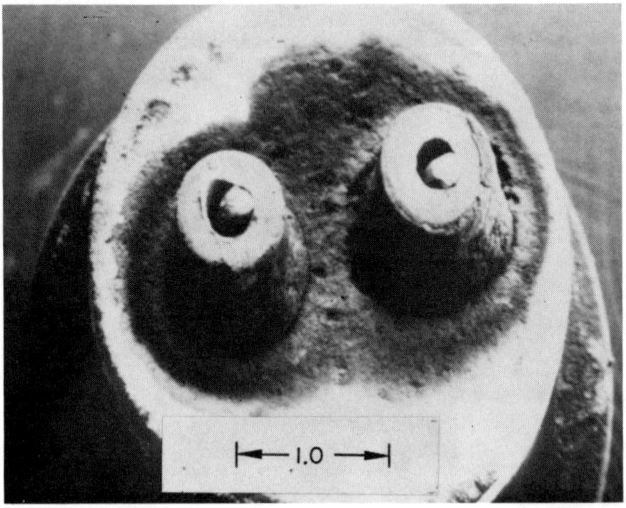

Fig. 4. SEM view of failed hermetic seal from pressure transducer (same specimen as seen in Fig. 1). Dimension is in mm.

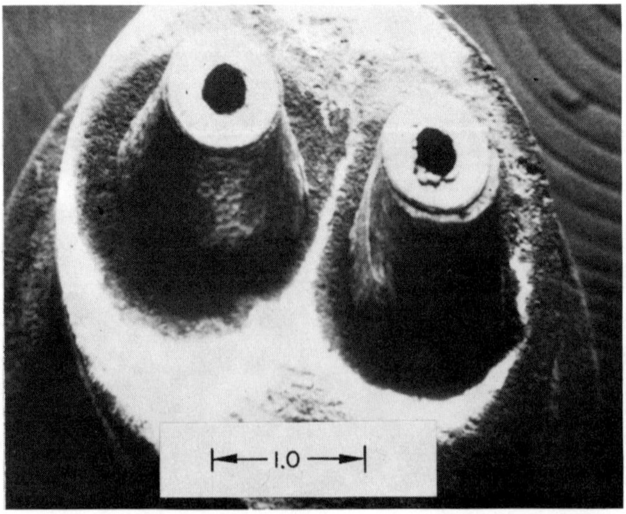

Fig. 5. SEM view of unfailed hermetic seal from pressure transducer. Dark coating on surface of insulator is cupric oxide. Dimension is in mm.

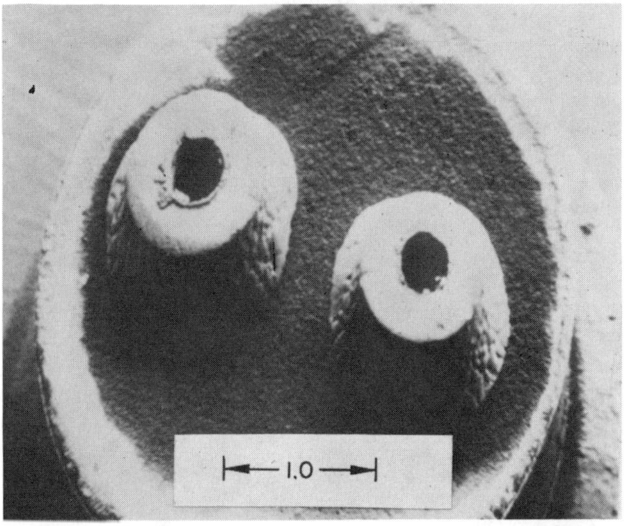

Fig. 6. SEM view of new, thermally-cycled hermetic seal. Dark coating on surface of insulator is very thin layer of cupric oxide. Dimension is in mm.

Fig. 7. SEM view of failed hermetic seal from pressure transducer. Dimension is in mm.

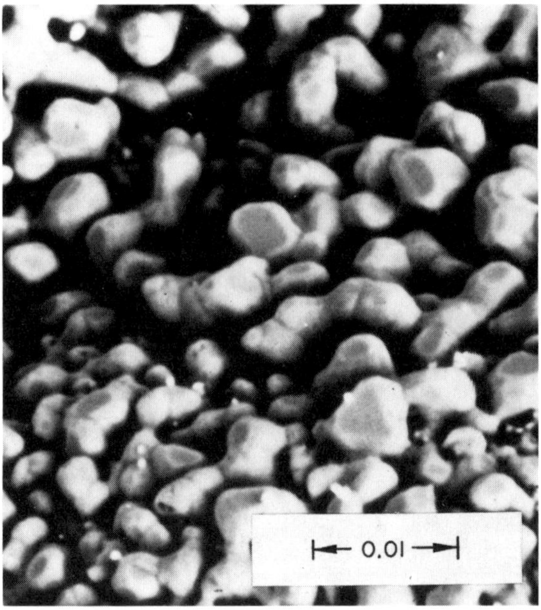

Fig. 8. SEM view of failed hermetic seal from pressure transducer, showing detail of copper coating on the surface of the braze alloy. (Same specimen as seen in Fig. 7). Dimension is in mm.

The failed specimen was sectioned and examined metallographically. It appeared that the braze alloy had decomposed on the surface and that copper was deposited as a thin film around the platinum leads, as shown in Fig. 9–11. It was the bridging of the copper film between the platinum leads that caused the electrical short in the pressure transducer circuitry.

It was difficult to understand what could have caused this to happen at 1150°F (621°C) within the pressure transducer, particularly since the hermetic seal was outgassed at 1400°F (760°C) prior to installation in the pressure transducer. It was conjectured that perhaps the brazing alloy was reacting with some residual material in the pressure transducer sensing element.

Investigation of the manufacturer's fabrication procedures revealed that there is one stage after the hermetic seal is added to the sensing element and prior to the final closure weld, in which the internal components of the pressure transducers are heated to 1000°F (538°C) under vacuum, and alternately in an oxygen atmosphere, in 15 min cycles for a total time of 8h. This procedure was a carryover from the fabrication procedure of other transducers, and was originally done to oxidize any residual organic materials. The oxygen cycle was not necessary for this particular design, and was later discontinued.

After examining the failed and unfailed hermetic seals metallographically, and observing that the copper which bridged the leads appeared as if it were plated-out or condensed from vapor, it was hypothesized that the cause of failure was a reaction between the gold-copper brazing alloy and residual oxygen which caused an exothermal reaction that drove the temperature up locally, and caused copper to diffuse from the alloy and deposit out between the surfaces of the platinum leads, with impurities of lead and niobium being deposited on the outer shell. See Figs. 9–11. The deposit might also contain some cuprous oxide (Cu_2O) which is also reddish in color

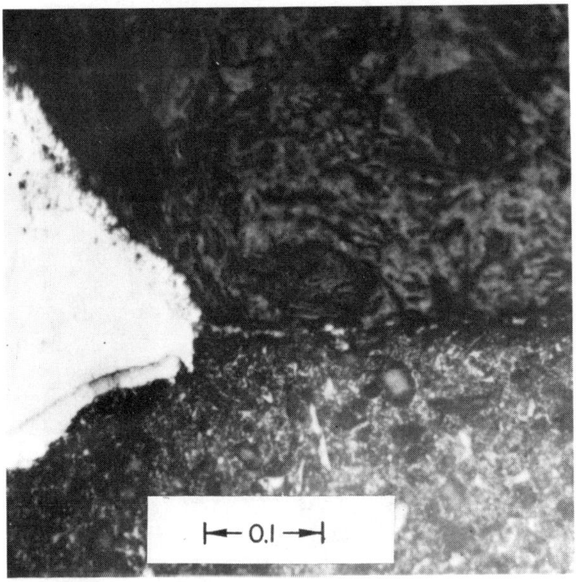

Fig. 9. Photomicrograph of as-polished section through failed hermetic seal. Braze alloy is at the left, alumina at bottom, and Bakelite at top. The thin film through the center is the copper coating on the alumina insulator. Dimension is in mm.

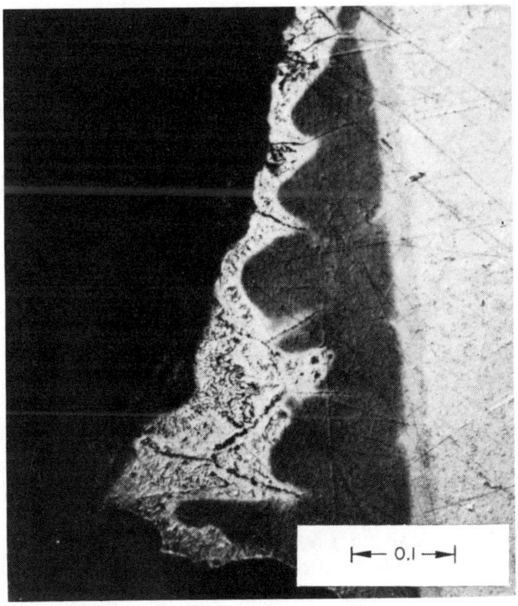

Fig. 10. Photomicrograph of etched section of failed hermetic seal showing copper-depleted region of braze alloy and copper coating on surface of braze alloy. Platinum lead is on right. Photograph taken with differential interference contrast. Dimension is in mm.

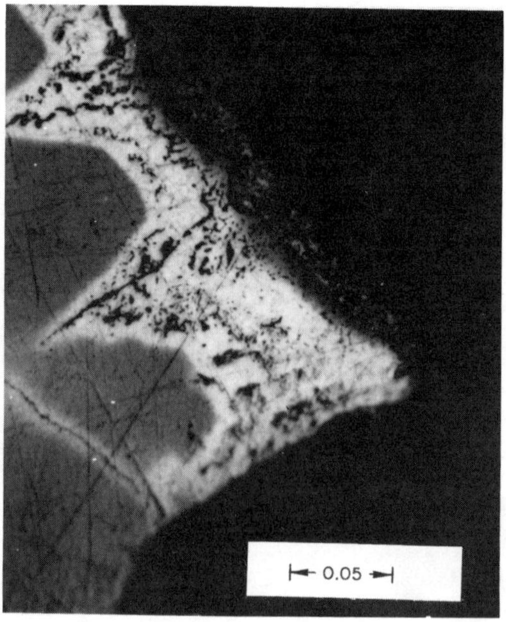

Fig. 11. Enlarged view of copper-depleted region of braze alloy and copper surface coating, as-etched. Dimension is in mm.

and is a p-type semiconductor whose conductivity increases with temperature. This reaction did not occur at 1000°F (538°C), but apparently began somewhere near 1150°F (621°C). When the hermetic seals failed, after being removed from the pressure transducers and heated at 1000°F (538°C) in oxygen at atmospheric pressure, the reddish deposit was oxidized to a black deposit (cupric oxide, CuO), and was no longer electrically conductive.

Although there may be other possible mechanisms for the failures, such as a reaction between the braze alloy and the alumina, or the braze alloy, alumina, and the manganese-molybdenum metallizing compound, time did not permit further investigation. When the heating-in-oxygen cycle was deleted from the fabrication procedures, no further failures occurred.

One puzzling question remained; why only some of the pressure transducers failed, and some did not, presumably under identical conditions of fabrication and testing. This was probably because gas flow rates during the pump-down and pressurization with oxygen cycles were governed by the very close-fitting annulus between the hermetic seal case and the pressure sensor body. In some instances where the fit was closer than others, not all of the oxygen may have been removed.

Because these custom-built pressure transducers are quite expensive to fabricate, an attempt was made to salvage the remaining failed pressure transducers. This was done by drilling a tiny hole in the case, heating the pressure transducers at 1000°F (538°C) in oxygen, evacuating, and welding the hole shut. The procedure successfully eliminated the electrical short by converting the copper and cuprous oxide to cupric oxide. These pressure transducers are being further tested to see if they can be used reliably up to 1300°F (704°C).

Stevens, Vander Voort, McCall, eds. Microstructural Science, Volume 8

METALLURGICAL ANALYSIS OF FIRE-DAMAGED PIPING FROM A U.S. STRATEGIC PETROLEUM RESERVE SUPPLY FACILITY

R.J. Gray*, G.M. Slaughter*, J.C. Griess, Jr.,*
and C.W. Houck*

ABSTRACT

On September 21, 1978, a fire seriously disrupted operations at the U.S. Strategic Petroleum Reserve site at Well 6 of Cavern 6 at West Hackberry, Louisiana. This facility stores strategic supplies of crude oil and is one of several subterranean salt domes in this area with a combined capacity of up to 750 million barrels. Members of the Metals and Ceramics Division at Oak Ridge National Laboratory were asked to investigate portions of the damage and to serve as consultants to the investigation committee assigned to this project from the Department of Energy (DOE). The objectives of the investigation were: (1) to provide information concerning the origin of a seeping petroleum leak in the wellhead and (2) to assess the condition of the brine casing before fire damage.

A definite leak path was found to exist through the welds in the wellhead. Furthermore, another leak path through a fracture in the casing seemed likely. Gross attack of several sections of the brine pipe was attributed to hot corrosion during the fire.

Presented in this paper are photographs of this awesome conflagration and a pictorial assessment of the damage after the fire. The investigative procedures that were followed in the careful sectioning and metallographic study of the wellhead and the metallographic survey of the brine pipe are also included.

INTRODUCTION

The U.S. now uses nearly 18 million barrels (MMB) of crude oil a day.[1] With such a demand on this vital commodity, there is a crucial need for very large storage facilities, particularly when various news media report that 45 to 50% of the crude oil used in this country now must come from nondomestic sources. A tremendous inventory is required for this important crude oil, and the containments must be geographically convenient to unloading from supertankers. Some of these tankers are larger than the biggest aircraft carriers and are capable of transporting from 1.02 to 2.04 MMB of oil. Although these supertankers, classified as Very Large Crude Carriers (VLCCs), are the current transports, the newest deep draft tankers, classified as Ultra Large Crude Carriers (ULCCs), have twice the capacity of the VLCCs, which must be taken into account in plans for future transportation and storage. A prime obstacle in the use of these huge petroleum carriers is that they are barred from all American ports – no port is deep enough to accommodate the draft. For years the VLCCs have had to unload the crude oil at terminals in the Caribbean or to transfer the oil at sea to smaller tankers for delivery to U.S. ports. The problems of transporting oil and storing a critical supply have been resolved by the U.S. Department of Energy (DOE) in cooperation with several oil companies through the use of huge hollowed-out subterranean salt domes sprinkled across large areas of Louisiana and parts of Texas, Mississippi, and Florida.[2] These salt deposits begin approximately 4600 m (1500 ft) below the ground and extend to a

*Metals and Ceramics Division, Oak Ridge National Laboratory, Oak Ridge, Tennessee 37830 USA.
ORNL is operated by Union Carbide Corporation under Contract W–7405–eng–26 with the U.S. Department of Energy.

depth of at least 12,000 m (40,000 ft). Oil has been stored in former salt mines in France and Germany. The West German government has had crude oil stored in salt caverns for over eight years. Cavities in the southern U.S. were formed by leaching or solution mining. The salt was removed by drilling a well into the salt formation. Fresh water or sea water was pumped down the well, forming a cavity at the bottom where the salt was leached to form a brine. The brine was pumped out, leaving a huge salt-enclosed chamber. In addition to the solution-mined cavities, additional caverns are being formed specifically for oil storage. The DOE acquired a total of 19 salt caverns in the 4 sites of Bayou Choctaw, Louisiana; Bryan Mound, Texas; West Hackberry, Louisiana; and Weeks Island, Louisiana. These sites have a combined storage capacity of 261 MMB. Ultimately, plans call for a total storage capacity of 750 MMB. An artist's concept of an unloading and storage facility is shown in Fig. 1.

This report is concerned with a failure that developed at Well 6 at West Hackberry, Louisiana. A drawing of this well is shown in Fig. 2. From this drawing it can be seen that the oil is transported to the subterranean cavern via a 0.24-m-ID (9 5/8-in.) casing, and the displaced brine is transported to the surface via a 0.14-m-ID (5½-in.) center casing. By reversing the pumping process, that is, by pumping water or brine into the cavern, oil can be removed.

On September 21, 1978, an explosion and fire occurred at Well 6 of Cavern 6 at West Hackberry, Louisiana. This cavern had been formed by solution mining. An aerial photograph of the site taken before the explosion is shown in Fig. 3. This incident developed during a remedial operation while 14 joints of 0.14-m (5½-in.) casing were being withdrawn (one joint of casing = 12-m or 40-ft). Excessive pressure in the cavern caused an uncontrolled flow and an airborne mist of oil. The oil mist was drawn into the intake of a diesel engine on the rig and into other nearby engines causing them to overspeed and ignite the oil. Local volunteer firemen under the overall direction of the Boots and Coots firefighter team were credited with control and prevention of the spreading of this conflagration, which burned for five days. The intensity of the fire is clear from photographs that are shown in Fig. 4. Of the original 7 MMB of oil in the salt dome, 67,510 barrels were expelled. Approximately half of this released crude oil was assumed to have burned.

Before the fire some oil leaks had been observed at the wellhead and attempts had been taken in the field to stop the leaks with a split clamp-type patch. After the fire Metals and Ceramics Division personnel from Oak Ridge National Laboratory (ORNL) were requested to examine the wellhead to provide information about this leak.. The request included an assessment of the condition of the 0.14-m(5½-in.) brine-carrying casing before and after the fire.

A detailed account of this accident has been reported elsewhere.[3]

WELLHEAD INVESTIGATION

The relative location of a wellhead to the casings and to a subterranean cavity is shown in Figs. 1 and 2. An on-site photograph of the Wellhead 6 after the fire is shown in Fig. 5. One of our objectives was to determine the origin of the leak that was reported by the well personnel before the fire.

The wellhead was sectioned on a bandsaw, and an internal bore weld was dye checked, as shown in Fig. 6. The dye checking clued us where additional cuts might be made for further study. The split clamp-type patch and elastomer are identified in Fig. 6. The microstructure of a failed bore weld joining the 0.24-m-ID (9 5/8-in.) interior casing to the heavy-walled casing head is shown in Fig. 7. A flame-cut spacer ring [identified in Fig. 6(a)] had been welded to position the outer 0.32-m-ID (12 3/4-in.) casing to the heavy-walled casing head. In the examination this spacer ring was cut from the wellhead and is shown in Fig. 8. Based on our investigation the wellhead was drawn (Fig. 9), locating the welds shown in Figs. 7 and 8; a proposed leak route for the crude oil can be traced.

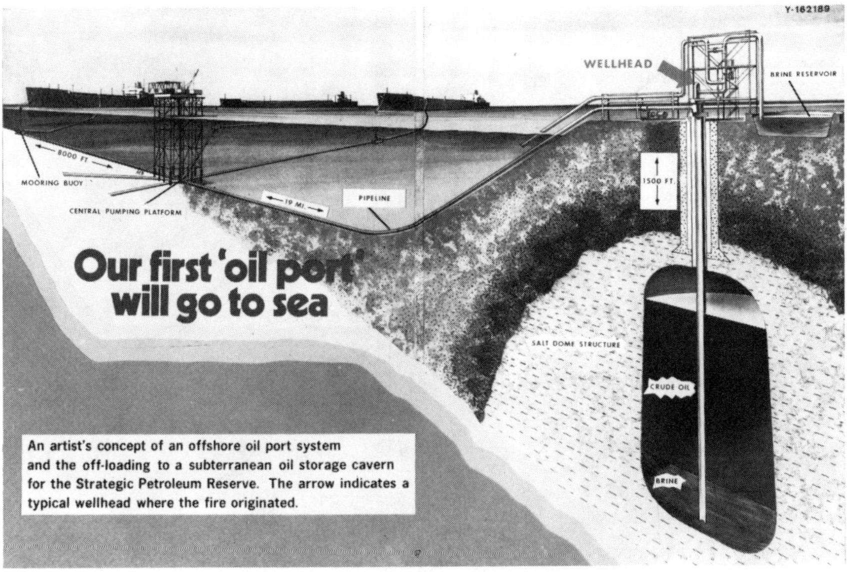

Fig. 1. Use of salt caverns for the strategic petroleum reserve. Reprinted by permission from *POPULAR MECHANICS.* Copyright 1978 The Hearst Corporation. All rights reserved. R. Petrow, "Our First 'Oil Port' Will Go To Sea", *Pop. Mech.* 150 (8); 74 (Aug. 1979).

The end of the 0.24-m (9 5/8-in.) casing in the wellhead, which transported the petroleum, had been fractured [see Fig. 10(a)]. The workers at the Strategic Petroleum Reserve Office (SPRO) site claimed that a crack might have developed before the fire while the well was in service and was a probable source for the oil leak. Further study of the fractured surface, particularly the chevron-type fracture pattern [Fig. 10(b)], indicated a possible fracture origin. The microstructure in this area revealed a fissure slightly below the plane of the fracture [Fig. 10(c)] that could have contributed to the fracture. We could find no other evidence that could cause the leak.

INVESTIGATION OF 0.14-m (5½-in.) BRINE PIPE

Several sections of the brine-carrying pipe were investigated. The pipe was identified as "5½-in., 15.5 lb/ft. K-55 tubing with an inside diameter of 4.950 in. The capacity was 0.0238 bbls/1in. ft. or 42.0 ft./bbl."[3] This pipe, shown in Fig. 2, transported the brine to and from the salt cavity.

We examined three conditions of pipe: (1) relatively undamaged pipe that was protected from the fire by overturned equipment, (2) some slightly fire damaged pipe, and (3) some severely damaged pipe (with massive holes through the wall). Several pipes in each condition were examined. Microstructures of pipe walls in the above conditions are compared in Fig. 11.

Relatively Undamaged Pipe (Pipe A)

This pipe did not show any evidence of wall thinning. Any corrosion evidence was minor. The microstructure of this pipe is shown in Fig. 11(a).

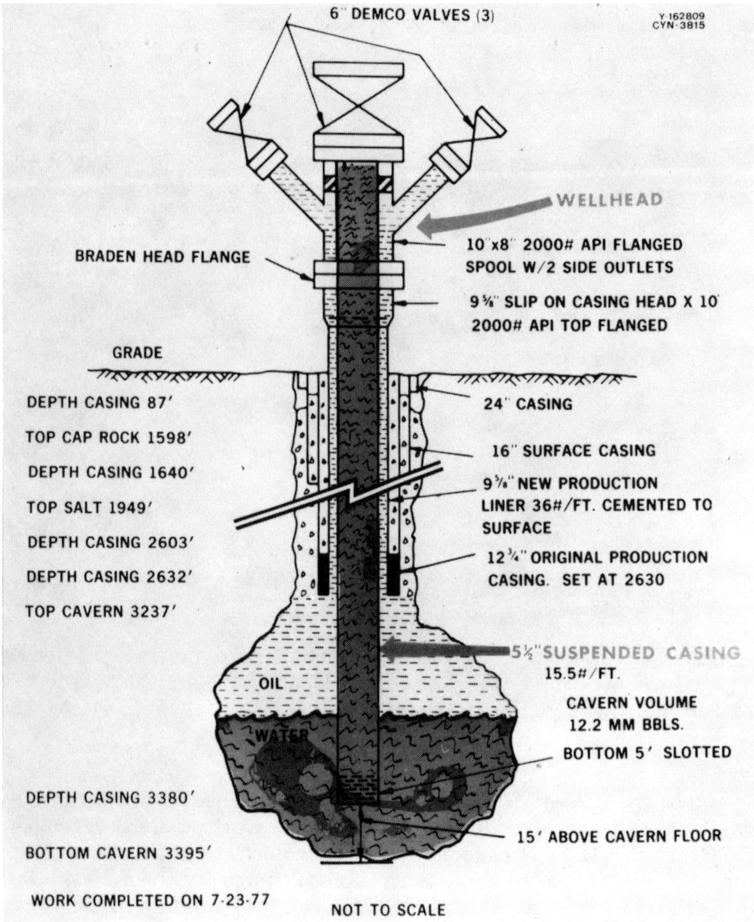

Fig. 2. Overall configuration of Well 6.

Relatively Sound but Fire-Damaged Pipe (Pipe B)

This pipe displayed some apparent fire damage. After cutting out a cross section of the pipe, some wall thinning could be seen. The pipe contained a reaction product that was a black, loosely adherent, lightweight, and partially magnetic powder. X-ray diffraction patterns showed this material to be mainly Fe_2O_3. Undoubtedly some magnetic Fe_3O_4 and/or metallic iron were present. Approximate chemical analysis of the powder showed 40% Fe, 17% C, and 660 ppm Cl. The microstructure of this pipe is shown in Fig. 11(b).

Severely Damaged Pipe with Massive Holes Completely Through The Pipe Wall (Pipe C)

This pipe showed severe burning (oxidation or corrosion) to the extent that 30- to 50-mm-diam holes penetrated the wall completely. X-ray diffraction patterns of the oxidation corrosion product revealed the presence of Fe_3O_4, Fe_2O_3, and FeO. In contrast with the analysis of the scale on Pipe B, the Pipe C scale showed a relatively small amount of carbon. The Pipe C scale contained 75% Fe, 870 ppm C and 230 ppm Cl.

Fig. 3. Aerial view of Cavern 6 site before explosion and fire.

A eutectic microstructure in the base metal just below the scale-metal interface is shown in Fig. 12(a). Areas labeled (X), (Y), and (Z) are of particular interest. Using energy dispersive X-ray analysis, we qualitatively studied the sound base metal (X), the scale-metal interface where the eutectic structure was found (Y), and the area completely composed of corrosion product (Z). Sulfur was pronounced in the scale-metal surface.

By chemical analysis sulfate ions were present in the brine from the storage cavity. In addition, some sulfur was present in the crude oil.

DISCUSSION AND CONCLUSION

Wellhead Investigation

Two leak paths were possibly present and would permit the crude oil to seep from inside the 0.24-m (9 5/8-in.) inner casing to the outside of the wellhead. One path involves two cracked welds (Fig. 9), and the other possible path is via a fracture that may have been caused by a fissure in the 0.24-m (9 5/8-in.) pipe, as shown in Fig. 10. We recommend that welding practices for this material be investigated. Furthermore, nondestructive procedures for evaluating the integrity of welds should be established if such steps are not included in field welding practices.

Investigation of 0.14-m (5½-in.) Brine Pipe

It is impossible to develop a positive set of conclusions because of the extensive fire damage to some of the pipe. Gross evidence of scaling and corrosion on the inside of some pipes and the side variations in carbonaceous products related to exposure to the fire point to several variables that contributed to the diverse conditions of the pipes.

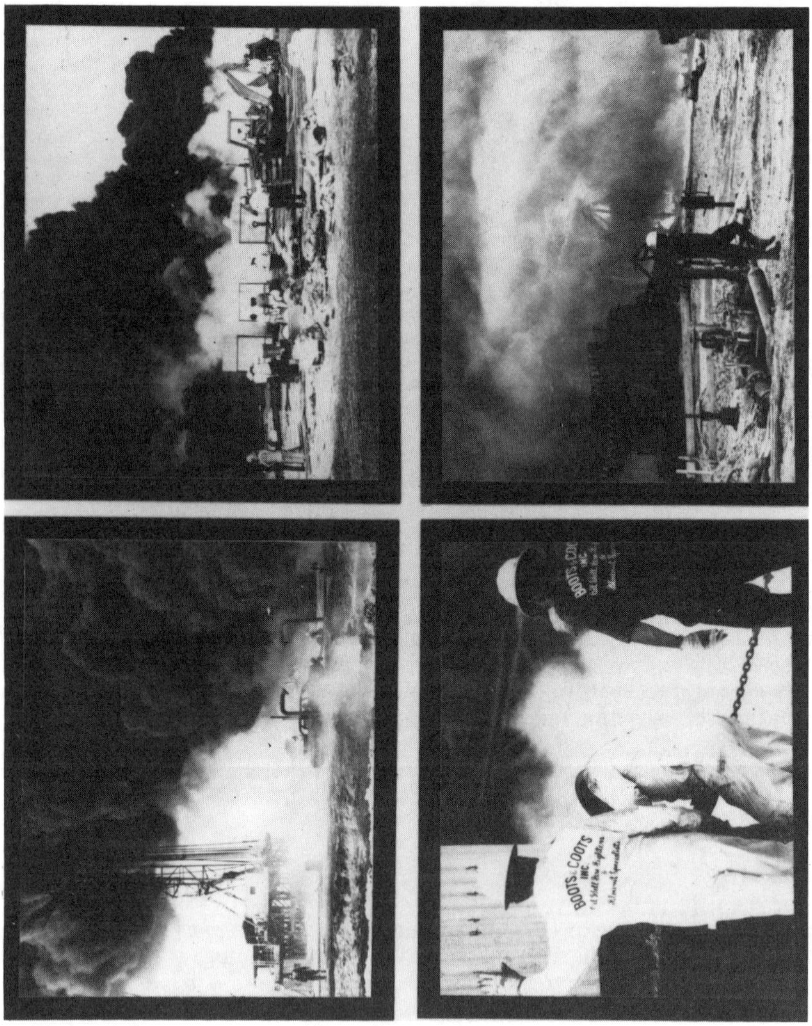

Fig. 4. Fire fighting.

Fig. 5. Wellhead 6 after the fire.

Some pipes were subjected to sufficient temperature and carbonaceous material to have developed considerable carburization of the interior surface, while other pipes were so severely attacked that the walls were perforated. During exposure to the fire, carbon from the petroleum and residual chloride from the brine played principal roles in the oxidation-corrosion reaction. A major contributor to the reduction of the pipe wall was the sulfur (sulfate) in the brine. Sulfur is a noted promoter of hot corrosion. There is well-documented evidence that rapid oxidation of mild- and low-alloy steels can occur in furnace atmospheres containing sulfur dioxide, and that this rapid oxidation is accompanied by intergranular penetration of the metal by iron-iron sulfide eutectic.[4] It has been reported[5] that pipe scales produced during the burning of fuel oil containing sulfur resulted in oxy-sulfide complex microstructures not unlike what we have presented here. Another related accounting describes the rapid penetration of steel by iron sulfide during the carburization of steel.[6]

Although positive conclusions cannot be made because of the many variables associated with this incident, the corrosion-oxidation of the pipe can be primarily related to the fire.

ACKNOWLEDGMENTS

The authors wish to acknowledge the extensive cooperation of many individuals at the SPRO site in identifying and providing the samples for our investigation. We also acknowledge the significant information obtained from and the interest shown by the DOE Accident Investigation Committee. Finally, we recognize the major efforts of several members of the Metals and Ceramics Division, especially R.S. Crouse, C.P. Haltom, K.F. Russell, T.L. Marlar, O.B. Cavin, T.G. Godfrey and J.R. Keiser, without whose enthusiasm and hard work this study could not have been conducted. We also offer our thanks to Barbara Ashdown for editorial suggestions and to Brenda Hatchett for the preparation of this report.

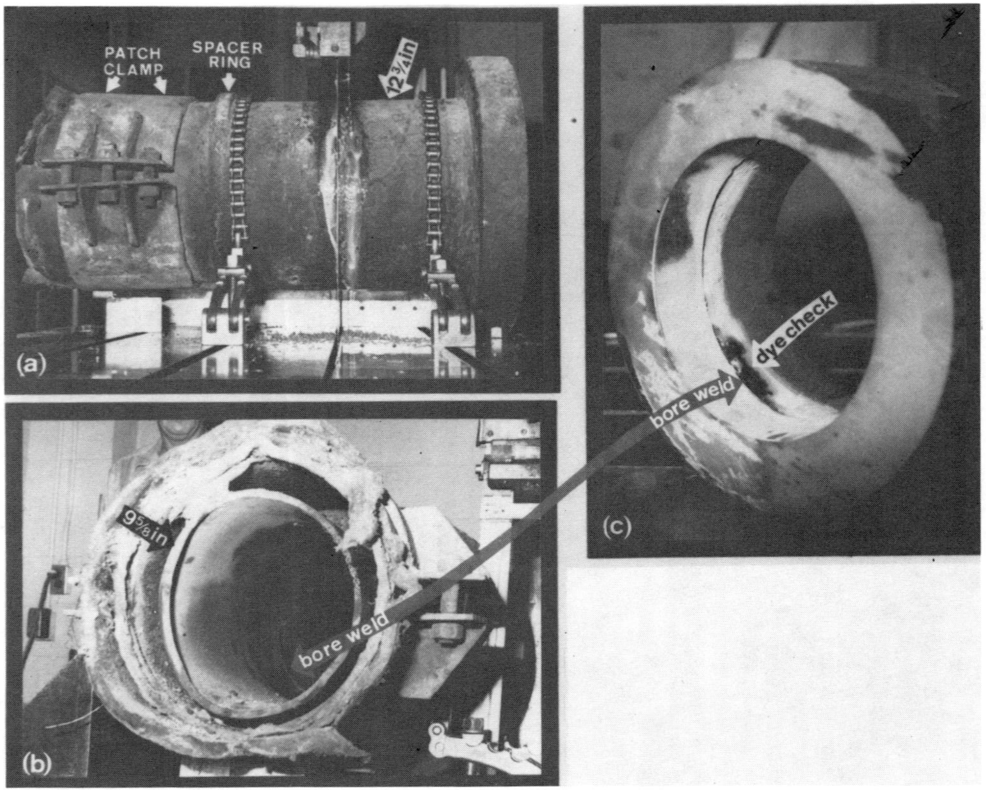

Fig. 6. Cutting and Dye Checking of Wellhead to Locate Leak in Internal Bore Weld. (a) Side view of wellhead during cutting operation. (b) Fractured end view of wellhead during cutting operation. (c) Dye check of internal bore weld. Arrow indicates dye seepage and probable leak.

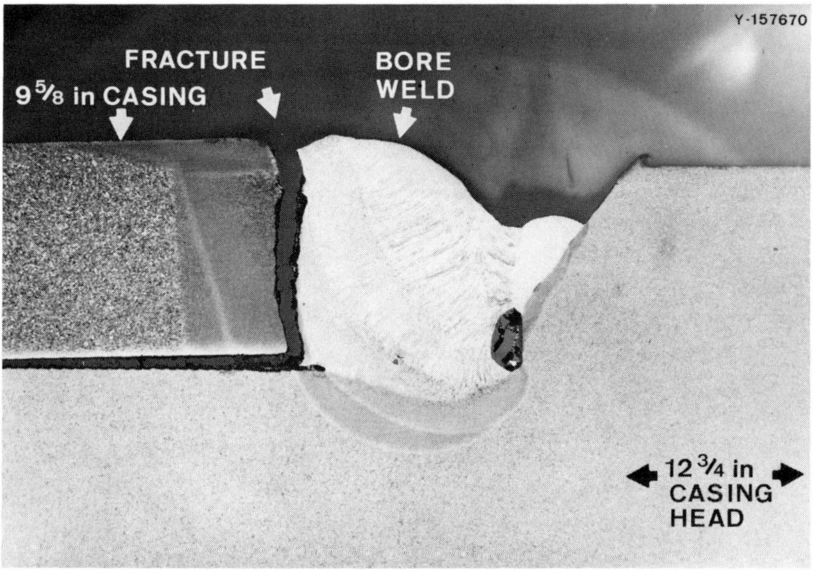

Fig. 7. Microstructure of failed bore weld. Note that fracture occurred in fusion line of weld.

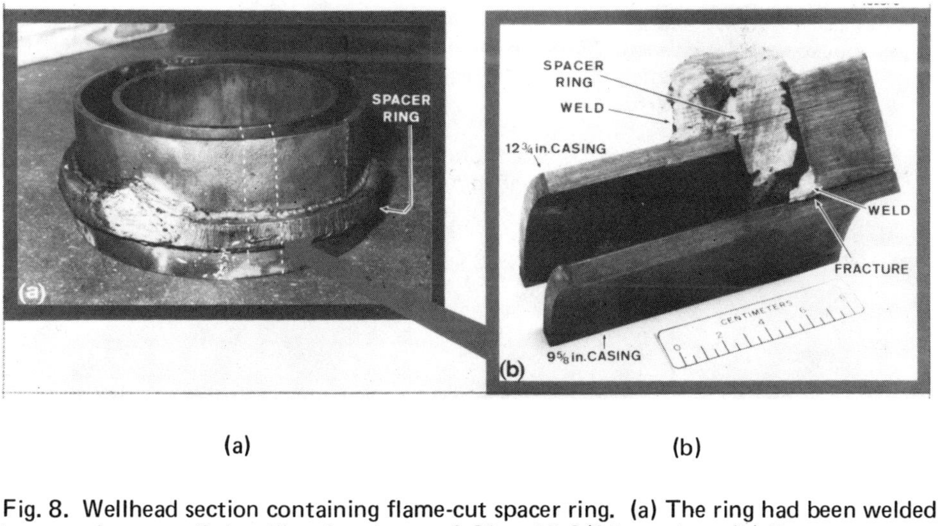

(a) (b)

Fig. 8. Wellhead section containing flame-cut spacer ring. (a) The ring had been welded between heavy-walled wellhead and outer 0.32-m 12 3/4-in. casing. (b) Dye penetrant tests show through-leaks in this area. Complete fracture of a weld is indicated.

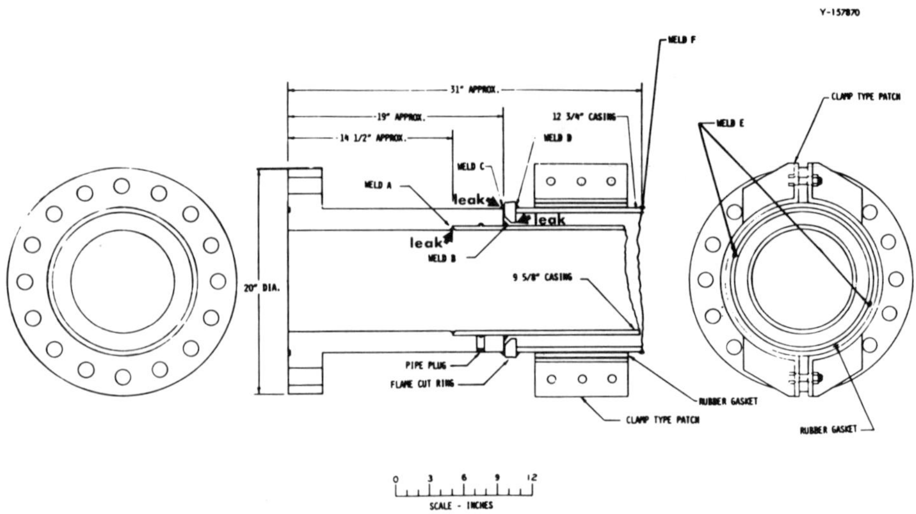

Fig. 9. Wellhead from West Hackberry Well 6 as received at ORNL. Drawing locates defective welds that could cause petroleum leak.

REFERENCES

1. K.A.D. Inglis, Ed., *British Statistical Review of the World Oil Industry,* 1977, Public Affairs and Information Dept., The British Petroleum Co., Ltd. London.
2. R. Petrow, "Our First 'Oil Port' Will Go To Sea", *Pop. Mech.* 150(8): 74 (Aug. 1978).
3. U.S. Department of Energy, *Report on the Explosion, Fire and Oil Spill Resulting in One Fatality and Injury on September 21, 1978 at Well 6 of Cavern 6 at the West Hackberry, Louisiana Oil Storage Site of the Strategic Petroleum Reserve,* DOE/EV-0032, U.S. Government Printing Office, Washington, D.C. 20402, Stock No. 061–000–00202–2 (1978).
4. J. Stringer, "Hot Corrosion in Gas Furnaces", MCIC Report 72–08, Battelle Columbus Laboratories, Columbus, Ohio (June 1972).
5. A. Preece, *Scaling of Gas-Turbine Alloys,* Spec. Rep. 43, Iron and Steel Institute, London, (1951).
6. A. Preece and K.J. Irwin, "Sulphide Penetration in the Carburization of Steel", *J. Iron Steel Inst., London* 157: 336–43 (Nov. 1947).

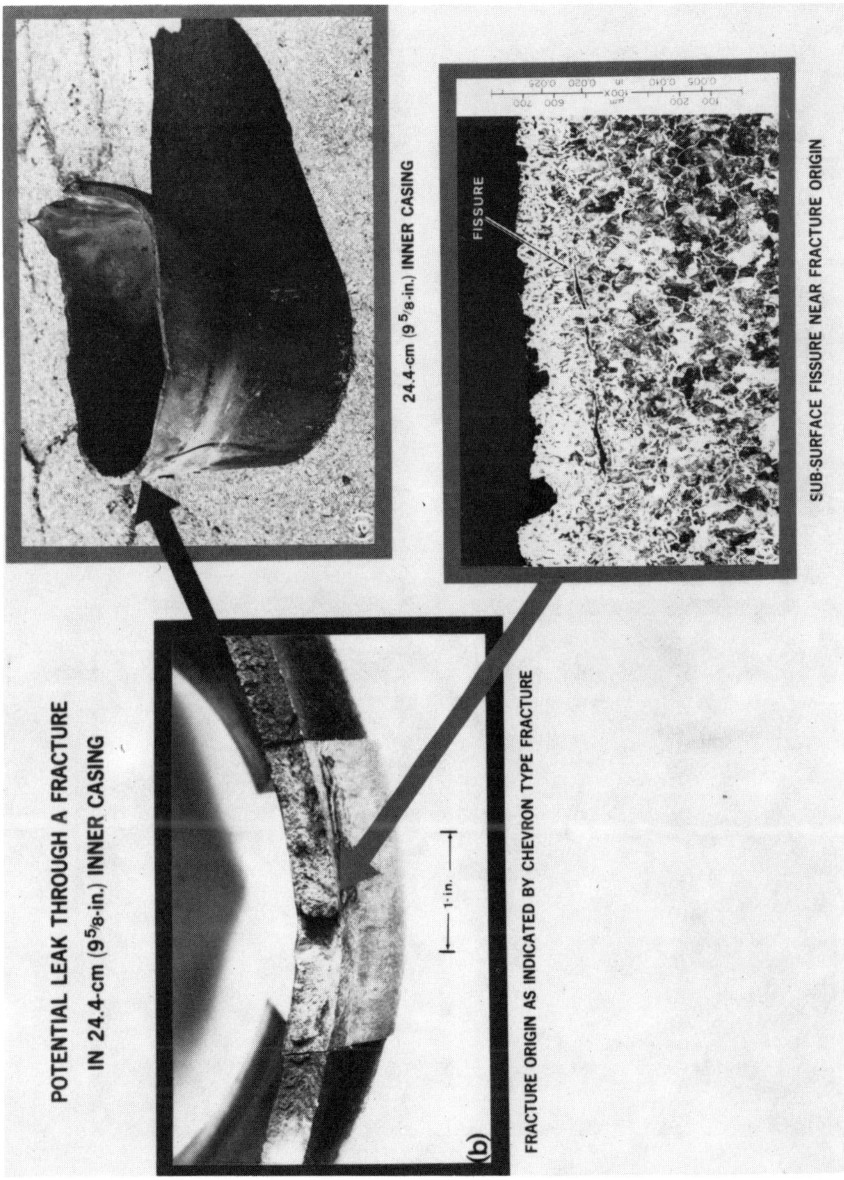

Fig. 10. Fractured end of inner casing where possible leak could occur.

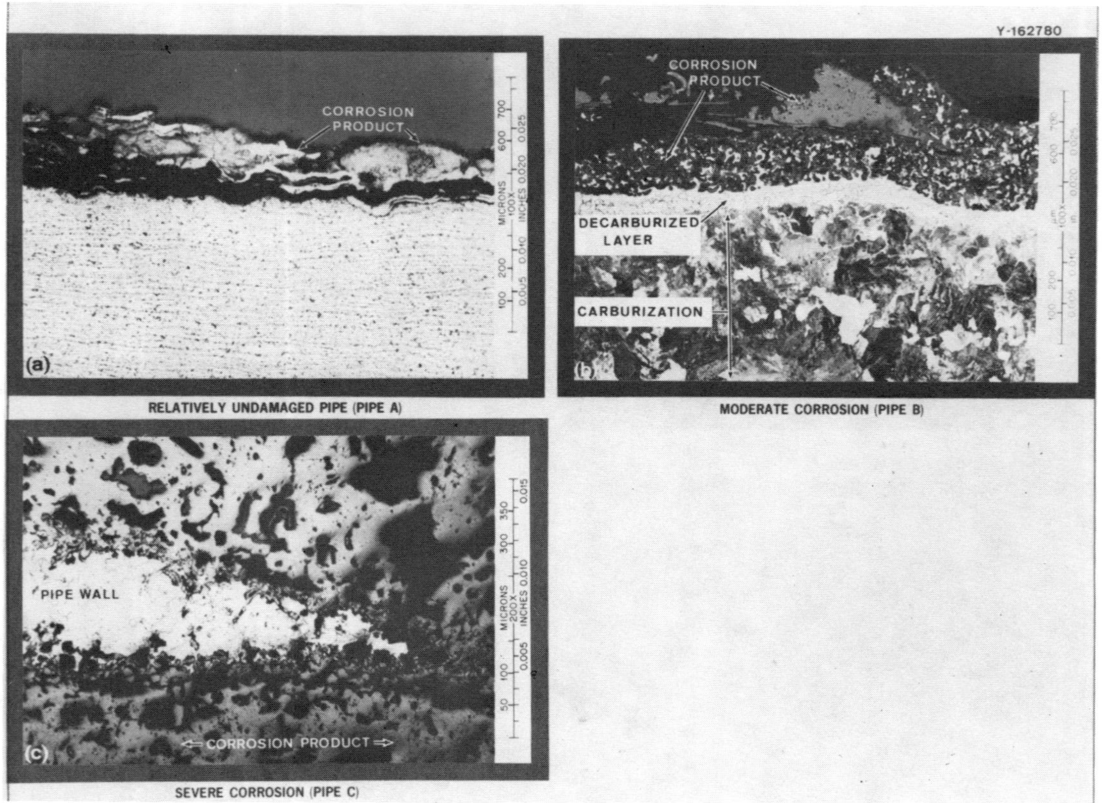

Fig. 11. Microstructural assessment of three 0.14-m (5½-in.) pipes illustrating corrosion conditions after fire. (a) Sound pipe. Corrosion probably present in as-fabricated pipe. (b) Moderate corrosion. Microstructure shows heavy carburization and decarburization near metal-corrosion interface. (c) Severe corrosion. Large holes in pipe wall.

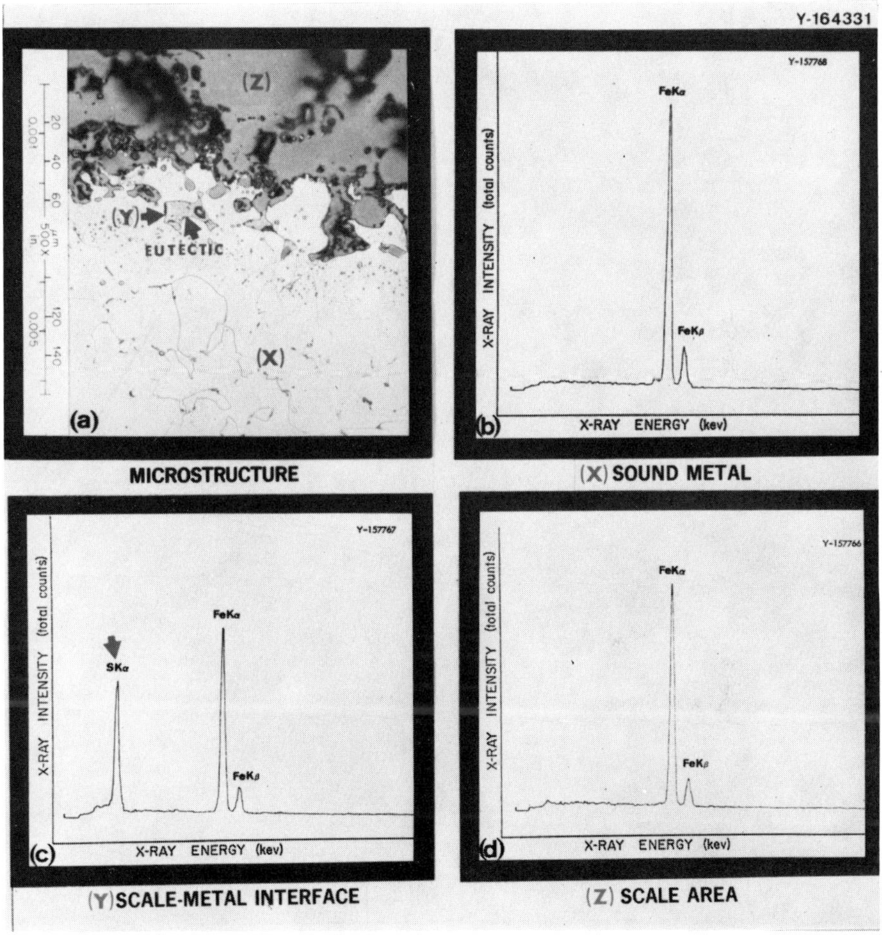

Fig. 12. Backscattered electron image and energy dispersive X-ray analyses of (X) sound metal, (Y) scale-metal interface, and (Z) scale area (Pipe C).

Stevens, Vander Voort, McCall, eds. Microstructural Science, Volume 8

A FRACTOGRAPHIC STUDY OF A THICK WALL PRESSURE VESSEL FAILURE

D.A. Canonico*, R.S. Crouse* and T.J. Henson*

ABSTRACT

A Heavy Section Steel Technology (HSST) Program sponsored by the Nuclear Regulatory Commission (NRC) and administered by the Oak Ridge National Laboratory** (ORNL) is concerned with the safety and reliability of nuclear pressure vessels. Some of the testing involves pressurizing to failure simulated pressure vessels containing large fatigue sharpened flaws. These vessels are fabricated from SA533 grade B, Class 1 plate or SA508, Class 2 forgings.

The vessel described in this paper is identified as Intermediate Test Vessel 1 (ITV–1) and was fabricated of SA508, Class 2 Steel. It was tested to failure at 54°C (130°F). The gross failure appeared to be a brittle fracture although accompanied by a measured strain of 0.9%. Seven regions of the fracture were examined in detail and the observed surfaces were compared to Charpy V-notch (C_v) specimens of SA508, Class 2 steel broken at temperatures above and below the ductile to brittle transition temperature. Three samples from the vessel were taken in the region around the fatigue notch and four from areas well removed from the notch. All these were carefully examined both optically and by scanning electron microscopy (SEM).

It was established that early crack extension was by ductile mode until a large flaw approximately 500 mm long by 83 mm wide was developed. At this point the vessel could no longer contain the internal pressure and final rupture was by brittle fracture.

INTRODUCTION

The Oak Ridge National Laboratory has managed the Heavy Section Steel Technology (HSST) program and participated in the experimental work associated with it since its inception in 1966. The HSST program is concerned with the safety and reliability of nuclear pressure vessels. One facet of the program involves pressurizing a 1 m dia. by 2.5 m tall intermediate test vessels (ITV) to failure. Typically these vessels are fabricated from SA 533 Grade B Class 1 plate or SA 508 Class 2 forgings that are 150-mm thick, and each vessel is provided with a large fatigue sharpened (or hydrogen-charged electron beam) crack. Fig. 1 depicts such ITVs.

To date seven vessels have been tested. The test conditions and results of these tests are provided in Table 1. This series of ITV tests has provided a great deal of engineering and design information as well as affording the opportunity to thoroughly assess the failure mode of a vessel whose wall thickness approximates that in commercial nuclear pressure vessels. This paper will describe the procedure employed in the post-test examination of one ITV which was tested to failure at 54°C.

*Oak Ridge National Laboratory, Oak Ridge, Tennessee 37830 USA.

**Research sponsored by U.S. Department of Energy and Union Carbide Corporation under contract W–7405–eng.26.

ORNL-DWG 70-320

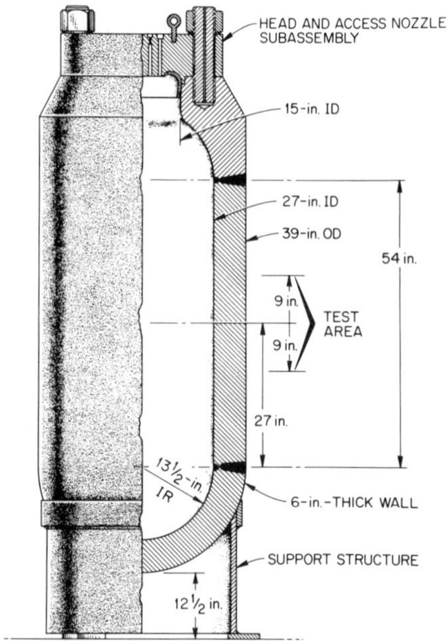

Intermediate Test Vessels.

Fig. 1. A typical Intermediate Test Vessel (ITV). From test to test, the flaw location and configuration varied. (1 in. = 0.0254 m)

FRACTURE PROPAGATION

ITV–1 was fabricated from ASTM A 508 Class 2 steel.[1] Vessel testing was conducted in a facility designed and built for these specific tests (see Fig. 2). The facility and testing procedure is described in an ORNL document.[2] The ITV–1 test resulted in the complete rupture of the vessel as shown in Fig. 3. The fracture originated from an 203 mm long by 66 mm deep semi-elliptical fatigue sharpened flaw and propagated to the top head in the upward direction and around the bottom hemihead in the downward direction. The upward propagation of the crack ran through the circumferential weld (shown in Fig. 1) that joins the transition piece to the cylindrical test section and terminated at the interface between the transition piece and the bolted-on cover plate. The downward propagation path was through the circumferential weld that joins the bottom hemihead to the center cylindrical test section. The crack continued around the bottom and terminated near the circumferential weld approximately 180° from where it first entered the bottom hemihead.

This well-documented test allowed the complete evaluation of the vessel failure. The test temperature, 54°C, is 56°C above the drop weight nil ductility temperature (NDT) for the SA 508 Class 2 heat of steel from which ITV–1 was fabricated. Moreover, 54°C is approximately the temperature of the onset of upper shelf for this heat of steel. Figure 4 contains the Charpy V-notch (C_V) curve for the 3/8 T and 5/8 T location for the ITV–1 steel.[3]

Table 1
Test conditions and results for the intermediate test vessels

Vessel	Flaw Size, mm Depth	Flaw Size, mm Length	Test Temperature, °C	Charpy V. Notch Toughness at Failure site, J	Load Factor at Failure[a]	Remarks
2	64.3	211	0	54	2.87	Flaw on outside of SA508 Class 2 vessel. Flat mode of failure.
4	76.2	209	24	135	2.75	Flaw on outside of weld. Mixed mode of failure.
9	30.5	(nozzle)	24	100	2.77	Flaw on inside of SA508 Class 2 vessel at nozzle. Flat mode of failure.
1	65	208	54	108	2.96	Flaw on outside of SA508 Class 2 vessel. Mixed mode of failure.
3	53.6	216	54	170[b]	3.19	Flaw outside of weld. Mixed mode of failure.
6	47.5	133	88	170[b]	3.28	Flaw outside of weld. Shear mode of failure.
5	30.5	(nozzle)	88	NA	2.74	Flaw on inside of SA508 Class 2 vessel at nozzle. Vessel leaked.
7	135	472	196	190[b]	2.20	Flaw on outside of SA533 Grade B Class 1 vessel. Largest flaw in all tests. Vessel leaked.

[a] Ratio of failure pressure to design pressure. Design pressure based on requirements of Section III Class 1 vessels in ASME Code.

[b] Upper shelf energy values.

NA Not available – Test temperature is on Charpy V notch upper shelf (135 J)

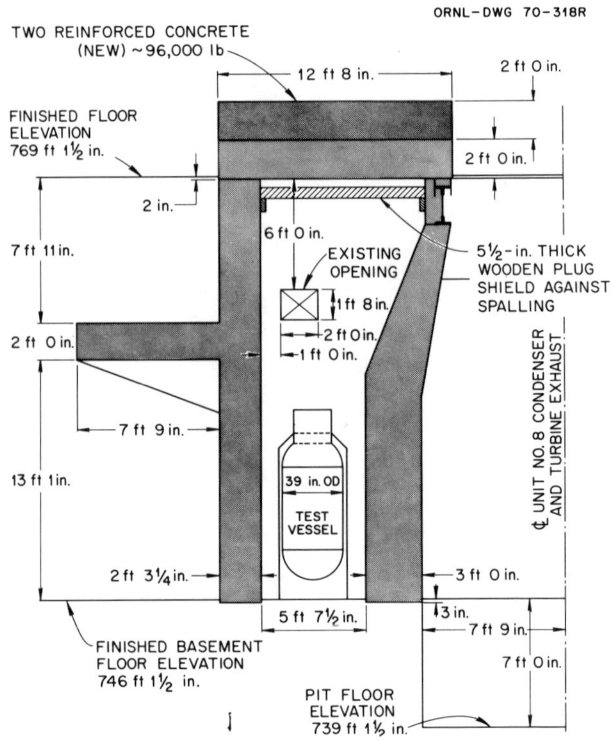

Fig. 2. Pressure Vessel Testing Facility. The vessel was surrounded by reinforced concrete. (1 pound = 0.4536 kilograms; 1 foot = 0.3048 meters; 1 inch = 0.0254 meters)

Although the vessel did undergo a strain of nearly 1%, the final fracture as observed immediately after vessel failure did not visually exhibit large shear lips and the initial observation was to conclude that the vessel had failed in a frangible mode. The "flat fracture" accompanied by a measured strain of nearly 1% was somewhat contradictory and it was decided that a fractographic study of the fracture surface was warranted.

Initially, an effort was made to extract fracture surface replicas with RTV rubber without disturbing the failed vessel. The rough texture of the fracture did not allow this nondestructive experimental approach and it was necessary to remove the fracture area from the vessel and subsequently saw it into pieces that could be handled in the scanning electron microscope. After considerable discussion concerning methods for removing the fracture surface from the failed vessel, it was concluded that oxyacetylene burning was the most expedient and least costly technique available. The flame cut was made approximately 8 in. away from the fracture surface. This distance is more than ample to assure no heat effect due to the flame cutting operation.[4] The fracture surface was protected during the flame-cutting. A photograph of the fracture surface prior to sawing is shown in Fig. 5. The original fatigued flaw is quite evident. The circumferential weld that joined the bottom hemihead to the cylindrical test section is easily seen at the left side (A) of the photograph. (It is located at the change in contour.) The circumferential weld at the upper end of the cylindrical test section is not as evident; however, it, too, is located at the geometric change in section size on the right side of the photograph.

Fig. 3. Intermediate Test Vessel No. 1 (ITV–1) after hydraulic testing to failure.

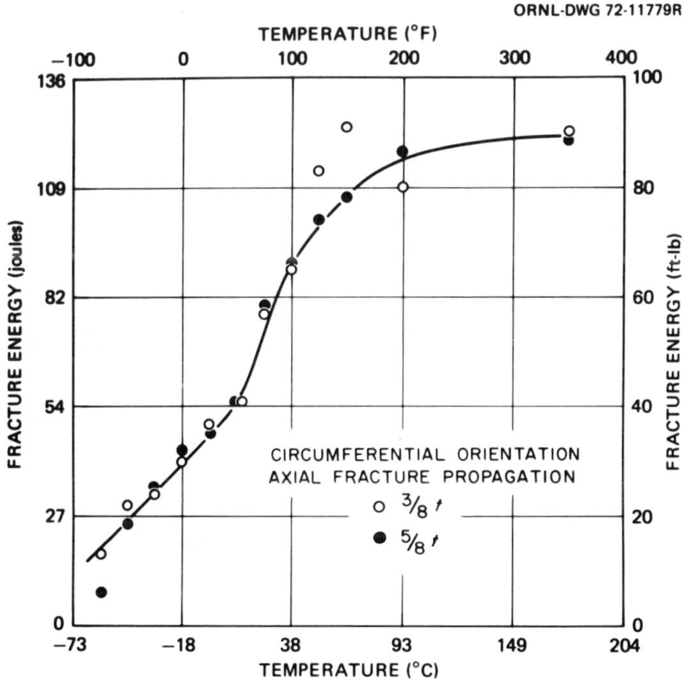

Fig. 4. Charpy V-notch impact test results obtained from intermediate vessel V–1, 6 in. thick, ASTM A508, class 2 forging.

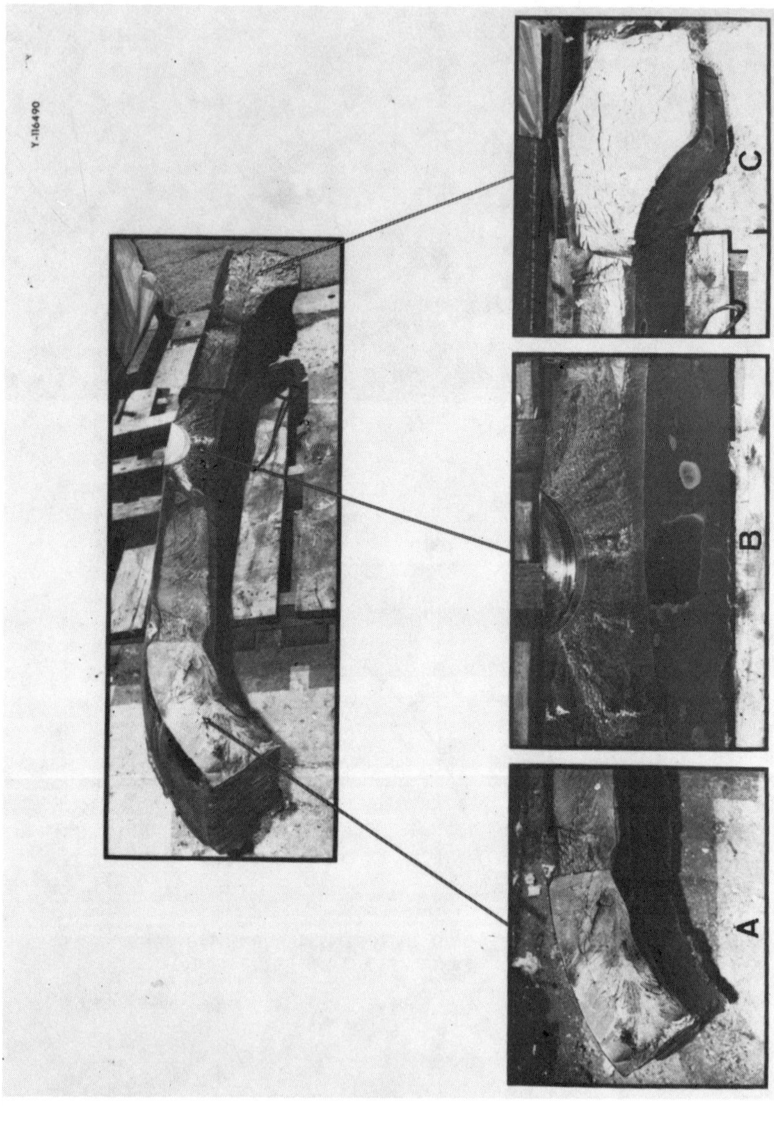

Fig. 5. One half of the final fracture removed from the rest of the vessel. Areas A, B and C were chosen from which to remove samples for scanning electron microscopy.

FRACTOGRAPHY STUDY

Seven regions of the fracture surface shown in Fig. 5 were studied. One is located in the transition piece, five are located in the cylindrical test section, and one is located in the hemihead. In order to relate the fracture morphology in ITV–1 to that exhibited by the same steel at the lower and upper Charpy V-notch (C_V) energy levels, a SEM examination of C_V specimens was conducted. Figure 6 contains the results of the SEM study on C_V specimens whose precise test conditions are known. The steel from which the C_V specimens were prepared was SA-508 and was heat treated in precisely the same manner as the cylindrical test section. Specimens from the C_V curve shown in Fig. 4 were used. A specimen tested at –59° C which absorbed 18 J of energy was used to illustrate brittle fracture appearance. A specimen tested at 50° C (4° C below the ITV–1 test temperature) which absorbed 112 J of energy was used to depict ductile behavior. Figure 6 shows the difference in appearance of a brittle and ductile fracture surface. Note the ribbon-like pocked appearance of the ductile failure. It is characterized by a dull appearance with surface contour somewhat akin to a classical cup-cone fracture of a ductile tensile failure. In contrast, the brittle fracture surface is faceted and both possess distinct characteristics that make each easily discernible.

The first ITV–1 specimen examined was from the region below the deepest point of the fatigue crack. Figure 7 shows two locations where specimens were taken and also illustrates the results of scanning electron microscopy studies of these fracture surfaces. Note the appearance of the surface of the two specimens is quite similar to that shown for the tough specimen (C_V energy of 112J) in Fig. 6. This region of the cylindrical test section exhibited a dimple fracture mode.

Figure 8 shows the results of the fractographic study from the region just adjacent to the fatigue crack. Two areas within this region were studied. (The corner of the flaw can be seen.) They, too, are similar in appearance to the ductile failures shown in Fig. 6. This region also exhibited dimple (tough) fracture mode.

The third region investigated is just under the steel rope shown in Fig. 5. On a macro scale this location exhibited a change in appearance and coloration. The dark "ductile" coloration seems to terminate in a V, with chevron lines, somewhat lighter in color, originating from this apex. This is usually indicative of an initiation site for fracture propagation. A fractographic study of this region (see Fig. 9) was carried out. The first area studied was located close to the center of the fracture surface and was within an area that was visually similar to the regions shown in Figs. 7 and 8. It has a fracture appearance (dimple mode) nearly identical to those previously shown. Therefore, this area is ductile (tough). The second area studied was outside the V-shaped area and in the more shiny, bright area. The SEM examination of this area revealed a cleavage mode of failure similar to those seen for the low temperature (frangible) specimens seen in Fig. 6. This area has undergone brittle failure.

Additional regions were studied that included regions approximately 76 mm from the circumferential welds in the top transition piece and the bottom hemihead. These regions exhibited cleavage fracture modes indicative of brittle failure.

INTERPRETATION OF RESULTS

An inspection of the exposed fracture surfaces showed that a shear lip of approximately 25–38 mm exists at both surfaces of the cylindrical test section. A sketch (Fig. 10) was prepared showing the dimensional relationship of the various topographical aspects of the region around the manmade flaw. The dark tough region (dimple fracture mode, see Figs. 7, 8, 9) is approximately 520 mm long and 76 to 83 mm wide. Beyond this region the surface exhibits a cleavage fracture mode.

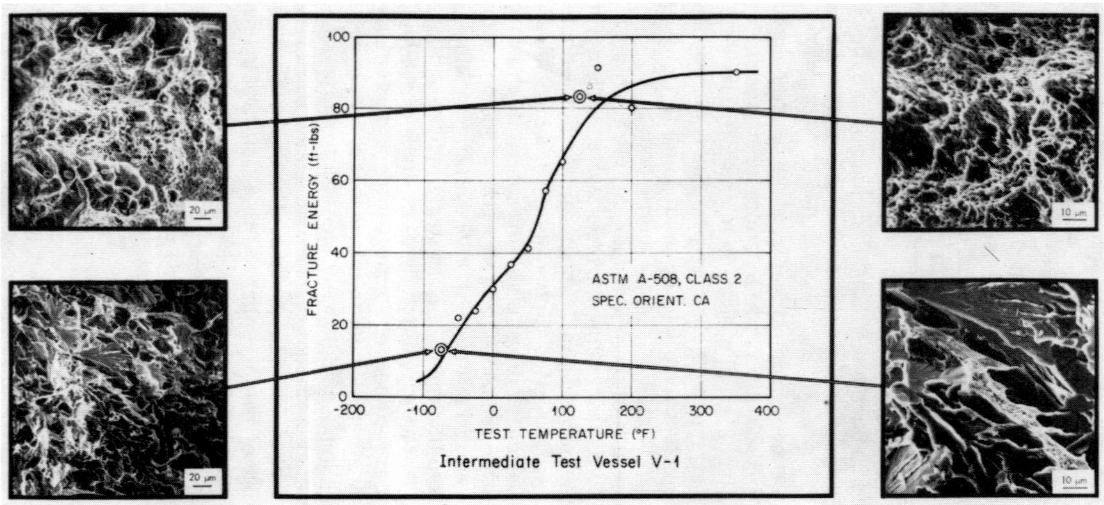

Fig. 6. Charpy V-notch test results. Carefully controlled toughness tests of ITV–1 steel provided base line data for evaluation of the vessel test. (1 foot pound = 1.36 joules; $t_c = (t_f - 32)/1.8$).

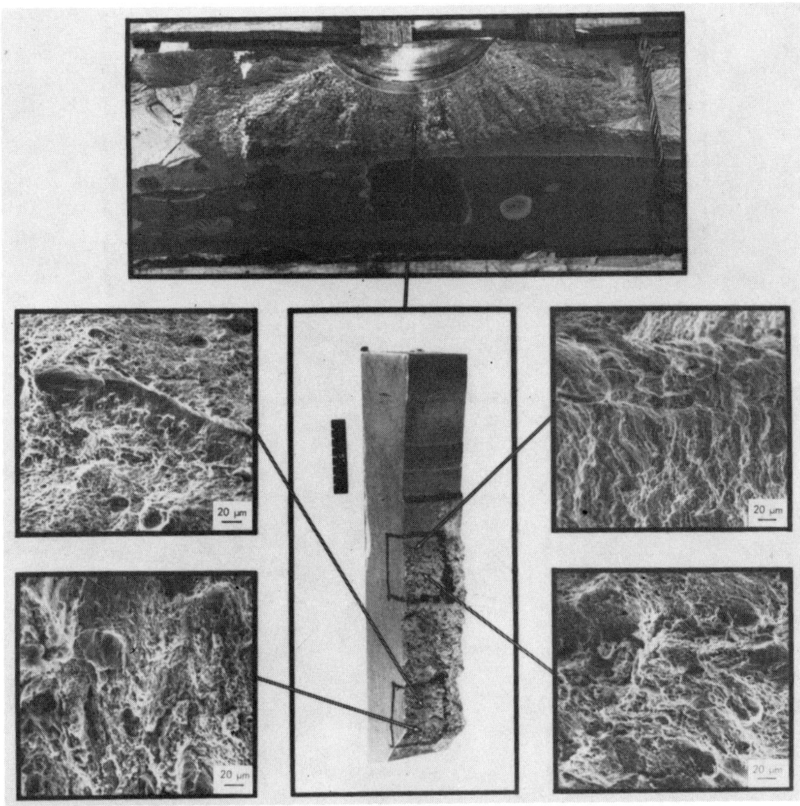

Fig. 7. Fractographic samples were taken from below the deepest penetration of the flaw. The failure was in the ductile mode.

This SEM study proved that the fatigue crack grew by ductile tearing until the flaw approached 500 mm in length and 83 mm in width. This extension, because of the stronger and perhaps tougher at 54°C surface material, was by internal tunneling. Slow crack propagation continued until the flaw became quite large and could no longer sustain the internal pressure which imposed a stress on the remaining vessel section that approached limit load. At this point fast fracture was initiated and crack extension was by cleavage mode.

SUMMARY

A study of the fracture surface of ITV–1 revealed a mixed mode of fracture. Initial crack growth is by ductile tearing. The growth was by tunneling and continued until the flaw size was such that unstable (plane strain) crack growth could occur. This results in a transition from ductile slow crack growth to brittle catastrophic failure. In summary, ITV–1 initially underwent ductile tearing. Final failure was rapid and fracture was by cleavage mode.

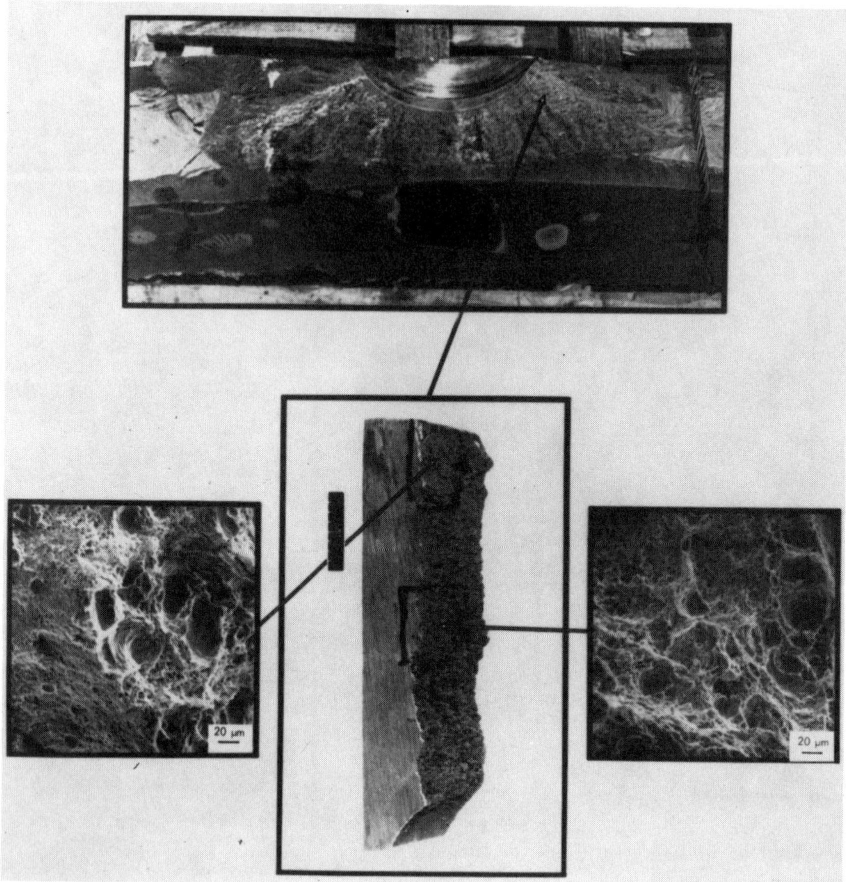

Fig. 8. The region adjacent to the fatigue crack was studied in the SEM. This area exhibited all ductile failure.

REFERENCES

1. *1964 Annual Book of ASTM Standards,* Part 4, pp. 696–704, American Society for Testing and Materials, A508–68, (1964).
2. R.W. Derby, J.G. Merkle, G.C. Robinson, G.D. Whitman and F.J. Witt, "Test of Six Inch Thick Pressure Vessels, Series 1: Intermediate Test Vessels V1 and V2," ORNL 4895.
3. F.J. Witt, "Heavy Section Steel Technology Program Semi-Annual Progress Report," ORNL–4855 (Sept. 1973), p. 39.
4. D.A. Canonico, "Technical Note: Depth of Heat-Affected Zone in Thick Pressure Vessel Steel Plate Due to Flame Cutting," *Welding J.* (N.Y.), 47 (9): 410-s and 419-s (Sept. 1968).

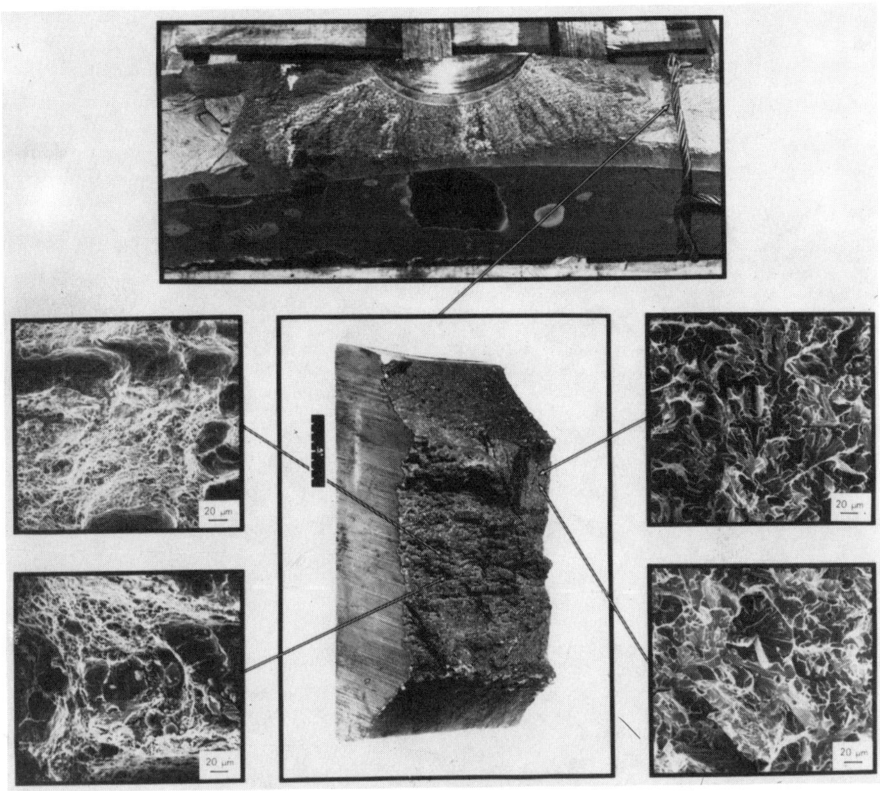

Fig. 9. Fractographic samples at the point where the surface appearance of the fracture changed showed a transition from ductile to brittle fracture.

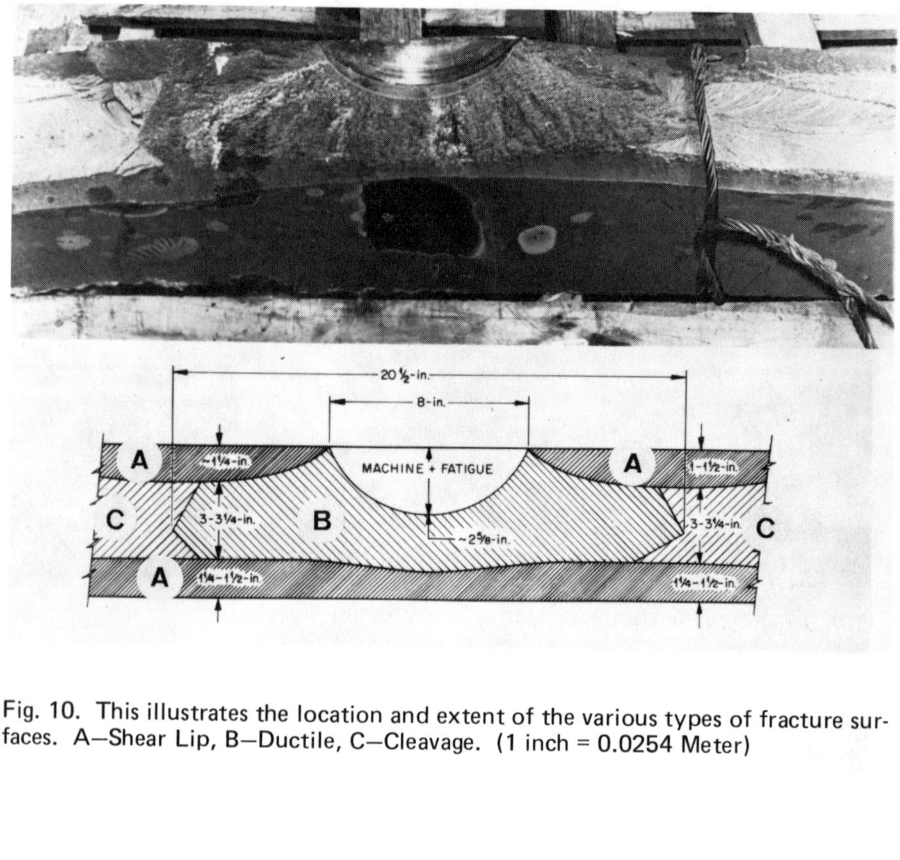

Fig. 10. This illustrates the location and extent of the various types of fracture surfaces. A—Shear Lip, B—Ductile, C—Cleavage. (1 inch = 0.0254 Meter)

STRUCTURE–PROPERTY–BEHAVIOR CORRELATIONS: III, QUANTIFYING STRUCTURE

Chairpersons: G.F. Vander Voort and J.E. Bennett

Stevens, Vander Voort, McCall, eds. Microstructural Science, Volume 8

METALLURGICAL CHARACTERIZATION OF TITANIUM ALLOY POWDERS

R.E. Omlor*, G.L. Teeters*, R.J. Bacon*, D. Eylon** and F.H. Froes***

ABSTRACT

The compactability and the final properties of consolidated titanium alloy powder products is strongly dependent on the morphology, microstructure, integrity and chemistry of the powder. To define the relationships between the powder and the properties of the final products, it is necessary to establish the complete characterization of the alloy powders. The present paper describes methods to identify the shape, size, microstructure, existence of pores or inclusions, as well as chemistry, of the various types of titanium alloy powders currently available in the market.

INTRODUCTION

For many aerospace applications, titanium alloys are the best design choice based on attractive mechanical properties, corrosion resistance and low density. However, the high fabrication cost of titanium alloy components negates full utilization. This high cost results both from workability problems and difficulty of machining. Substantial effort has been expended in recent years to develop new manufacturing methods of net-shape titanium alloy components to circumvent this problem[1]. Net-shape or near net-shape production lowers the working requirements and can eliminate a large proportion of the machining costs.

Hot Isostatic Pressing (HIP) of powders is one of the most promising net-shape production methods for complicated geometry titanium alloy components. In this method, the titanium alloy powder is held in a shaped canister, die or mold and subjected at temperatures (typically 150°F below the beta transus) to 15 ksi hydrostatic pressure for several hours[2]. The powder is consolidated to 100% density to the general shape of the mold. The tensile properties of HIPed components are comparable to wrought products, and in production scale quantities, their price can be lower than that of conventionally produced components.

POWDER MAKING

There are several methods by which titanium alloy powders can be produced. The Rotating Electrode Process (REP)[2,3], developed by Nuclear Metals, Inc., (NMI), is the only method currently used on a production scale. In this method, molten droplets are ejected from a rapidly rotating titanium alloy electrode and solidify in an inert atmosphere into the spherical shapes shown in Fig. 1a. In this and the following methods, the powder is made out of cast and wrought alloy billets.

*Systems Research Laboratories, Dayton, Ohio 45440 USA.
**Department of Materials Science and Metallurgical Engineering, University of Cincinnati, Cincinnati, Ohio 45221 USA.
***Air Force Materials Laboratory, Metals and Ceramics Division, Wright-Patterson Air Force Base, Ohio 45433 USA.

In another method developed by Crucible Research Center (designated C–T) [2], molten droplets are produced by electron beam (EB) melting of hydrogenated titanium alloy billet. The droplets are thrown away from the bar during the evolution of hydrogen. These powder particles have a basically spherical shape, similar to the REP powder (Fig. 1a), but a slightly different size distribution[2]. This method generally requires dehydrogenation of the powder prior to subsequent compaction.

Another method is the hydride/dehydride (HDH) method in which a titanium alloy billet embrittled by the addition of hydrogen is crushed to powder. This produces an angular powder as shown in Fig. 1b. As for the C–T powder, the HDH powder generally requires dehydrogenation prior to the compaction step.

The REP, C–T and HDH are currently the methods providing most of the titanium powder available in the U.S. market on a commercial or semi-commercial basis. Other methods, such as the PSV powder produced by CEN in France and the EB melting method by Leybold-Heraeus in Germany, will not be covered here.

An alternate compaction technique to HIP is to Cold Isostatically Press (CIP). The cold compactability of the powder for green strength depends in part on the particle share characteristics. Angular powder (such as the HDH, Fig. 1b) provides more contact area between the particles and requires less energy to cold compact the particles. For HIP compacts, the spherical powders (such as REP and C–T, Fig. 1a) are more desirable since they provide better flow characteristics and a higher "tap" density; both requirements for reproducible high accuracy shape making[2].

To increase the contact area of the spherical REP powder and to provide strain energy for subsequent recrystallization, it is possible to flatten the particle by passing them through a rolling mill, Fig. 1c, in a method known as the Strain Energizing Process (SEP) developed by Kelsey-Hayes[3].

All the powders shown in Fig. 1a through 1c are made of the Ti-6Al-4V alloy, which currently accounts for the majority of alloyed titanium in the aerospace industry. However, other titanium alloy powders have been produced successfully by the techniques described above.

CHARACTERIZATION

The metallurgical characteristics of titanium alloy powders is a powerful tool for the understanding and study of their compactability and subsequent mechanical properties. The following table illustrates the various steps typically followed in characterization, and properties that can be related to each item:

CHARACTERIZATION STEP	STUDIED PROPERTIES
Particle Morphology	Compactability
Microstructure	Mechanical Properties (general)
Defects	Fatigue Properties
Chemistry	Mechanical Properties & Stability

For proper evaluation of the different powders utilization, it is suggested that the steps indicated in the table above be carried out. The present work describes the results of using this procedure at the Air Force Materials Laboratory Characterization Facility to evaluate and compare titanium alloy powders obtained from different sources.

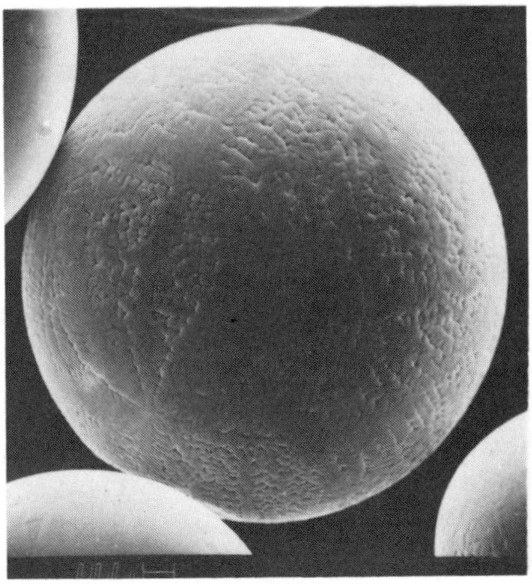

(a)

(b)

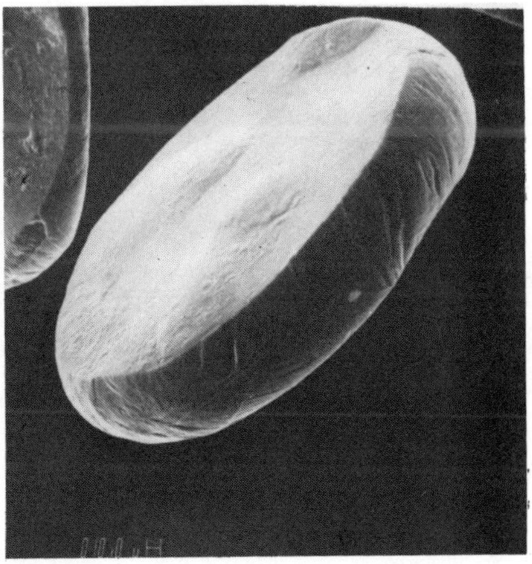

(c)

Fig. 1. Ti-6Al-4V powders made by the: a) Rotating Electrode Process (REP); b) Hydride/Dehydride Process (HDH); c) REP + Strain Energizing Process (SEP).

Particle Morphology

The various powders (Fig. 2) were obtained from their makers or users. The particle morphology was established (Fig. 1) by SEM which reveals the following features:

a. Shape
b. Size distribution
c. Particle surface features

Surface features, such as dendrite structure (Fig. 1a), can yield information about the cooling rates from the dendrite arm spacing. Other features like surface facets (Fig. 1b) may have some influence on the compactability.

Microstructure

The mechanical properties of all titanium alloys, such as fracture toughness[6], creep[7], and fatigue[8,9], are greatly influenced by the microstructure. Although the powder particle microstructure may change during the consolidation the initial microstructure may have considerable influence on final properties since the HIP consolidation is carried out below the beta transus temperature.

Powder particles were metallographically mounted and ground, polished and etched using standard specimen preparation procedures.

Figures 4a through 4d show the various microstructures obtained from alloys with an increasing degree of beta phase stabilization where Ti-6Al-4V is the least stabilized and Beta III the most highly stabilized.

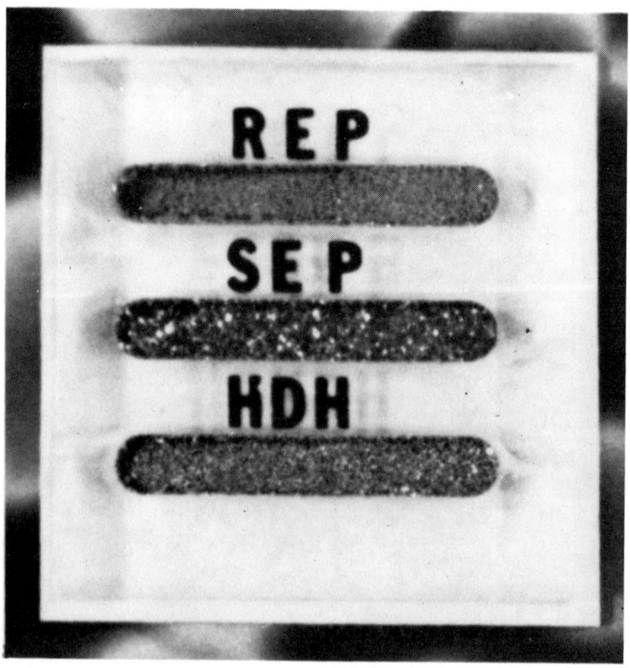

Fig. 2. The REP, SEP and HDH powders in capsules.

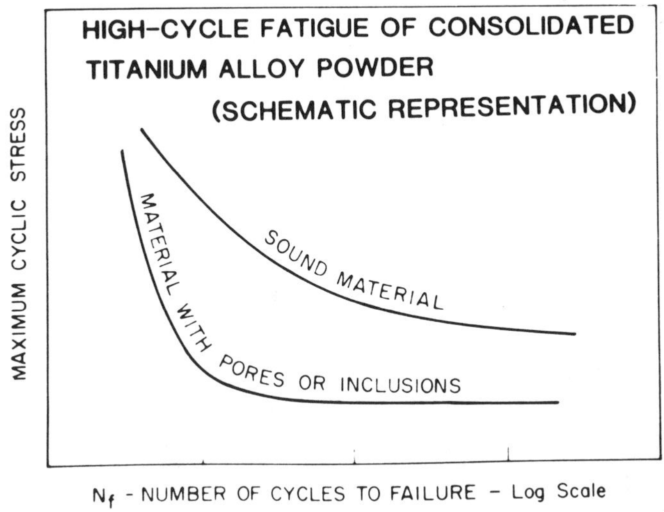

Fig. 3. Fatigue life schematic representation of consolidated P/M material with and without defects.

Features of Special Interest

As demonstrated in the schematic representation in Fig. 3, pores or hard inclusions generally reduce the fatigue strength due to premature crack initiation. The identification of such features is thus of prime importance for components where fatigue resistance is a major consideration.

Figures 5a through 5d show gas pores, shrinkage pores, embedded tungsten inclusions and loose tungsten inclusions in Ti-6Al-4V REP powder. Pores and inclusions are found also in powders made by other methods, while the tungsten is an inherent feature of the REP process as currently practiced.

It is possible through microprobe chemical analysis to identify inclusions (Fig. 5e) and, in some cases, to minimize their influence on mechanical properties by process modification.

CONCLUSION

A complete characterization of loose titanium alloy powders holds the initial key to the understanding of the structure and properties of consolidated titanium alloy P/M products. To be able to predict both the compactability and mechanical behavior of those powder products, morphological microstructure and chemical analysis have to be performed.

ACKNOWLEDGEMENT

Parts of this work were carried out under U.S.A.F. Contract Numbers F33615-77-C-5008 and F33615-76-C-5227. Useful discussions with Mr. C. Kelto of the AFML Manufacturing Technology Division are acknowledged.

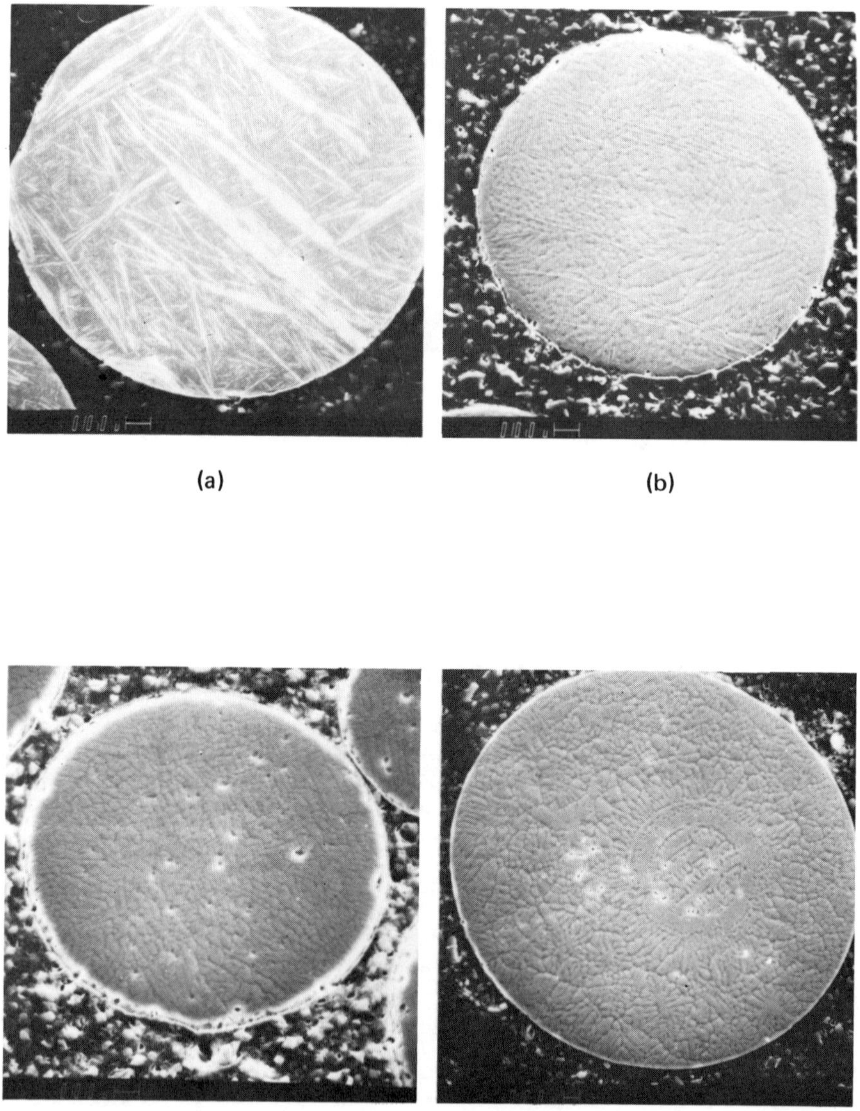

(a) (b)

(c) (d)

Fig. 4. Microstructure of sectioned and etched powder particles of the alloys: a) Ti-6Al-4V made by the REP method; b) Corona 5 (4.5Al-5Mo-1.5Cr) – REP; c) Ti-17 (5Al-2Sn-2Zr-4Mo-4Cr) – REP; d) Beta-III (11.5Mo-6Zr-4.5Sn) – REP.

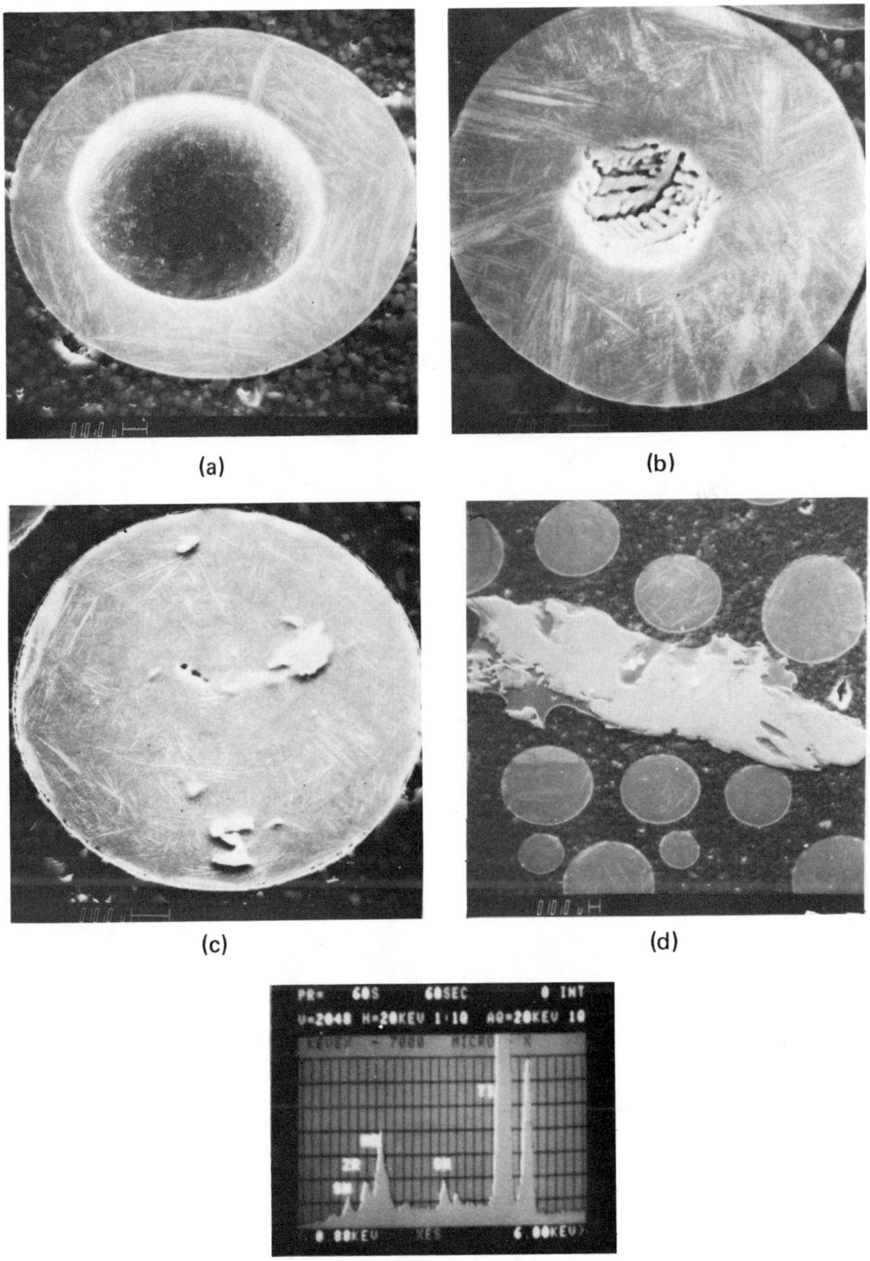

(e)

Fig. 5. Features ot special interest in REP Ti-6Al-4V powder particles: a) gas porosity; b) shrinkage porosity; c) tungsten inclusions; d) tungsten particles; e) chemistry of areas of interest.

REFERENCES

1. N.G. Tupper, J.K. Elbaum and H.M. Burte, *J. Metals,* pp. 7–13 (September 1978).
2. E.J. Dulis, V.K. Chandhok, F.H. Froes and L.P. Clark: *Proceedings of 10th National SAMPE Meeting,* pp. 316–329 (1978).
3. T.E. Miles and J.F. Rhodes: *Proceedings of Rapid Solidification Processing Conference,* pp. 347–361 (1977).
4. A. Lawley: *The Int. J. of Powder Met. & Powder Tech.* 13, pp. 173–188 (1977).
5. L.P. Clark, AGARD Report #627, Technical Editing and Reproduction Ltd.; London, (March 1975).
6. J.A. Hall, C.M. Pierce, D.L. Ruckle, and R.A. Sprague: *Mater. Sci. Eng.,* Vol. 9, pp. 197–210 (1972).
7. D. Eylon, J.A. Hall, C.M. Pierce and D.L. Ruckle: *Met. Trans. A,* Vol. 7A, pp. 1817–76 (1976).
8. M.S. Greenfield, C.M. Pierce, and J.A. Hall: *Titanium Science and Technology,* Vol. 3, pp. 1731–1943, Plenum Press, New York, (1973).
9. D. Eylon and C.M. Pierce: *Met. Trans.A,* Vol. 7A, pp. 111–121 (1976).

Stevens, Vander Voort, McCall, eds. Microstructural Science, Volume 8

MICROSTRUCTURAL ANALYSIS BY NEW IMAGE ANALYSIS TECHNIQUES

James F. Golden* and Werner Hunn*

ABSTRACT

The rapid technical development of high-speed digital electronics and microcomputers dramatically increases the power of television-based image analyzers. New hardware features include a 100 level densitometer, multiple binary image planes connected by an arithmetic logic unit, and a versatile individual structure analyzer. Under the control of a microcomputer and BASIC-like operating system, these modules perform complex image operations such as removal of frame intersected particles and structure reconstruction. Applications include free and locked phase analysis, separation of inter- and intra-grain porosity, grain sizing distributions and directional analysis.

INTRODUCTION

Early television-based image analyzers were designed to provide rapid analyses such as volume fraction and grain sizing by mean intercept length over many microscope fields. With the increased performance of computers and digital circuitry, it has become feasible to couple imaging devices and computers, resulting in systems with greater power and flexibility. One basic approach taken by systems manufacturers is to use the television system as a viewing device, digitize the image, and store the digitized image in a dedicated computer, where analyses are performed by software algorithms. A second method utilizes special high-speed hardware under control of a micro- or minicomputer. The LEITZ Texture Analyzing System (TAS) with Individual Analysis is of this latter type.

INSTRUMENTATION

A block diagram of the TAS is shown in Fig. 1. It consists of a number of hardware modules, operating at TV speeds, coupled to a DEC LSI 11 microcomputer with peripherals. The software operating system resident in the LSI 11 is similar to BASIC, but includes many machine-specific statements for controlling the various hardware modules, thus providing flexibility in terms of application.

The signal from the plumbicon camera is passed through the densitometer module. The densitometer divides the full amplitude range (from black to white) into 100 parts and then sets an amplitude window of selectable position and width. As the video signal is passed through the module, picture points whose amplitude falls within the window are detected (assigned a binary value of 1), all others are not detected (binary value 0). The result is a binary "thresholded" image. The densitometer can be set, for instance, so that phases in metal are detected. With the high resolution (100 levels) of the densitometer, phases differing only slightly in reflectance can be separately detected and analyzed. Alternately, the densitometer can be scanned through its entire range to produce reflectograms or mean reflectance values.

*E. LEITZ, INC., Rockleigh, New Jersey 07647 USA.

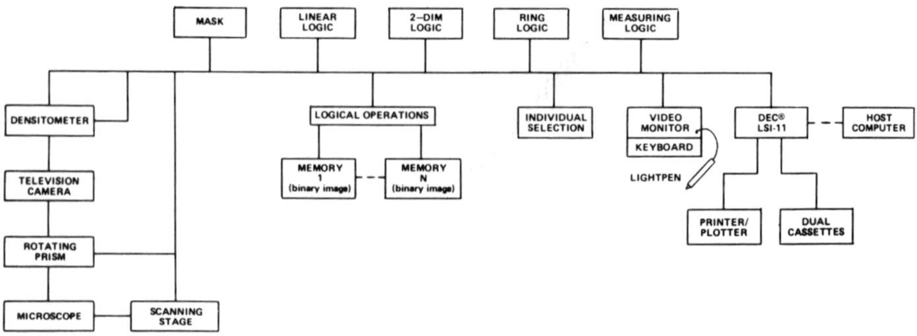

Fig. 1. Block diagram of LEITZ Texture Analyzing System (TAS).

The detected image can either be analyzed directly ("live") or can be stored in a binary image memory. Since the TAS contains several memories, different images can be stored and compared; for instance, images produced by different densitometer settings or different illumination methods. The memories are interconnected through an arithmetic logic unit, which can perform logical operations on one or two memories. The logical operations can be combined to produce powerful algorithms. For instance, particles which intersect the edge of the measuring field can be instantly removed. Also, particles stored in one memory can be reconstructed from markers stored in another memory.

At times, due to the nature of the specimen, the image as detected is not an accurate representation of the material. For example, some grain boundaries might not be detected due to low contrast or inadequate etching. In these instances, operator interaction with the image is necessary. In the TAS, the content of an image memory can be modified, using a light pen, by adding or erasing detail and drawing acceptance or rejection regions.

In many cases, such as volume percent or grain sizing applications, measurements taken over the entire measuring field are sufficient. However, for other applications such as individual particle size or shape distributions, measurements on individual structures are required. The TAS includes individual analysis circuitry that allows individual features to be selected, either manually by touch of the light pen, or sequentially under computer control. Once selected, the particle can be analyzed using the full range of operations available for whole field analysis. Based on these analyses, the particle can then be sorted by storing in other image memories.

In addition to the basic densitometric detection and measurement operations, the TAS incorporates both linear and two-dimensional transformations. The two basic transformations, erosion and dilation, can produce particle size and proximity distribution histograms, in either one or two dimensions. Under software control, the sizing distribution program automatically handles touching particles. Using the linear transformation, covariance analysis can be performed to detect periodicity in the material.

The detected image from the densitometer can be passed through a mask prior to analysis. One type of available mask is the circular mask. This mask, in combination with an image rotating prism, makes accurate directional analysis possible. Also, any image stored in memory, including an individual structure, can be set as a mask. For instance, the mean reflectance of a particular grain can be easily determined. Masking can also be used to study fine detail within a single structure.

GENERAL APPLICATIONS

The above described combination of high-speed hardware and program control permits solutions to problems not possible by conventional techniques. While it is, for example, very easy to measure a volume fraction of a given phase with most image analyzers, it is difficult to proportion the volume between the free part of a component and the one interlocked to a second phase in, for example, an alloy or mineral. We define two phases to be interlocked whenever they have a common surface. With the help of the image planes and two-dimensional analyses, the TAS separates the free and locked portion for volume fraction measurement, as shown in Fig. 2, or for subsequent analyses such as sizing of the respective group of grains.

In sizing of grains in single phase systems with boundaries, the problem of incomplete boundaries always arises. When determining the mean grain size dimensions by intercept methods, this effect can very often be neglected. However, when actual sizing distributions by area or dimension are required, the broken boundaries lead to erroneous results. These can be overcome with a program called "ultimate erosion". Any grain, even if just indicated by the start of a boundary, is reduced to a point in repetitive steps of shrinkage. This is exemplified in Fig. 3. The program also determines the diameter of the largest hexagon which can be inscribed in each grain, which is easily printed out as a histogram for all the grains present.

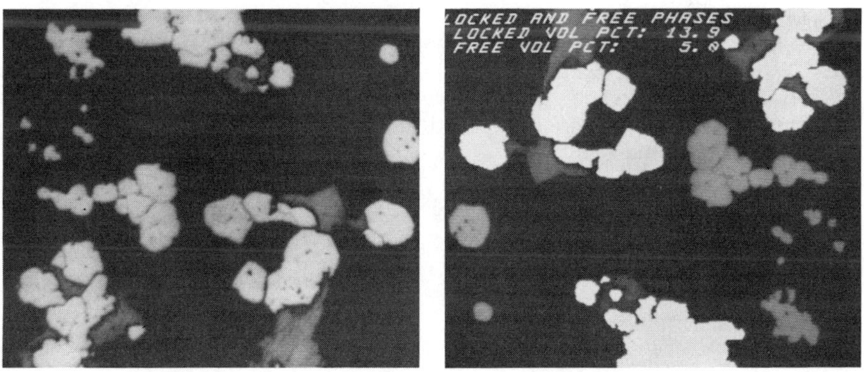

(a) (b)

Fig. 2. Locked and free phase analysis. a) Television image showing phase of interest (light grey) and second phase (dark grey). b) Detection (white) of locked phase with volume percent of locked and free phase.

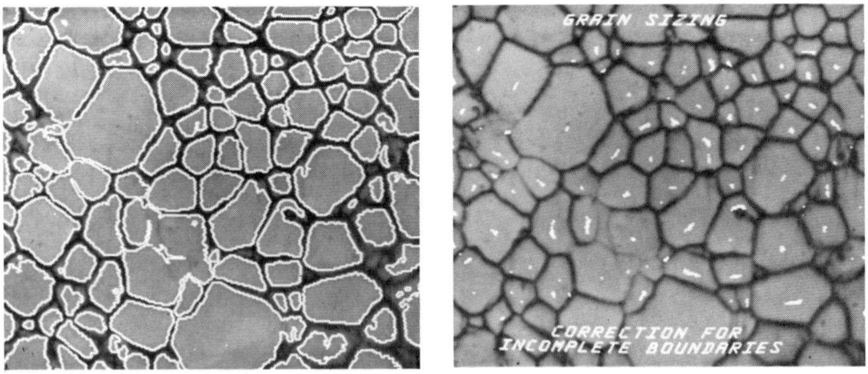

(a) (b)

Fig. 3. Grain sizing with incomplete boundaries. a) Perimeter display (white) of detected image showing incomplete grain boundary detection. b) Result of ultimate erosion, where each detected grain is marked. Note: detection of grains even where boundaries are incomplete.

Another example of considerable difficulty is the separate measurement of intra- and inter-grain porosity. This separation may require image editing with the light pen to complete grain boundaries. Once this editing is complete, the separation of intra- and inter-grain porosity is automatically accomplished, including suppression of the grain boundaries (see Fig. 4).

Directional analysis to study structural anisotropy, such as in grains in rolled steel, or preferred orientation, such as needles in aluminum alloys, can yield valuable information. A prism rotating the image in one degree steps is coupled with a circular electronic frame to always measure the same image content. As shown (Fig. 5) in the case of hydrides in steel tubing, such an analysis is performed with a direction sensitive parameter, such as the vertical projection, measured over 180 degrees. The results are displayed as a polar plot from which the tangential to radial ratio of the hydrides can be extracted.

In addition to these and other general applications, there are of course numerous problems specific to one industry or process. Some of these cannot be adequately addressed by image analysis. However, thanks to recent advances in image analysis technology, solutions are now practical which only a few years ago were well beyond the state-of-the-art.

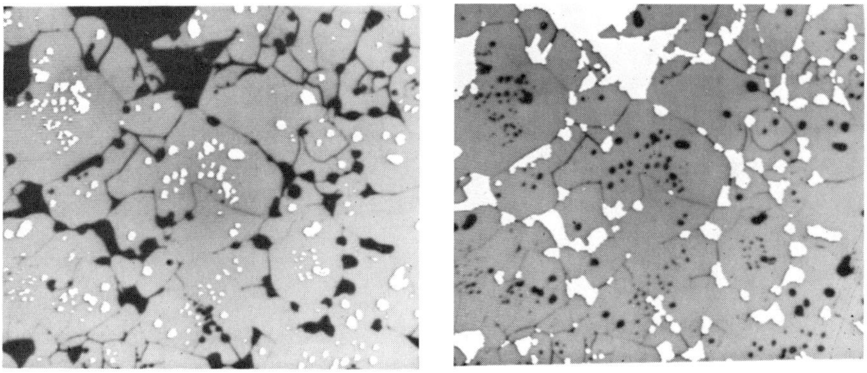

(a) (b)

Fig. 4. Separation of intra- and inter-grain porosity. a) Detected intra-grain porosity. b) Detected inter-grain porosity. Note: detection of grains even where boundaries are incomplete.

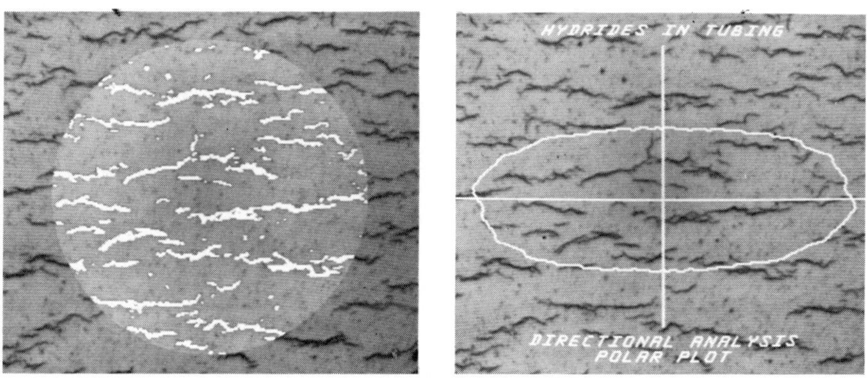

(a) (b)

Fig. 5. Directional analysis. a) Detection of hydrides in steel tubing, showing circular electronic mask. b) Polar plot of projected length of hydrides as a function of angle.

Stevens, Vander Voort, McCall, eds. Microstructural Science, Volume 8

A METHOD TO STUDY SINTERING OF A SUPPORTED METAL CATALYST

H.K. Kuo*, P. Ganesan*, and R.J. De Angelis*

ABSTRACT

An x-ray diffraction single profile analysis method has been used to monitor the changes in the average nickel particle sizes and particle size distributions (PSD) during sintering of a silica-supported nickel catalyst. Sintering was carried out in nitrogen and hydrogen atmospheres at temperatures from 500° to 800°C for times varying from 5 to 100 h. The particle size distribution functions determined both by the x-ray single order line profile analysis and by transmission electron microscopy were in excellent agreement. The sintering process occurred very rapidly initially and then proceeded more slowly at longer times at the higher temperatures. The effects of sintering temperature on the changes of PSD were found to be much more pronounced than the effects of sintering time. As sintering progresses, the PSD's developed long tails to the large diameter side. The changes in the PSD's tend to indicate that the probable sintering mechanism for the silica-supported nickel catalyst is associated with particle migration.

INTRODUCTION

Metal catalysts are generally employed in the form of small metal particles or crystallites dispersed on high surface area supports. In this form, a high ratio of metal atoms exists on the surface and an active catalyst results. The support also serves the function of physically separating the small metal particles, which tends to inhibit agglomeration of the small particles into larger particles. Strong interaction between the support and the metal particle also inhibits agglomeration. This agglomeration, known as sintering, occurs very rapidly in unsupported metal catalysts and leads to fewer metal atoms at the surface of the particle with consequent loss of catalyst activity. Sintering also occurs at high temperatures in supported metal catalysts and is the main source of catalyst thermal deactivation. The process of sintering, by definition, requires changes to occur in the particle size distribution. The sintering behavior of a metal catalyst depends on the initial particle size distribution, and knowledge of the progressive changes in the particle size distribution during sintering can assist in elucidating the mechanism of sintering in metal supported catalysts.[1] Therefore, a reliable determination of the initial particle size distribution (PSD) function and the changes occurring in the PSD during sintering becomes essential if the sintering process is to be understood and ultimately controlled.

The purpose of this paper is to present an x-ray diffraction method based on a single profile analysis technique which gives the diffracting particle size, lattice microstrains, particle size distribution (PSD) function. Also, comparisons are made for PSD's determined by x-ray diffraction and by electron microscopy. This comparison will be made on catalytic materials containing a metallic nickel particle in the range of 30 to 100 Å. The lower part of this range has until recently been considered to be a particle size range which could not be completely investigated using x-ray diffraction techniques. [2]

*Department of Metallurgical Engineering and Institute for Mining and Minerals Research, University of Kentucky, Lexington, Kentucky 40506 USA.

THEORY

Single Profile Analysis Technique

The single diffraction profile technique is based on the work of Gangulee[3] and Mignot.[4] The Stokes corrected cosine coefficients[5] from an (hkl) diffraction profile are composed of two components, a size coefficient A_L^S and a microstrain or distortion coefficient A_L^D; their relationship can be expressed as

$$A_L = A_L^S \cdot A_L^D \tag{1}$$

where A_L are the Stokes corrected cosine coefficients at a given L, where L is nδ. Here, n is the harmonic number and δ is a distance normal to the diffracting planes. Defining a variable $X = 1/D_e$, where D_e is the effective diffracting particle size for small values of n and δ, where the values of L are small and such that the number of diffracting domains in the specimen with this dimension is insignificant, the particle size term of the Fourier coefficients can be expanded as

$$A_L^S = (1 - LX) \tag{2}$$

The distortion coefficients can be expanded for small values of n, as

$$A_L^D = (1 - KL^2 \langle\epsilon_L^2\rangle) \tag{3}$$

where K is $2\pi^2/d^2$, d is the (hkl) planar spacing and $\langle\epsilon_L^2\rangle$ is mean square of the microstrain averaged over all distance in the diffracting specimen spaced L apart.

Letting $Y_L = K\langle\epsilon_L^2\rangle$, the small L value cosine coefficient can be written as

$$A_L = (1 - LX)\ (1 - Y_L L^2) \tag{4}$$

Eq. (4) can be solved for X if the functional form of Y_L is known. In this way the particle size can be separated from the microstrain terms. The problem then resolves to determination of the most suitable forms of Y_L. Assuming the form of the strain function to be [6]

$$Y_L = (C/L)\delta K \tag{5}$$

leads to an expression for A_L of

$$A_L = 1 - L(X + C\delta K) + L^2\ (XC\delta K) \tag{6}$$

The expression for A_L has the form of a second order polynomial in L. Now a polynomial of second degree is defined here as

$$R_L = a_o + a_1 L + a_2 L^2 \tag{7}$$

whose coefficients by comparing with Eq. (6) are

$$a_o = 1,\ a_1 = -(X + c\delta K),\ \text{and}\ a_2 = XC\delta K \tag{8}$$

From these relations it follows that,

$$X = 1/2\ (-a_1 \pm (a_1^2 - 4a_2)^{1/2})$$

$$D_e = 1/X$$

$$C = a_2/X\delta K \tag{9}$$

$$<\epsilon_L^2> = (C/L)\delta$$

The determination of the coefficients a_o, a_1, and a_2 can be made by solving the system of equations:

$$\sum_t A_L = a_o N + a_1 \sum_t L + a_2 \sum_t L^2$$

$$\sum_t LA_L = a_o \sum_t L + a_1 \sum_t L^2 + a_2 \sum_t L^3 \tag{10}$$

$$\sum_t L^2 A_L = a_o \sum_t L^2 + a_1 \sum_t L^3 + a_2 \sum_t L^4$$

Where N is the number of coefficients used in the evaluation, $N \geq 4$, t is an integer satisfying $t = (L_N - L_o) / (L_1 - L_o)$, and L_o is the distance corresponding to the initial coefficient used in the computation of Eqs. (10).

The best solution to Eqs. (10) are selected from four criteria:

(1) $a_o \approx 1$

(2) $a_2 > 0$, since a_2 from Eqs. (6) and (7) can be written as d^2A_L/dL^2 which must be positive.

(3) $\sum_t (R_L - A_L)^2 /N$ is a minimum

(4) $(dA_L/dL)_{L \to 0} = a_1 = -(X + C\delta K)$

After the best solution to Eq. (10) is determined, it is a simple process to calculate the average coherent diffracting particle size D_e and lattice mean square strain $<\epsilon_L^2>$ by employing Eq. (9).

Particle Size Distribution Function

While single profile analysis gives a convenient method to calculate lattice microstrain, it is important to note that the PSD function cannot be obtained by this method alone, since from Eq.(2), $d^2A_L^S/dL^2$ reduces to zero. However, by employing the usually accepted small (L/d) approximation it is possible to obtain the particle size distribution function by combining results from both the single profile and multiple order analysis methods. Although this approach does contain limitations, it is a major improvement over the zero strain approximation.

The expression for the relationship between the particle size and distortion coefficients in the Warren-Averbach double order method[7] is given by

$$A_L = A_L^S < \cos 2\pi\, L/d\, \epsilon_L > \tag{11}$$

For small values of L/d the argument of the microstrain term is small and using the approximations $< \cos X > = 1 - < X^2 > /2$ and $(1 - X) \cong \exp(-X)$, Eq. (11) can be written as

$$A_L = A_L^S \exp(-2\pi^2 L^2 / d^2 <\epsilon_L^2>) \tag{12}$$

This method of separation of particle size and microstrain term makes no assumption of the nature of strains; however, Eq. (12) is most exact at small values of L/d. It is easily shown that Eq.(12) is exact for all values of L/d if the microstrain distribution is Gaussian[8]. The distribution of strains at a given L has been shown to be close to Gaussian[9–11], but the variation of $<\epsilon_L^2 (hkl)>^{1/2}$ with L is not Gaussian.

Now it is possible to obtain the particle size coefficient A_L^S from Eq. (12) by substituting the values for mean square strain $<\epsilon_L^2>$ obtained from the single profile analysis (Eq.(9)). The particle size distribution function, P(L), can then be obtained from the second derivative of A_L^S with respect to L[8], that is

$$P(L) = \frac{d^2 A_L^S}{dL^2} \quad D_e \tag{13}$$

The particle size distribution (PSD) data can be employed to calculate the average particle size using the two relationships

$$\overline{D} = \sum_L LP(L)\ \Delta L \tag{14a}$$

$$\overline{D} = [\sum_L \frac{1}{L} P(L)\ \Delta L\ ]^{-1} \tag{14b}$$

where P(L) is the value of the distribution function at the size L normalized such that $\int P(L)dL = 1$, $\overline{D}$ is the volumetric mean particle size, and $\overline{D}_1$ is the average value of the thickness of the particle normal to the (hkl) reflecting planes.[12]

The complete x-ray diffraction profile shape analysis which involves the calculation of Stokes corrected coefficients A_L, particle size coefficients A_L^S and the particle size distribution function P(L) has been computerized.[13]

RESULTS AND DISCUSSION

The silica-supported catalytic material (C150–1–01) used in this investigation, supplied by United Catalysts Inc., Louisville, Kentucky, contained 51.7% Ni, 2.94% C and 0.06% S. It had a surface area of 211m^2/g, a pore volume of 0.34 cm^3/g and a density of 1.05 g/cm^3. The catalytic materials NiO/SiO_2 were produced by a co-precipitation process. This is one of the four catalysts reported in a previous investigation.[13]

The particle size distribution functions obtained by x-ray diffraction technique for the silica-supported nickel catalyst C150–1–01 on sintering up to 20 h at temperatures between 500 to 800°C in hydrogen atmosphere are given in the form of isochronal plots in Figs. 1–4. The sintering behavior observed in nitrogen atmosphere was very similar to the hydrogen atmosphere results. Sintering occurs very rapidly at short sintering times and proceeds much more slowly at longer times. The effects of sintering temperature on the PSD are much more pronounced than the effects of sintering time. There appears to be a limiting particle size which increases with increasing sintering temperature. Also, there is a definite formation of tails in the distribution functions to larger particle size side. This type of tail formation supports a sintering model based on a particle migration mechanism.[14,15] The lack of formation of particles smaller than the original particles tends to indicate that an atomic migration mechanism[16] was inoperative.

In order to independently check the validity of particle size distribution functions determined by the x-ray diffraction method, transmission electron microscopy work was done on specimens from five selected runs to obtain the PSD's by direct observation. TEM specimens were prepared by suspending the catalyst powder in methyl alcohol ultrasonically. A drop of the suspension solution was placed on copper grid which was coated with carbon on Formvar and the alcohol was allowed to evaporate. After drying, the specimens were observed in a Hitachi HU–11B Electron Microscope

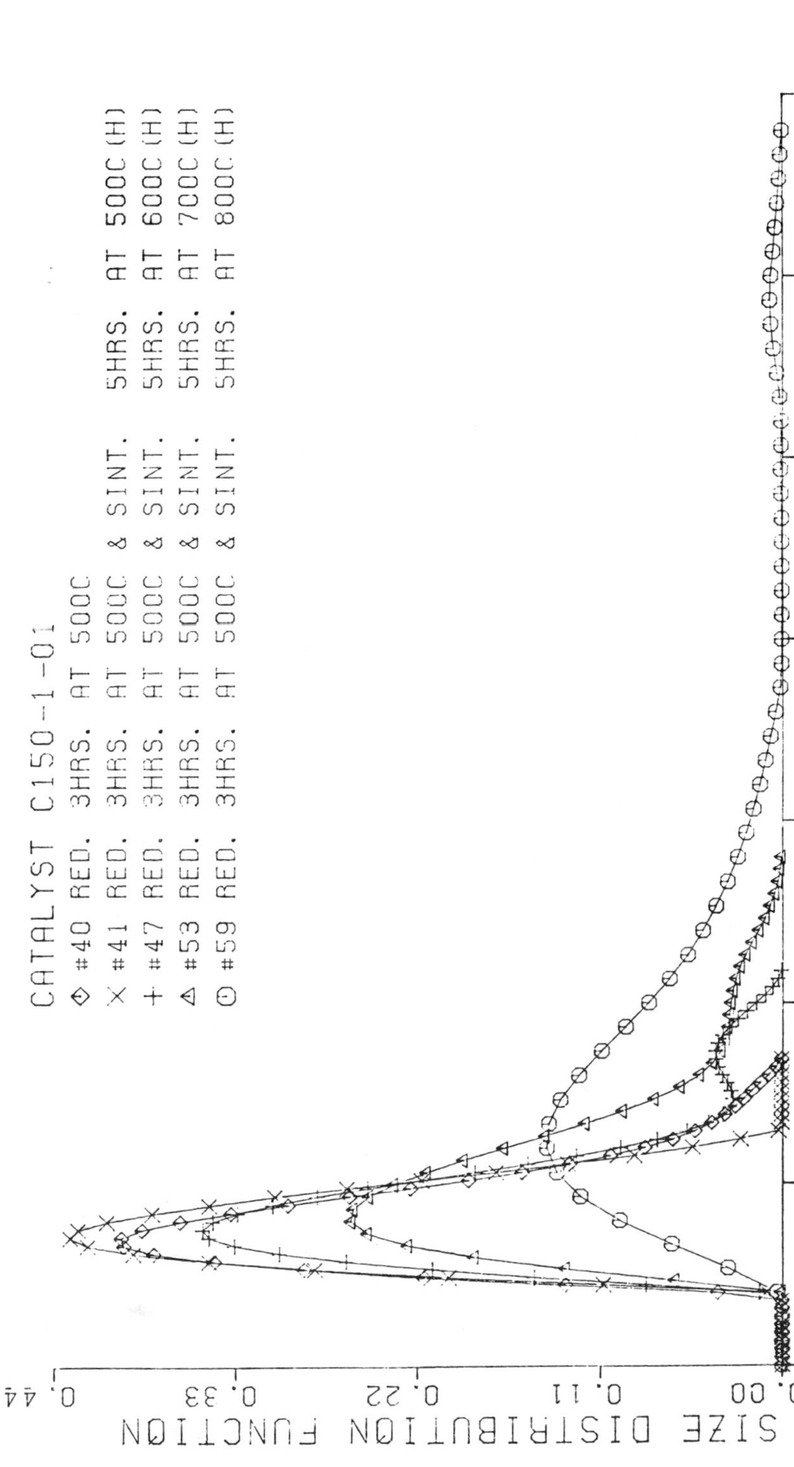

Fig. 1. Particle size distribution functions of catalyst C150–1–01 reduced at 500°C for 3 h and sintered at 500, 600, 700 and 800°C for 5 h in hydrogen atmosphere.

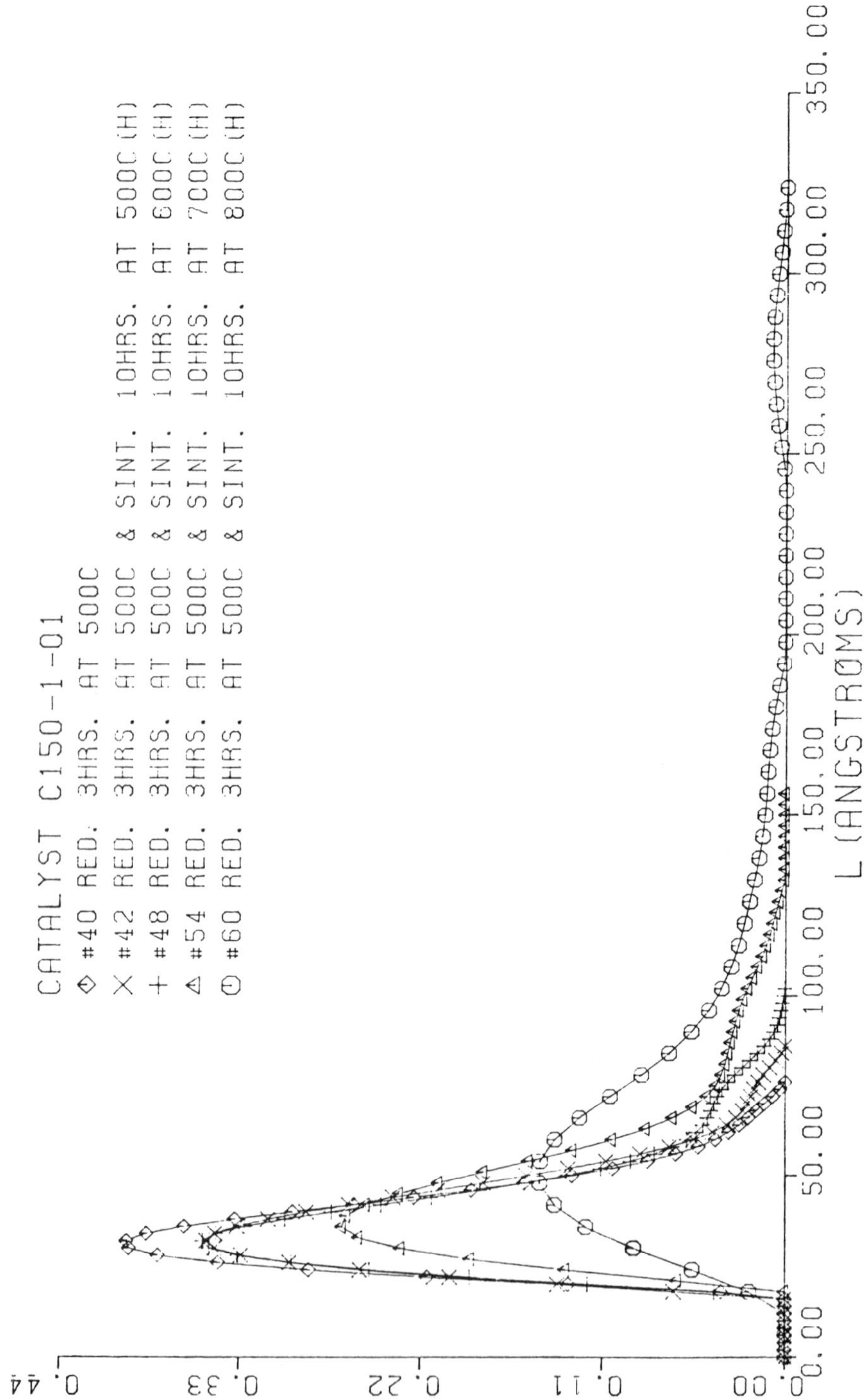

Fig. 2. Particle size distribution functions of catalyst C150–1–01 reduced at 500°C for 3 h and sintered at 500, 600, 700 and 800°C for 10 h in hydrogen atmosphere.

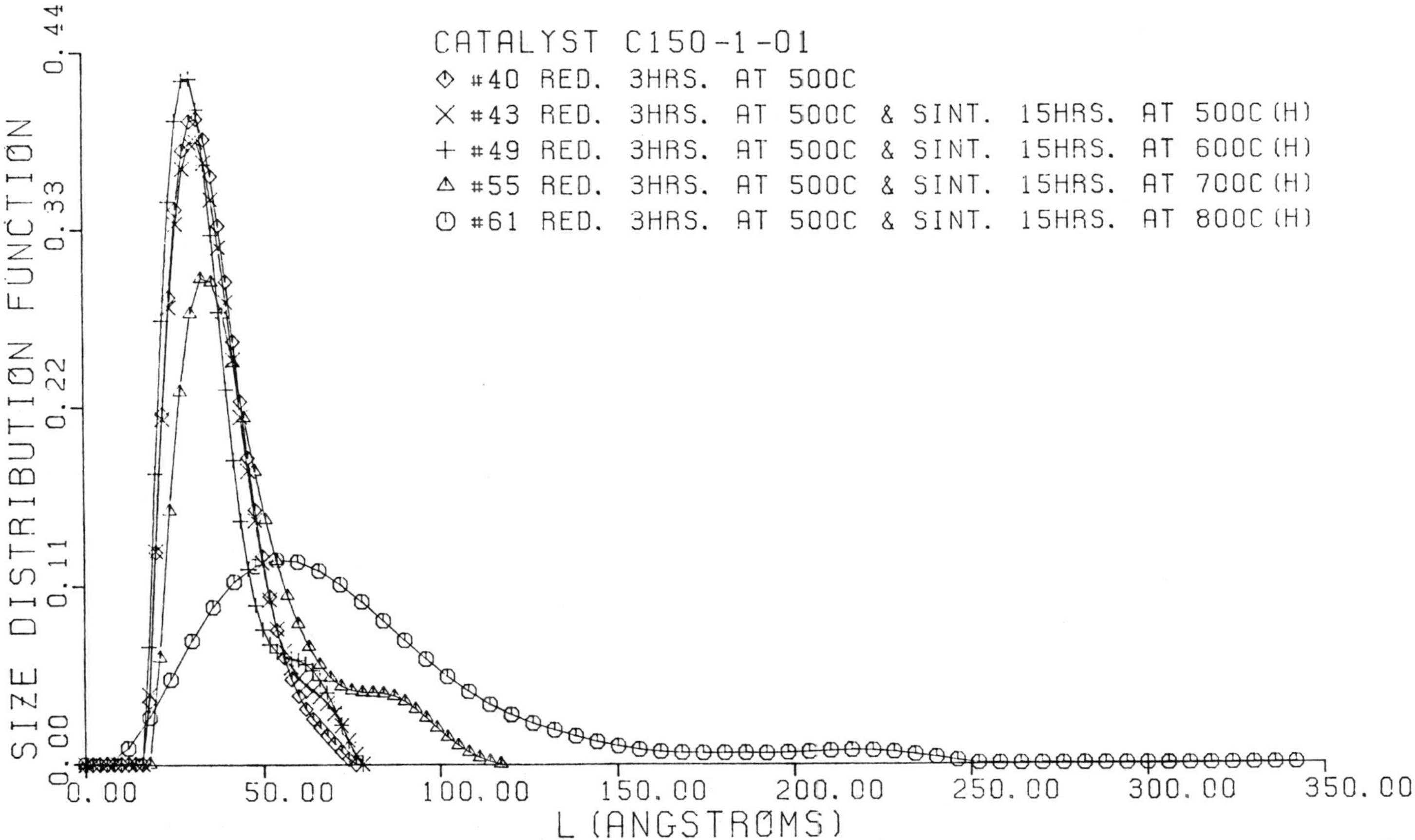

Fig. 3. Particle size distribution functions of catalyst C150–1–01 reduced at 500°C for 3 h and sintered at 500, 600, 700 and 800°C for 15 h in hydrogen atmosphere.

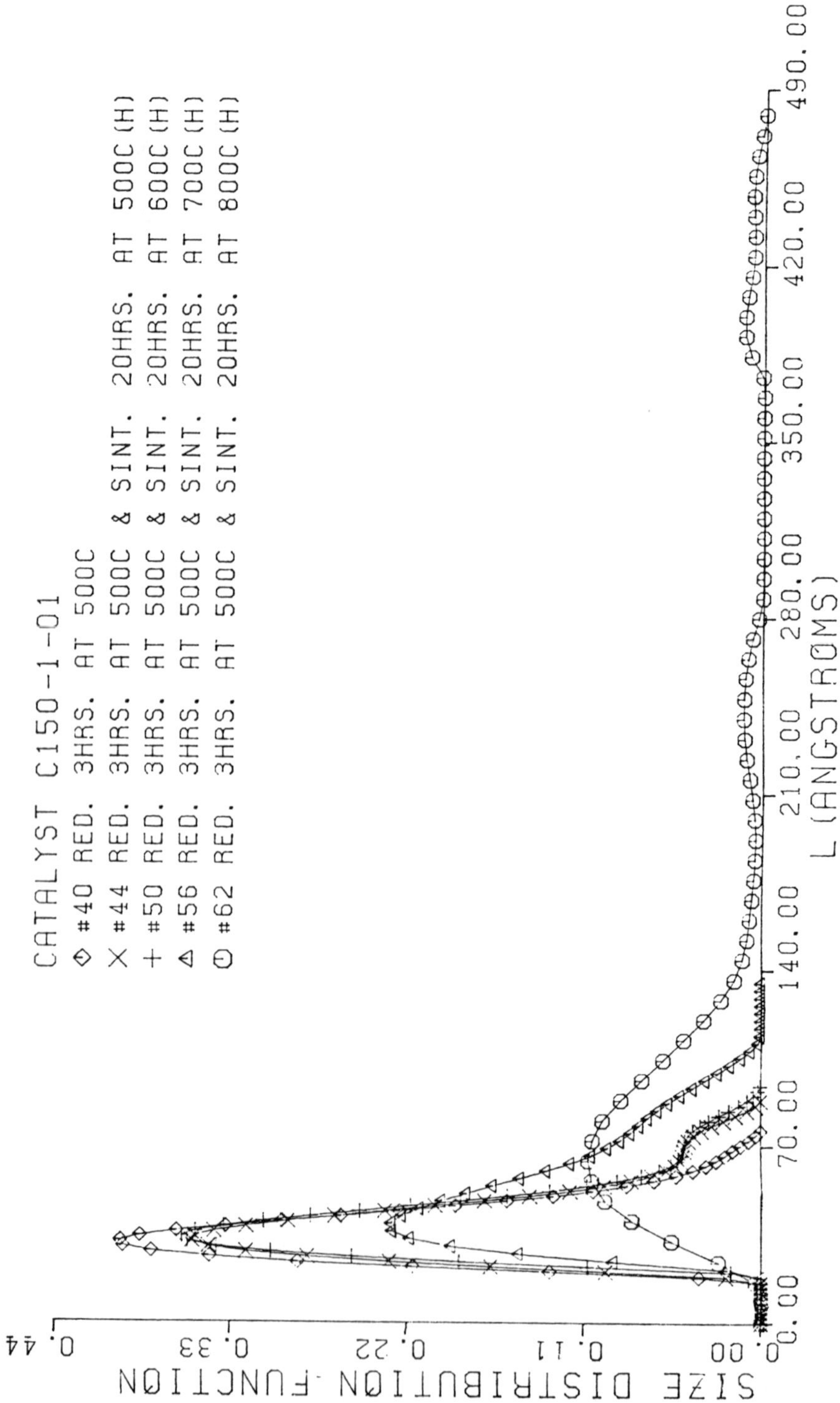

Fig. 4. Particle size distribution functions of catalyst C150–1–01 reduced at 500°C for 3 h and sintered at 500, 600, 700 and 800°C for 20 h in hydrogen atmosphere.

set up for high resolution observations. Photographs of each specimen were enlarged to greater than half a million magnification. Tedious and time consuming measurements of images of about one thousand particles were made from photomicrographs of each specimen to determine the PSD. Typical photomicrographs of the "as-reduced" (3 h at 500°C) specimen and specimens reduced and sintered; at 700°C for 100 h in nitrogen, at 700°C for 50 h in hydrogen, and at 800°C for 20 h in hydrogen atmosphere, are shown in Fig. 5.

Comparisons of the normalized PSD's determined by x-ray diffraction and transmission electron microscopy for catalysts that were; reduced 3 h at 500°C in hydrogen, reduced 3 h at 500°C in hydrogen and sintered via 50 h at 700°C in nitrogen and hydrogen, 100 h at 700°C in nitrogen, or 20 h at 800°C in hydrogen, are shown in Figs. 6–10.

The agreement between the results obtained from the two techniques is considered to be excellent, especially in the cases of the "as-reduced", sintered at 700°C for 50 h in nitrogen and hydrogen atmospheres (Figs. 6–8). The agreement is not as good in the case of the specimen sintered at 700°C for 100 h in nitrogen (Fig. 9). However, this specimen was stored in the atmosphere about three weeks after the x-ray work was completed and prior to making the TEM specimens. During this period of time a portion of the nickel metal possibly reoxidized, increasing the apparent size of the particles. But even for this worst specimen the agreement observed is still good. Similar results on domain size distribution functions in Cu_3Au have been obtained using the Warren double order x-ray technique;[17] however, these are the initial data reported which support the single profile x-ray technique of determining the PSD function.[13] In addition, although PSD determined by TEM have been compared with small angle x-ray results on a platinum catalyst,[18] this is the first time x-ray line profile analysis and TEM determined PSD functions have been performed on a supported nickel catalyst.

CONCLUSIONS

1. An x-ray technique based on the analysis of a single order profile which allows the reliable determination of the particle size distribution function has been developed. This technique has been successfully applied to monitoring of changes in the particle size distribution taking place during sintering of a silica-supported nickel catalyst.
2. Sintering of a silica supported nickel catalyst occurs rapidly at short sintering times and progresses very slowly at longer sintering times.
3. The extent of sintering observed is strongly dependent on the temperature of sintering and less dependent on the sintering time at a particular temperature.
4. The operative sintering mechanism is one which contains a particle migration step.

ACKNOWLEDGEMENTS

This research was sponsored by the Institute for Mining and Minerals Research through Contract No. PD 18 and by the Energy Research and Development Agency through Contract No. E(49–18)2229.

REFERENCES

1. Wanke, S.E., and Flynn, P.C., *Catal. Rev.* 12 (1), 93 (1975).
2. Cinneide, A.D.O. and Clarke, J.K.A., *Catal. Rev.* 7 (2), 213 (1972).
3. Gangulee, A., *J. Appl. Crystallogr.* 7, 434 (1974).
4. Mignot, J., and Rondot, D., *Acta Met.* 23, 1321 (1975).
5. Stokes, A.R., *Proc. Phys. Soc.* 61, 382 (1948).
6. Rothman, R.L., and Cohen, J.B., *Advan. X-Ray Anal.* 12, 208 (1969).

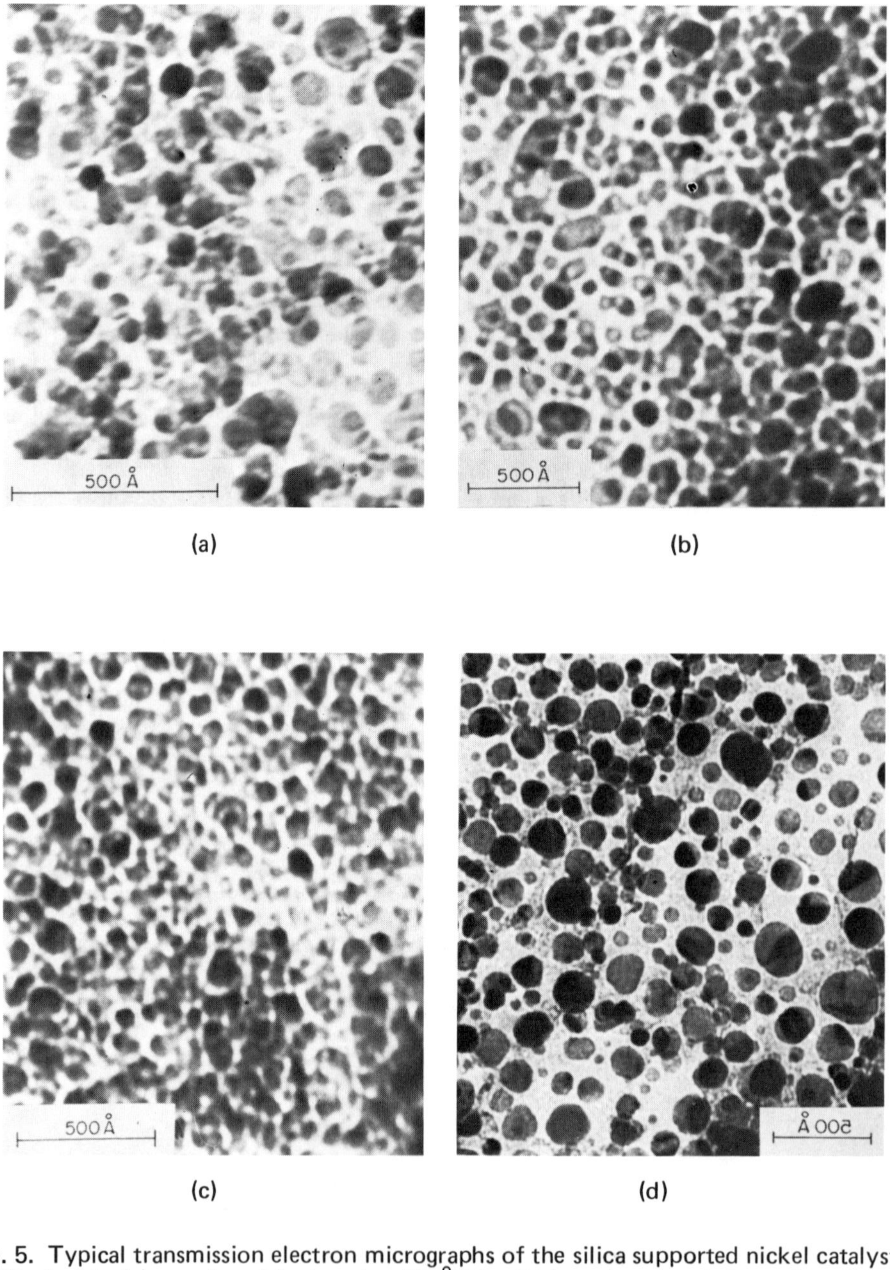

Fig. 5. Typical transmission electron micrographs of the silica supported nickel catalyst C150–1–01; (a) as reduced (reduced at 500°C for 3 h); reduced and sintered at: (b) 700°C for 100 h in N_2, (c) 700°C for 50 h in H_2, and (d) 800°C for 20 h in H_2.

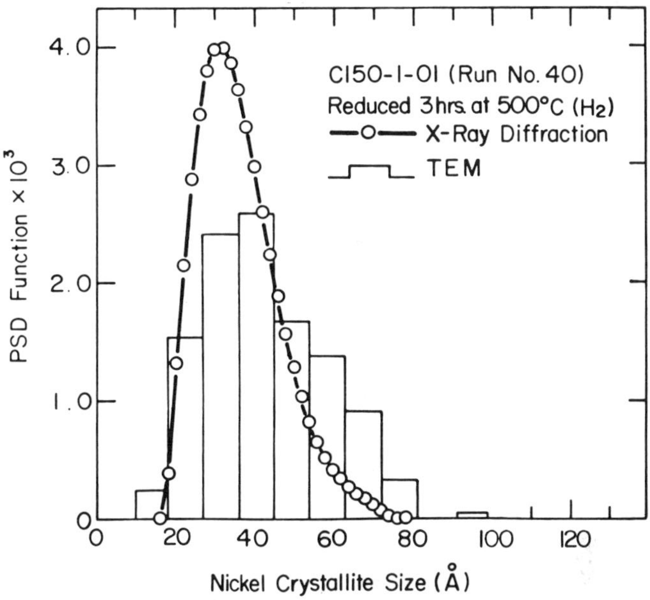

Fig. 6. Particle size distribution functions determined by transmission electron microscopy and x-ray diffraction for catalyst C150–1–01 in the as reduced condition.

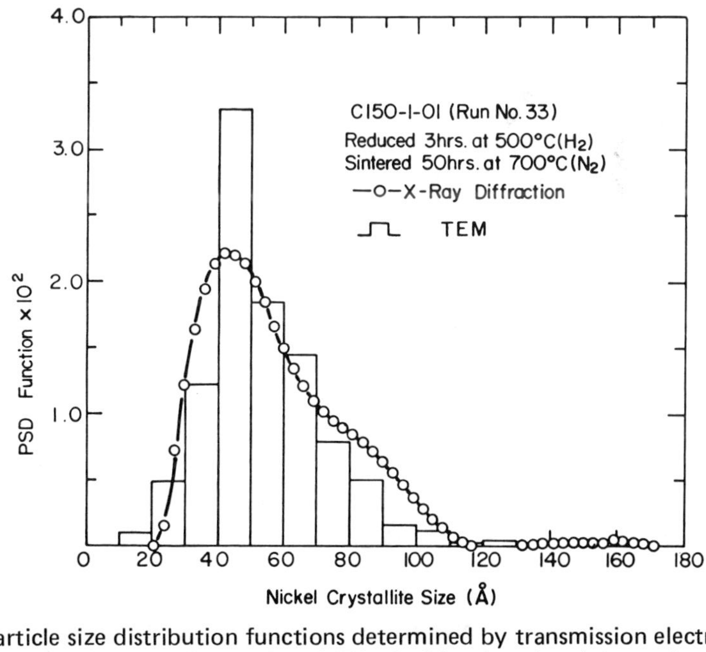

Fig. 7. Particle size distribution functions determined by transmission electron microscopy and x-ray diffraction for catalyst C150–1–01 reduced and sintered at 700°C for 50 h in nitrogen atmosphere.

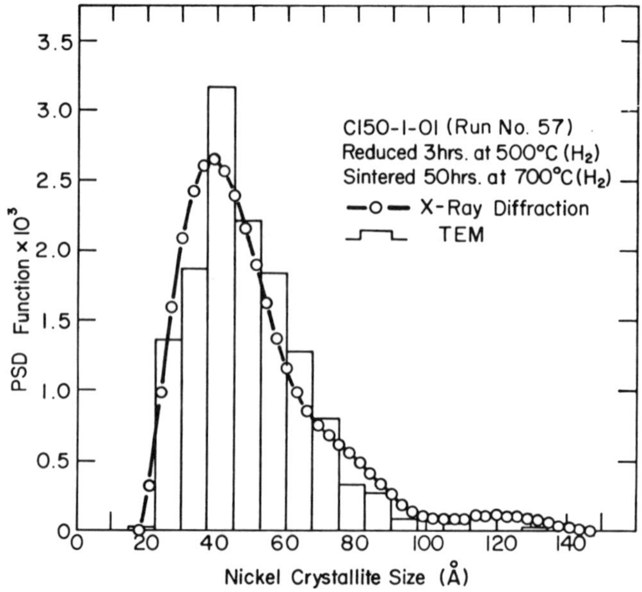

Fig. 8. Particle size distribution functions determined by transmission electron microscopy and x-ray diffraction for catalyst C150–1–01 reduced and sintered at 700°C for 50 h in hydrogen atmosphere.

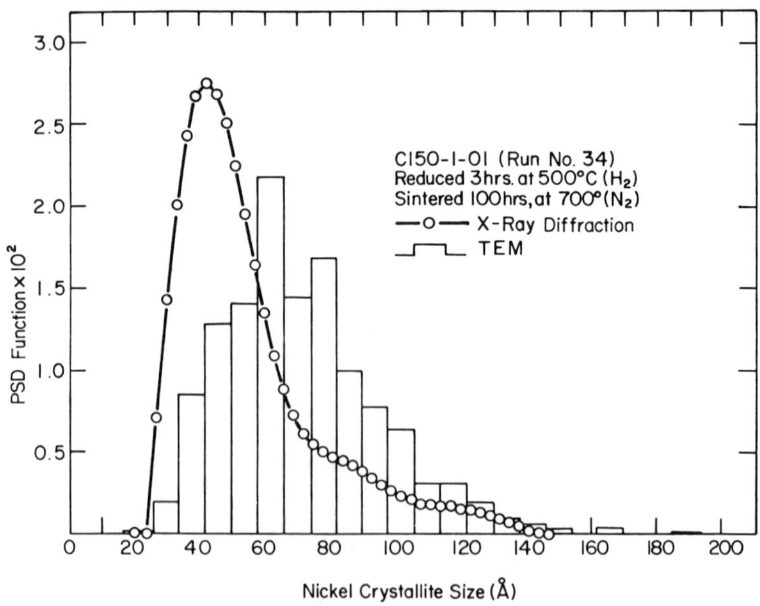

Fig. 9. Particle size distribution functions determined by transmission electron microscopy and x-ray diffraction for catalyst C150–1–01 reduced and sintered at 700°C for 100 h in nitrogen atmosphere.

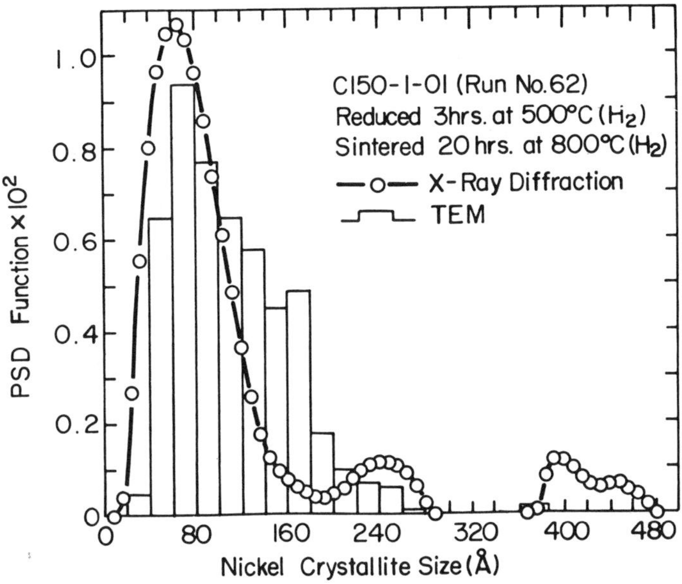

Fig. 10. Particle size distribution functions determined by transmission electron microscopy and x-ray diffraction for catalyst C150–1–01 reduced and sintered at 800°C for 20 h in hydrogen atmosphere.

7. Warren, B.E., and Averbach, B.L., *J. Appl. Phys.* 21, 595 (1950).
8. Warren, B.E., *Prog. Metal Phys.* 8, 152 (1959).
9. McKeehan, M., and Warren, B.E., *J. Appl. Phys.,* 24, 52 (1953).
10. Evans, W.P., Ricklefs, R.E., and Millan, J.H. in *Local Atomic Arrangements Studied by X-ray Diffraction* (J.B. Cohen and J.E. Hilliard, Eds.) p. 271, Gordon and Breach, New York (1966).
11. De Angelis, R.J., in *Local Atomic Arrangements Studied by X-ray Diffraction* (J.B. Cohen and J.E. Hilliard, Eds.), p. 271, Gordon and Breach, New York (1966).
12. Guinier, A., *X-ray Diffraction in Crystals, Imperfect Crystals and Amorphous Bodies,*Freeman, San Francisco, (1963).
13. Ganesan, P., Kuo, H.K., Saavedra, A., and De Angelis, R.J., *J. Catal.* 52, 310 (1978).
14. Ruckenstein, E., and Pulvermacher, B., *AICHE J.* 19, 356 (1973).
15. Ruckenstein, E., and Pulvermacher, B., *J. Catal.* 29, 224 (1973).
16. Flynn, P.C., and Wanke, S.E., *J. Catal.* 34, 390, 400 (1974).
17. Sakai, M., and Mikkola, D.E., *Met. Trans.* 2, 1635 (1971).
18. Renouprez, A., Hoang-Van, C., and Compagnon, P.A., *J. Catal.* 34, 411 (1974).

Stevens, Vander Voort, McCall, eds. Microstructural Science, Volume 8

METALLOGRAPHIC STUDY OF Ni–Mo–Al FIBER REINFORCED COMPOSITE SUPERALLOYS

V. Biss*

ABSTRACT

Metallographic studies have been conducted on sixteen Ni-Mo-Al-base directionally solidified eutectic superalloys containing variable additions of chromium, tantalum and tungsten, added singly and in combination. The effect of variation of the base composition, and of chromium, tantalum and tungsten additions on the microstructure and on the compositions and volume fractions of individual phases has been evaluated. The microstructures of the alloys generally consist of a γ/γ' matrix containing fibrous or lamellar αMo phase. Chromium, added in quantities up to 9.0%, partitions about equally between the αMo and the matrix. Tantalum dissolves preferentially in the γ' phase, and tungsten dissolves extensively in the αMo. The effects of composition on the morphology and distribution of αMo are illustrated and discussed.

INTRODUCTION

Directionally solidified eutectic superalloys, based on compositions located in the nickel-rich corner of the Ni-Mo-Al ternary system, have recently been explored as potential candidates for applications requiring increased high temperature strength, such as gas turbine blades. A partial isothermal section of the Ni-Mo-Al phase diagram at 1300 C is shown in Fig. 1. The dashed curves on the diagram indicate positions of projected monovariant liquidus troughs[1]. Alloys with compositions located on the eutectic trough in the $\gamma + \alpha$ phase field consist of αMo fibers in a γ phase matrix following directional solidification. On cooling below the eutectic temperature, γ' phase precipitates from γ. The final microstructure is a pseudo ternary eutectic consisting of ductile αMo fibers in a ductile γ/γ' matrix. The presence of the ductile αMo fibers in the γ/γ'–αMo composite system constitutes the major difference between this system and other fiber-reinforced composite systems which are characterized by brittle fiber/ductile matrix combinations, such as the systems strengthened by TaC and Ni_3Cb fibers. One of the advantages of the ductile αMo fiber strengthened system over other systems is the potential for increased rupture life.

The present paper summarizes the results of metallographic studies conducted on sixteen Ni-Mo-Al-base fiber-reinforced eutectic superalloys consisting of a series of ternary Ni-Mo-Al alloys plus several series of modifications with chromium, tantalum and tungsten, added singly and in combination. These studies were conducted as part of a cooperative project between United Technologies Research Center and Climax Molybdenum Research Laboratory to conduct metallographic, oxidation and hot corrosion resistance studies on Ni-Mo-Al-base fiber composite superalloys[2,3].

---

*Climax Molybdenum Company of Michigan, Ann Arbor, Michigan, USA., a Subsidiary of AMAX Inc.

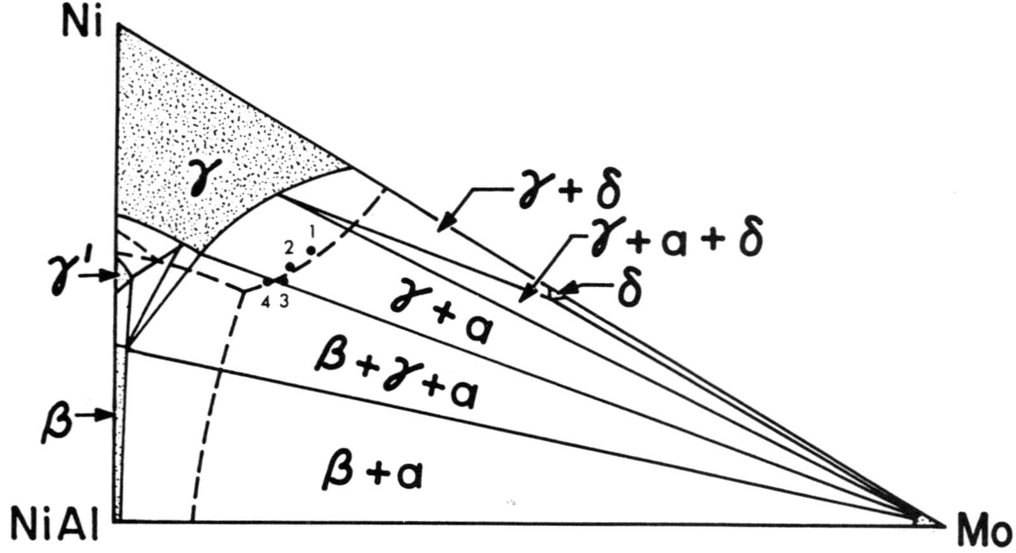

Fig. 1. Isothermal section at 1300 C of a nickel rich corner of Ni-Mo-Al phase diagram (At. 5). The dashed lines indicate positions of the projected monovariant liquidus troughs[1]. The points numbered 1 through 4 indicate the positions of Ni-Mo-Al ternary alloys No. 1 to 4.

EXPERIMENTAL PROCEDURES

Materials

The alloys subjected to metallographic study consisted of 9.5–10.0 mm diameter rods of directionally solidified eutectic superalloys prepared by the United Technologies Research Center. The alloys were solidified in a cylindrical alumina crucible with a growth rate of 3 cm/h (except two alloys which were prepared with a 2 cm/h growth rate) and a liquid phase temperature gradient of 300 C/cm. The nominal chemical compositions of the alloys are presented in Table 1.

Table 1

Nominal Compositions and Solidification Rates of Alloys

Alloy	Element, Wt. %						Growth Rate, cm/h
	Mo	Al	Cr	Ta	W	Ni	
1	34.6	4.7	-	-	-	Bal.	3
2	31.5	6.2	-	-	-	Bal.	3
3	31.0	7.0	-	-	-	Bal.	3
4	29.3	7.5	-	-	-	Bal.	2
5	31.0	6.2	2.0	-	-	Bal.	3
6	34.0	5.5	3.0	-	-	Bal.	3
7	29.0	6.2	6.0	-	-	Bal.	3
8	27.5	6.2	9.0	-	-	Bal.	3
9	31.5	5.5	-	3.0	-	Bal.	2
10	19.2	7.7	-	11.6	-	Bal.	3
11	32.0	5.5	3.0	2.0	-	Bal.	3
12	30.0	5.5	3.0	3.0	-	Bal.	3
13	31.0	5.5	3.0	-	3.0	Bal.	3
14	34.0	5.5	3.0	1.0	1.0	Bal.	3
15	31.0	5.5	3.0	2.0	2.0	Bal.	3
16	26.0	5.5	3.0	3.0	3.0	Bal	3

Metallographic Techniques

The microstructures of the alloys were examined using both scanning and transmission electron microscopy. Metallographic specimens were prepared by mounting transverse sections of the rods. The specimens were mechanically polished and etched electrolytically in an electrolyte consisting of 10 parts H_3PO_4, 50 parts H_2SO_4 and 40 parts HNO_3. The microstructures were examined in an AMR-1000 scanning electron microscope.

The volume percent of αMo phase was determined in selected alloys by point counting directly on the display screen of the SEM. On each alloy, a total of 1600 points was counted in 20 randomly selected areas.

For elemental analysis of the fibers, the specimens were re-etched electrolytically in a 10% HCl-methanol electrolyte for about 15 minutes at 4 volts to preferentially dissolve the γ/γ' matrix to a sufficient distance below the cross sectional plane of the fibers to permit reasonably accurate microanalysis of the fibers by energy dispersive X-ray (EDX) spectroscopy.

The specimens were then repolished and etched electrolytically in 50% aqueous HNO_3 at 2 volts using the specimen as a cathode to preferentially dissolve the α phase to a sufficient distance below the plane of polish necessary for EDX analysis of the γ' phase without interference from the α phase. The EDX analyses were performed employing an EDAX 711B Analyzer interfaced with the AMR-1000 scanning electron microscope.

Thin foils for transmission electron microscopy were prepared by first cutting 0.4 mm thick transverse slices from the rods using a Buehler Isomet low speed diamond cutoff wheel. The slices were ground from both sides on wet silicon carbide abrasive papers to a thickness of 0.1 mm and electrolytically thinned to 0.025 mm employing a double jet, hollow cathode electrolytic thinning apparatus and a 12.5% H_2SO_4-methanol electrolyte. The final thinning was performed in the same electrolyte using a conductive mask electrolytic thinning technique. The foils were examined in a 100 kV JEM-6AS transmission electron microscope.

RESULTS

Characterization of Microstructures

Base Ni-Mo-Al alloys

The alloys numbered 1 through 4 in Table 1 are ternary Ni-Mo-Al alloys containing 29.3 to 34.6% Mo and 4.7 to 7.5% Al. The locations of these compositions, as indicated by the respective numbers 1 through 4 on the Ni-Mo-Al phase diagram in Fig. 1, correspond with the projected eutectic trough in the $\alpha + \gamma$ field.

Typical microstructures of the base Ni-Mo-Al ternary alloys are illustrated by the scanning electron micrographs in Figs. 2, 3 and 4. The microstructure of the alloy with the lowest aluminum content, 4.7% Al, illustrated in Fig. 2a, contains γ/γ'—αMo eutectic as the major microconstituent with isolated areas of primary γ phase. The γ/γ'—αMo eutectic exhibits a slight tendency for cell formation and contains both lamellar and fibrous morphologies of αMo. The αMo fibers have faceted, square and rectangular cross sections, as illustrated in Fig. 2b. It is evident also that the fibers are completely surrounded by the γ' phase, indicating that the formation of the γ' on cooling below the γ—αMo eutectic temperature occurs preferentially at the interface between γ and αMo.

The ternary alloy containing 7.0% Al, illustrated in Fig. 3, is free of primary γ, and has a cellular microstructure. The majority of the αMo has a lamellar morphology, arranged within a given cell to present a radiating, spherulitic appearance.

The microstructure of the alloy containing 7.5% Al, illustrated in Fig. 4, is completely fibrous with uniform size and distribution of the αMo rods. The alloy illustrated in Fig. 4 has a relatively similar composition to that shown in Fig. 3, but was produced at a slower solidification rate, namely 2 cm/h instead of 3 cm/h as employed for most of the other alloys. The microstructural differences exhibited by these two alloys indicate that the morphology of αMo is sensitive to the solidification rate.

The microstructure of the ternary Ni-Mo-Al alloy containing 6.2% Al was completely eutectic and contained both fibrous and lamellar morphologies of αMo.

Chromium modified alloys

The chromium modified series of alloys contained 2.0, 3.0, 6.0 and 9.0% Cr at essentially constant aluminum content. A typical microstructure of the alloy containing 6.0% Cr is shown in Fig. 5. The microstructure consists of γ/γ'—αMo eutectic containing fibrous αMo with faceted square and rectangular cross sections. The γ phase contains a fine precipitate believed to be secondary γ' phase particles. The microstructural features of the γ/γ'—αMo eutectic present in the other chromium

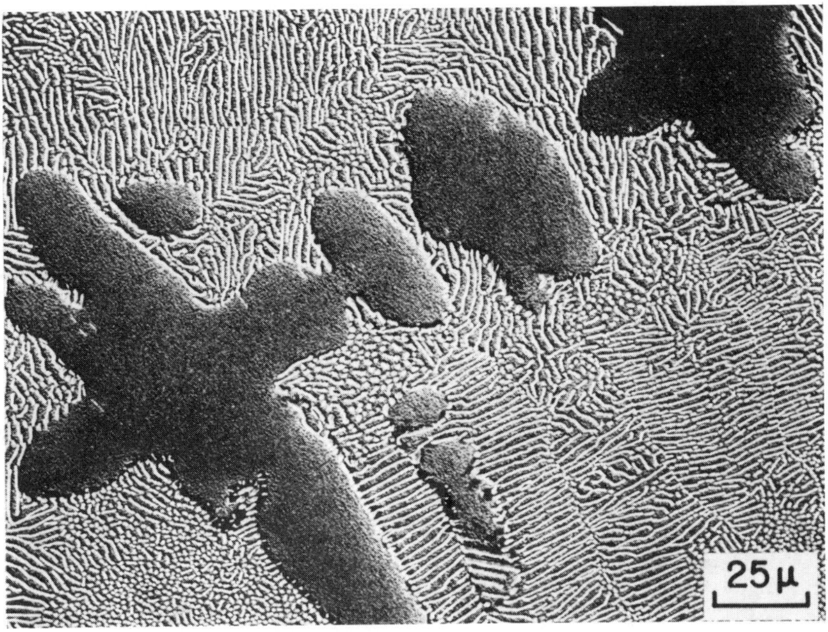

(a)

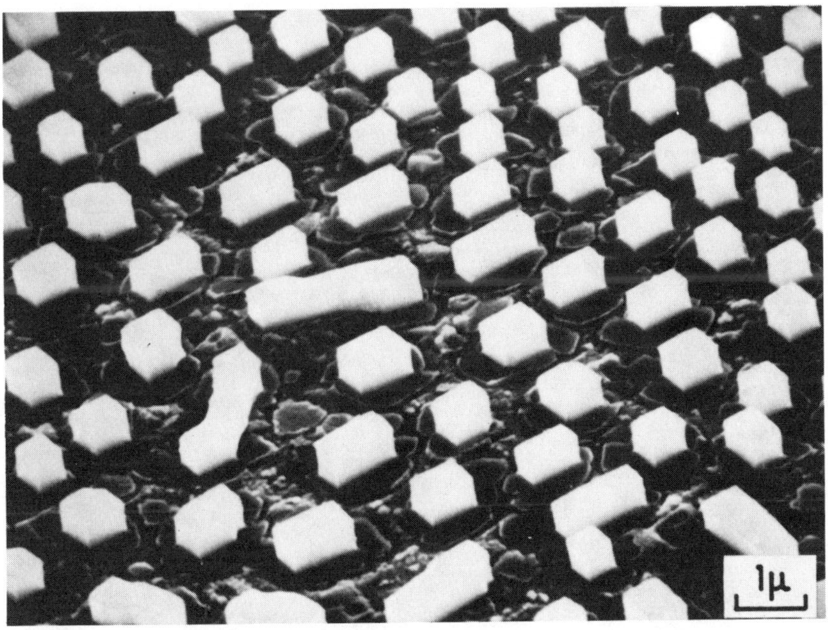

(b)

Fig. 2. Microstructure of Ni-34.6Mo-4.7Al alloy. a) General microstructure consisting of γ/γ'–αMo eutectic and islands of primary γ phase. b) Area containing faceted αMo fibers completely surrounded by γ' phase.

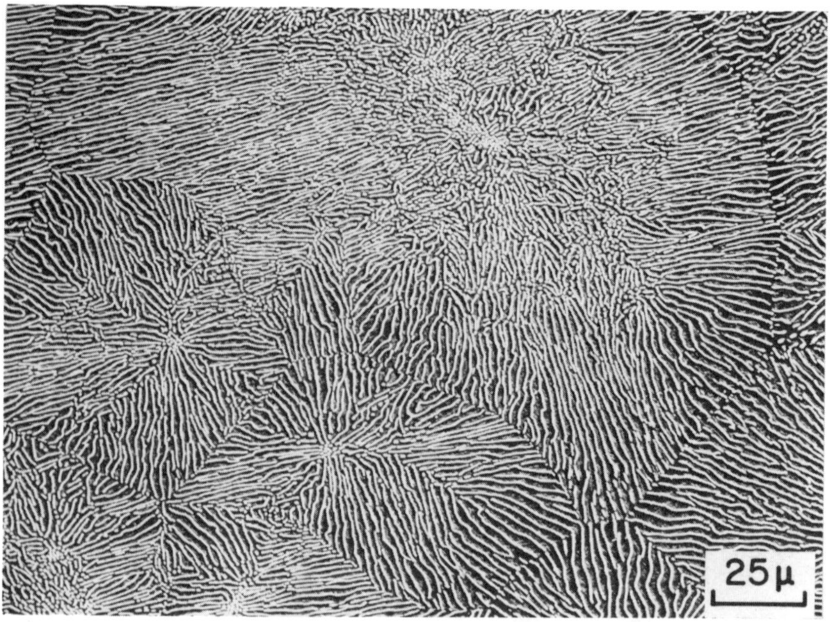

Fig. 3. Cellular microstructure of Ni-31.0Mo-7.0Al alloy.

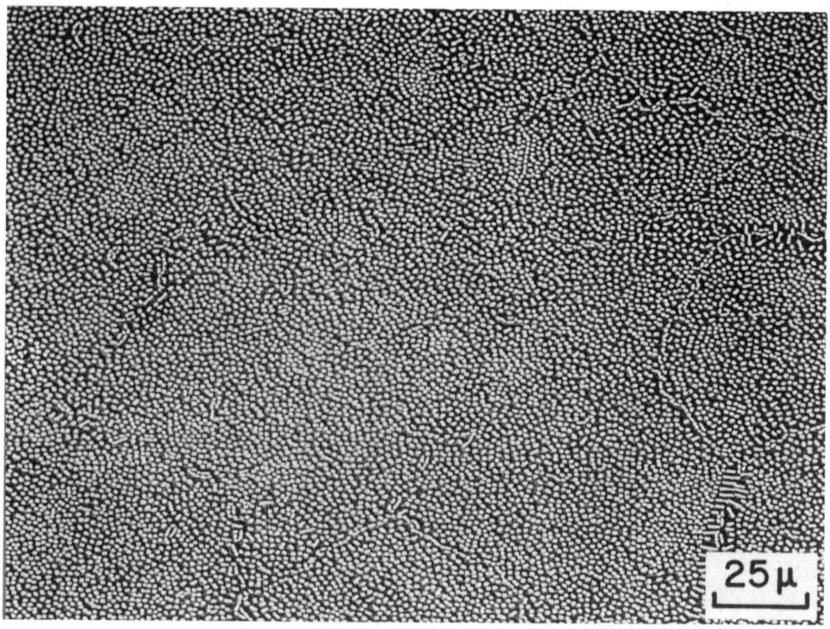

Fig. 4. Fibrous microstructure of Ni-29.3Mo-7.5Al alloy.

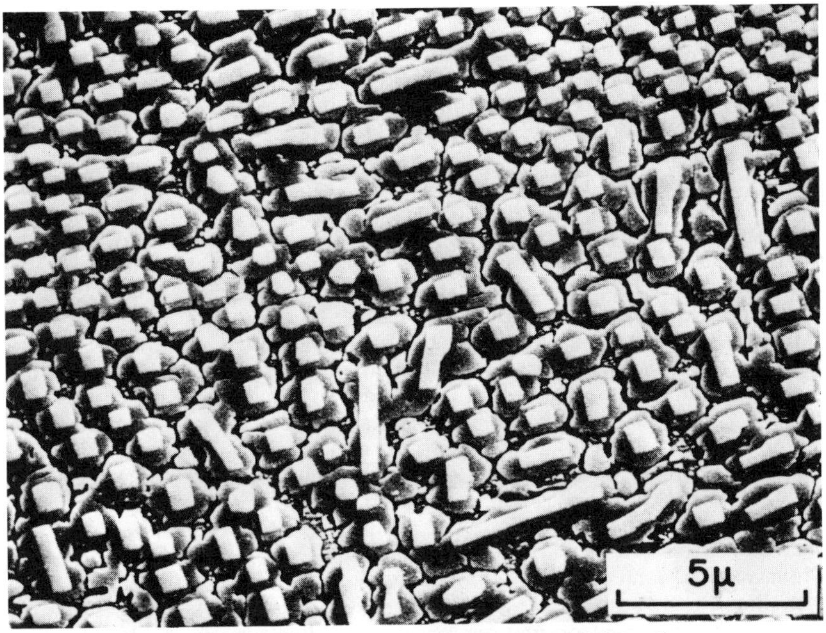

Fig. 5. Microstructure of Ni-29.0Mo-6.2Al-6.0Cr alloy.

modified alloys were found to be similar to that shown in Fig. 5. In the alloys with Cr+Mo content above 35%, namely Alloys No. 6 and No. 8, the microstructure contained scattered primary αMo dendrites typified by that illustrated in Fig. 6.

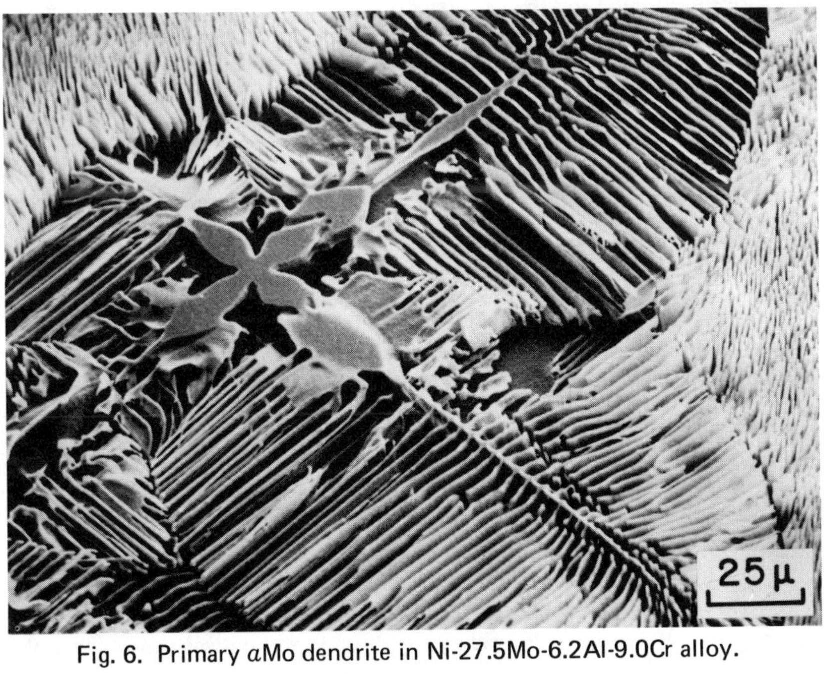

Fig. 6. Primary αMo dendrite in Ni-27.5Mo-6.2Al-9.0Cr alloy.

Tantalum modified alloys

The microstructures of alloys modified by 3.0 and 11.6% Ta are illustrated in Figs. 7a and 7b, respectively. The alloy containing 3.0% Ta was produced using a 2 cm/h solidification rate. The microstructure contains uniformly distributed αMo fibers with faceted, mostly square cross sections. The γ phase contains very fine secondary γ' phase particles.

The alloy containing 11.6% Ta was produced using a 3 cm/h solidification rate. The microstructure, illustrated in Fig. 7b, is composed of very fine, uniformly distributed αMo fibers in a γ' phase matrix. No evidence of the presence of γ phase was found in this alloy. The reduced αMo fiber size present in the 11.6% Ta alloy apparently reflects the reduced diffusion of elements in the liquid near the solid-liquid interface caused by the presence of larger, less mobile tantalum atoms in greater quantity than present in the previous alloy, plus the higher rate of solidification.

The internal microstructure of the tantalum-modified alloys is illustrated by the transmission electron micrographs in Figs. 8 and 9. The internal microstructure of the 3.0% Ta alloy, shown in Fig. 8, contains faceted, square cross sections of αMo fibers surrounded by essentially dislocation-free γ' phase. The dark mottled-appearing phase is γ containing fine secondary γ' phase particles. The γ–γ' interface is frequently oriented parallel to the diagonal of the αMo fiber cross sections, suggesting a definite crystallographic relationship between the three phases. By electron diffraction analysis conducted on the thin foil specimens, the crystallographic relationship of the interfaces between αMo and γ' phase was determined to be $(110)_{\alpha}//(100)_{\gamma'}$. The growth direction of both phases was found to be parallel to [100].

The transmission electron micrograph in Fig. 9 illustrates the multifaceted nature of the αMo fiber sections in the 11.6% Ta alloy. The γ' matrix of this alloy is free of dislocations and has a mottled appearance, except in the areas around the fibers.

Chromium plus tantalum modified alloys

The alloys modified by 3.0% Cr plus 2.0 and 3.0% Ta were found to have similar microstructures, containing both fibrous and lamellar morphologies of αMo. A typical microstructure illustrating the arrangement of the αMo morphological varieties in these two alloys is shown in Fig. 10.

Chromium plus tungsten modified alloy

The alloy modified by 3.0% Cr plus 3.0% W has a completely cellular microstructure, as illustrated in Fig. 11. The morphology of the αMo is lamellar with no evidence for the presence of rod-shaped fibers.

Chromium, tantalum and tungsten modified alloys

This series consisted of three alloys containing 3% chromium plus 1, 2 and 3% each of tantalum and tungsten. The microstructure of the alloy containing 1% each of tantalum and tungsten, illustrated in Fig. 12, consists of fine, uniformly distributed αMo fibers in a γ/γ' matrix. Increasing the tungsten content to 2.0 and 3.0% resulted in a gradual transition from fibrous to cellular microstructure and the formation of primary αMo dendrites. A typical primary αMo dendrite, present in the microstructure of the alloy containing 2.0% each of tantalum and tungsten, is shown in Fig. 13.

Volume Percent of αMo Phase

The effect of the alloying elements chromium, tantalum and tungsten on the volume percent of αMo was determined in the alloys with relatively constant molybdenum

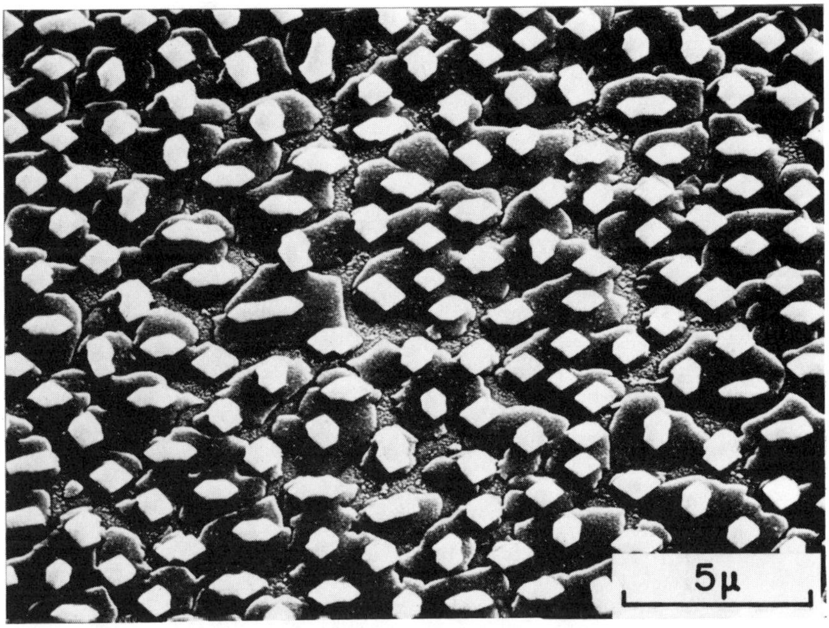

(a)

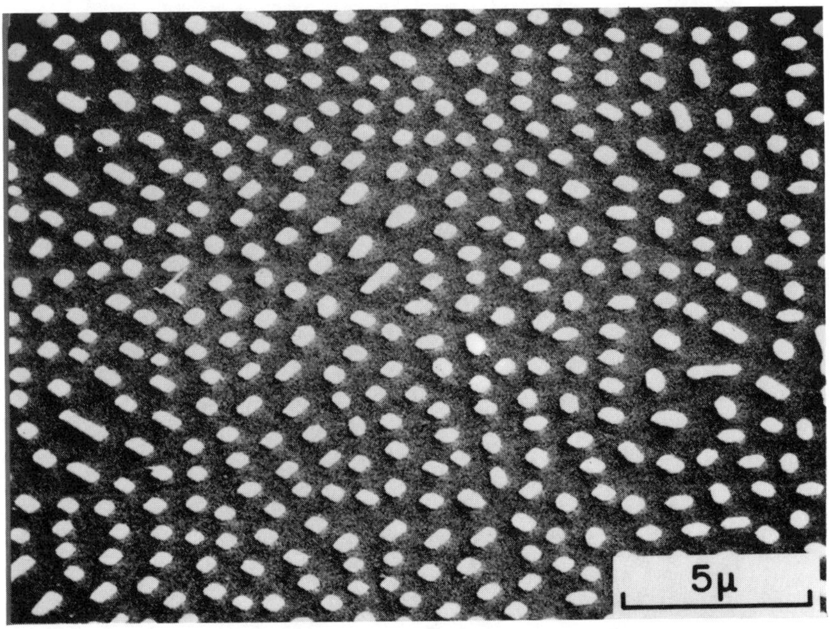

(b)

Fig. 7. Microstructures of tantalum modified Ni-Mo-Al-base alloys. a) Ni-31.5Mo-5.5Al-3.0Ta alloy. b) Ni-19.2Mo-7.7Al-11.6Ta alloy.

Fig. 8. Internal microstructure of Ni-31.5Mo-5.5Al-3.0Ta alloy.

Fig. 9. Internal microstructure of Ni-19.2Mo-7.7Al-11.6Ta alloy.

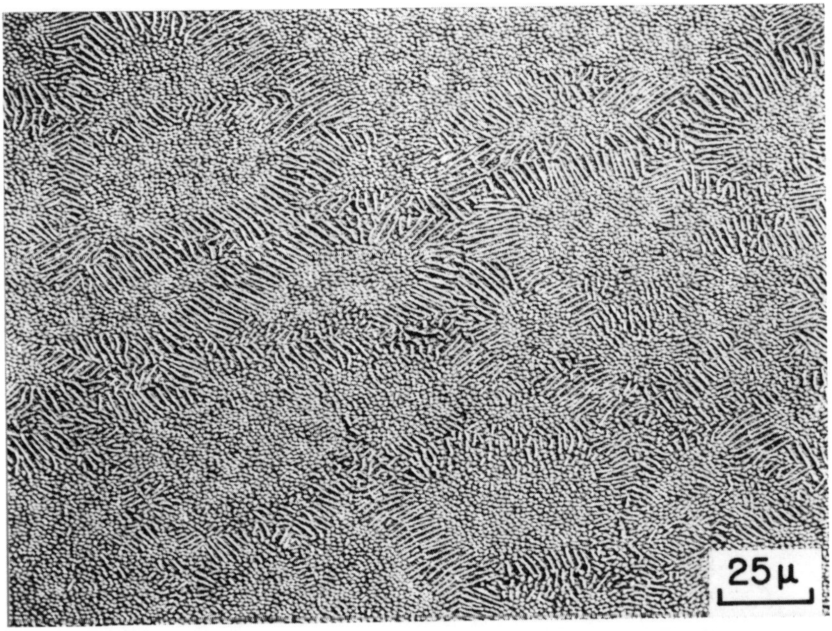

Fig. 10. Microstructure of Ni-32.0Mo-5.5Al-3.0Cr-2.0Ta alloy.

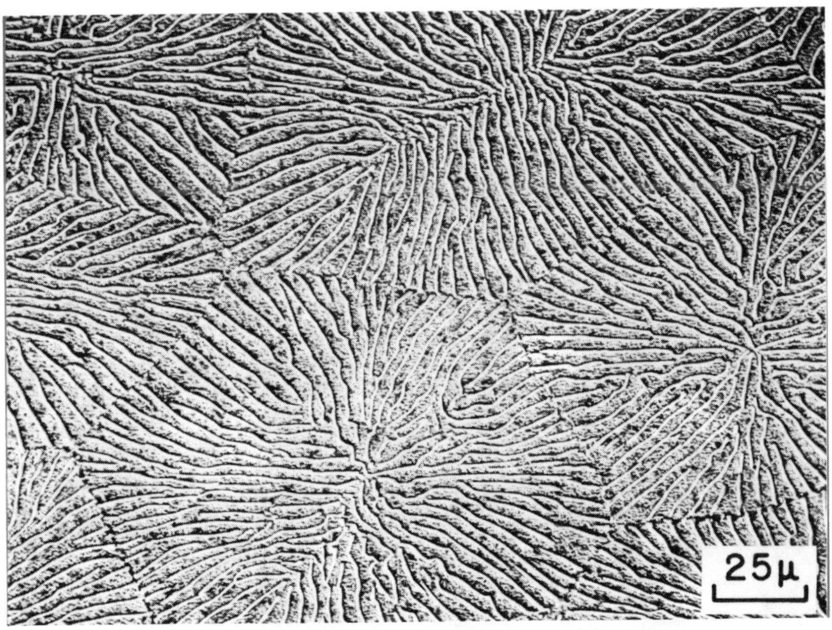

Fig. 11. Microstructure of Ni-31.0Mo-5.5Al-3.0Cr-3.0W alloy.

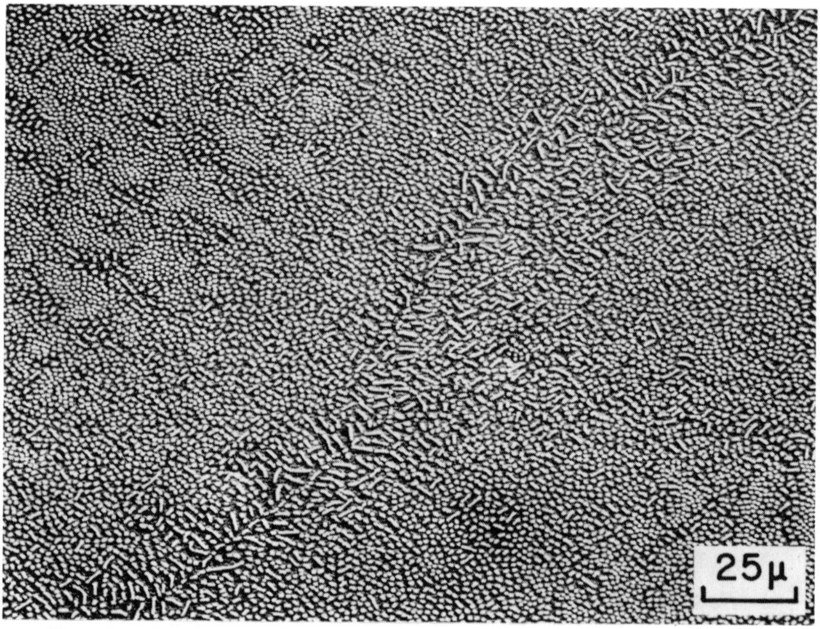

Fig. 12. Microstructure of Ni-34.0Mo-5.5Al-3.0Cr-1.0Ta-1.0W alloy.

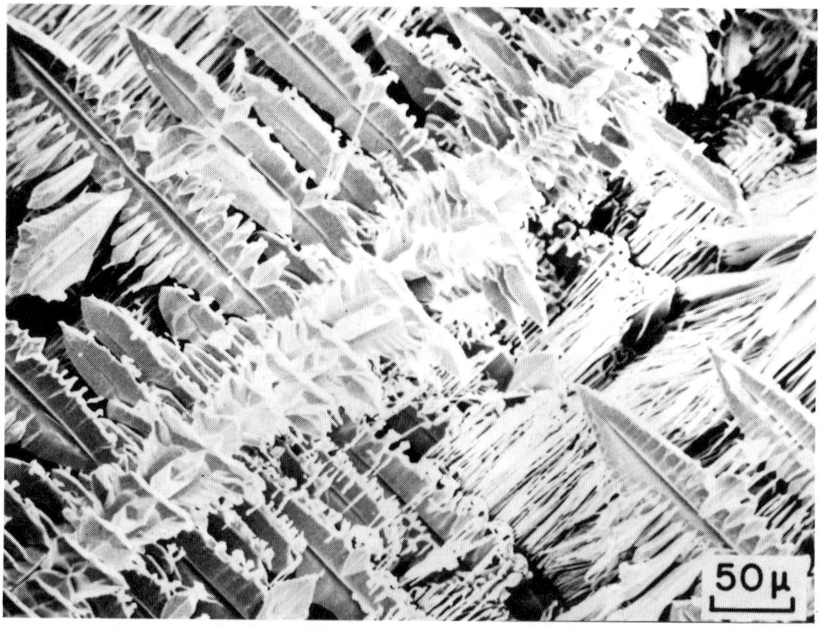

Fig. 13. Primary αMo dendrites in Ni-31.0Mo-5.5Al-3.0Cr-2.0Ta-2.0W alloy.

content of 31.0–32.0%. The results are summarized in Table 2. It can be seen that the addition of these elements in moderate quantities, up to 3% of each, resulted in a moderate increase of the αMo volume percent. The addition of 2.0% Cr to the base alloy increased the volume percent of αMo from 19.8 to 22.1%. Less effective is the addition of 3.0% Ta, resulting in the increase of the volume percent of αMo from 19.8 to 21.3%. Part of this increase may be due to the slightly higher molybdenum content of the tantalum-containing Alloy No. 9, as compared to the base Alloy No. 3 (31.5 vs 31.0% Mo). The highest volume percent of αMo (22.9%) was found to be present in Alloy No. 11 containing 3.0% Cr plus 2.0% Ta. This increase apparently reflects the combined effects produced by the higher molybdenum content of Alloy No. 11 (32.0% Mo), plus the presence of alloying elements. Alloy No. 13, containing 3% each of chromium and tungsten, was found to have the same volume percent of αMo as base Alloy No. 3, namely 19.8%. It can be seen from Fig. 11 that the morphology of the αMo phase in this alloy is completely lamellar. It cannot be stated with certainty, however, whether or not the morphological variation has any effect on the volume percent of αMo.

Table 2

Volume Percent of αMo Phase

No.	Alloy	Vol. % αMo
3	Ni-31.0Mo-7.0Al	19.8
5	Ni-31.0Mo-6.2Al-2.0Cr	22.1
9	Ni-31.5Mo-5.5Al-3.0Ta	21.3
11	Ni-32.0Mo-5.5Al-3.0Cr-2.0Ta	22.9
13	Ni-31.0Mo-5.5Al-3.0Cr-3.0W	19.8

Composition of αMo Phase

The results of EDX analyses conducted on αMo fibers of the alloys are summarized in Table 3. Nickel is present in the αMo phase of all alloys in quantities ranging from 0.5 to 3.4%. In the majority of the alloys examined, the nickel content of the αMo is between 2.0 and 2.9%, or 2.4% average. Aluminum was not detected by EDX analysis of the αMo phase in any of the alloys examined, although small quantities of aluminum are expected to be present. Of the alloying elements chromium, tantalum and tungsten, the highest solubility is exhibited by tungsten, which dissolves in the αMo phase preferentially. Chromium appears to be partitioning between the αMo and matrix phase about equally. Tantalum dissolves in the αMo phase to a limited degree. The solubility of tantalum in αMo phase appears to be increased, however, by the presence of chromium. This can be seen by comparing the analyses of the αMo in Alloy No. 9, containing 3.0% Ta, and Alloy No. 12, containing 3.0% Ta plus 3.0% Cr. The αMo of Alloy No. 9 contains no tantalum, while that of Alloy No. 12 contains 1.1% Ta. The effect, if any, of tungsten on the solubility of tantalum in the αMo could not be evaluated because the X-ray energy peaks of tantalum and tungsten overlap. It seems reasonable to assume, however, that the αMo phase in Alloys No. 15 and 16, in Table 4, containing 2.0 and 3.0% of tungsten and tantalum, respectively, may have contained small amounts of tantalum, contributing to the size of the energy peak of tungsten, since the tungsten analyses of αMo phase in these two alloys are higher than would be expected.

Table 3

Compositions of αMo Phase

No.	Alloy	EDX Analysis, Wt. %				
		Ni	Cr	Ta	W	Mo
1	Ni-34.6Mo-4.7Al	2.0				Bal.
2	Ni-31.5Mo-6.2Al	3.4				Bal.
3	Ni-31.0Mo-7.0Al	2.3				Bal.
4	Ni-29.3Mo-7.5Al	2.7				Bal.
5	Ni-31.0Mo-6.2Al-2.0Cr	2.0	0.8			Bal.
6	Ni-34.0Mo-5.5Al-3.0Cr	2.1	1.2			Bal.
7	Ni-29.0Mo-6.2Al-6.0Cr	2.9	3.6			Bal.
8	Ni-27.5Mo-6.2Al-9.0Cr	2.8	5.5			Bal.
9	Ni-31.5Mo-5.5Al-3.0Ta	2.5				Bal.
10	Ni-19.2Mo-7.7Al-11.6Ta	3.3		3.5		Bal.
11	Ni-32.0Mo-5.5Al-3.0Cr-2.0Ta	2.2	1.4			Bal.
12	Ni-30.0Mo-5.5Al-3.0Cr-3.0Ta	1.6	1.4	1.1		Bal.
13	Ni-31.0Mo-5.5Al-3.0Cr-2.0W	1.4	1.6		8.7	Bal.
14	Ni-34.0Mo-5.5Al-3.0Cr-1.0Ta-1.0W	2.5	1.1		3.4	Bal.
15	Ni-31.0Mo-5.5Al-3.0Cr-2.0Ta-2.0W	0.5	1.0		11.6	Bal.
16	Ni-26.0Mo-5.5Al-3.0Cr-3.0Ta-3.0W	1.2	1.2		16.7	Bal.

Table 4

Effect of Tantalum on the Composition of γ' Phase

No.	Alloy	EDX Analysis, wt. %			
		Al	Mo	Ta	Ni
4	Ni-29.3Mo-7.5Al	12.6	8.0	-	Bal.
9	Ni-31.5Mo-5.5Al-3.0Ta	10.4	7.5	6.6	Bal.
10	Ni-19.2Mo-7.7Al-11.6Ta	10.9	2.3	17.2	Bal.

The elemental composition of the γ' phase in selected base alloys and tantalum containing alloys, as determined by EDX analysis, is presented in Table 4. The aluminum content of the γ' phase in all three alloys remains reasonably constant, ranging from 10.4 to 12.6%. Molybdenum and tantalum contents of the γ' phase are strongly affected by the alloy composition. As tantalum is added in increasing quantities to the alloy, the molybdenum content of the γ' phase is gradually reduced from 8.0% in the tantalum-free alloy, to 2.37% in the 11.6% tantalum alloy. The tantalum content of the γ' phase increases from 6.6 to 17.2% as the tantalum content of the alloy is raised from 3.0 to 11.6%.

Effect of Tantalum on the Lattice Parameters of αMo and γ' Phases

The relative effect of tantalum on the lattice parameters of αMo and γ' phases can be seen from the results of electron diffraction analyses conducted on thin foil specimens and presented in Table 5. It can be seen from the data given in Table 5, that the addition of 3% Ta to the ternary Ni-Mo-Al alloy produces no change in the lattice parameter of αMo. When 11.6% Ta is added to the alloy, the lattice parameter of αMo is increased from 3.05 to 3.13Å, or 2.6%. The lattice parameter of γ' phase is increased slightly with each addition of tantalum.

Table 5

Effect of Tantalum on the Lattice Parameters of αMo and γ' Phases

No.	Alloy	Lattice Parameter	
		αMo, Å	γ' Phase, Å
4	Ni-29.3Mo-7.5Al	3.05	3.57
9	Ni-31.5Mo-5.5Al-3.0Ta	3.05	3.58
10	Ni-19.2Mo-7.7Al-11.6Ta	3.13	3.59

DISCUSSION

The sensitivity of the morphology of the γ/γ'–αMo eutectic to the solidification rate, as reflected in a transition from fibrous to cellular morphology with increasing solidification rate, is consistent with observations by other investigators[4,5].

The presence of primary γ in the Ni-34.6Mo-4.7Al alloy may be caused by compositional differences due to gravity-induced segregation or by the specimen being taken from the end of the bar away from the zone of steady state solidification. The nominal composition of the alloy being located slightly off the eutectic trough toward the γ phase field on the Ni-Mo-Al phase diagram, in Fig. 1, may also have contributed to the formation of γ.

The observation that the γ/γ' interface in the Ni-31.5Mo-5.5Al alloy is frequently oriented parallel to the diagonal of the αMo fiber cross section is consistent with the data of Lemkey[6] and Sprenger, *et al.*,[4] who observed similar orientation relationship in various directionally solidified Ni-Mo-Al-base alloys. Also consistent with the results of studies by Sprenger, *et al.*, and by Henry[5] is the crystallographic orientation relationship between the α and γ' phases determined in the present study.

The increase in lattice parameter of the αMo phase from 3.05 to 3.13Å, associated with increasing the tantalum content of the alloy from 3.0 to 11.6%, apparently reflects the effect of the presence of larger tantalum atoms dissolved in the α phase (3.5%) of the 11.6% Ta alloy. Similarly, the gradual increase of the lattice parameter of the γ' phase from 3.57 to 3.59Å, associated with the increased additions of tantalum to the alloy composition, is believed to be reflecting the observed gradual replacement of molybdenum with larger tantalum atoms in the γ' phase.

The absence of γ phase in the 11.6% Ta alloy apparently reflects a strong tendency for tantalum to promote γ' formation.[7]

The transition from fibrous to cellular microstructures associated with increasing tungsten content above 1% apparently reflects the morphological instability of the eutectic microstructure controlled by the diffusion of elements in the liquid at the solid-liquid interface. At slower solidification rates, it may be possible to suppress the formation of the cellular microstructure at a tungsten content above 1%.

SUMMARY OF RESULTS

The results of metallographic studies conducted on Ni-Mo-Al base directionally solidified eutectic superalloys containing additions of chromium, tantalum and tungsten may be summarized as follows:

1. The microstructures in general consist of γ/γ' matrix containing fibrous or lamellar αMo phase.
2. Chromium added in quantities up to 9.0% partitions about equally between αMo and matrix.
3. Tantalum dissolves preferentially in the matrix and replaces molybdenum in the γ' producing a slight increase of the γ' lattice parameter. With increasing concentration in the alloy, tantalum dissolves in αMo to a moderate extent, increasing the lattice parameter of the α phase.
4. Tungsten dissolves preferentially in αMo phase.
5. The growth direction of the αMo fibers and of the γ' phase is parallel to [100], and the crystallographic relationship of the interfaces between the two phases is $(110)_{\alpha}//(100)_{\gamma'}$.

ACKNOWLEDGEMENTS

The author is grateful to Mr. F.D. Lemkey and Mr. D.D. Pearson of United Technologies Research Center for preparing the materials used in this study and for their useful comments and discussions during the course of the investigation.

REFERENCES

1. D.D. Pearson and F.D. Lemkey, "Solidification and Properties of γ/γ'–αMo Ductile/Ductile Eutectic Superalloy" *Proc. of International Conf. for Solidification,* Sheffield, England, (18–21 July, 1979), pp. 526–532.
2. E.P. Whelan, W.C. Hagel, and F.D. Lemkey, "Oxidation of Ni-Mo-Al-Cr Eutectic Alloys at 1000 C," to be published.
3. E.P. Whelan, W.C. Hagel, and F.D. Lemkey, "Na_2So_4–Induced Oxidation of Ni-Mo-Al-Cr Eutectic Alloys," to be published.
4. H. Sprenger, H. Richter, and J.J. Nickl, "Unidirectional Solidification of Ni-Mo-Al Eutectic Alloys," *Journal of Materials Science,* 11 (1976), pp. 2075–2081.
5. M.F. Henry, "Evaluation of D.S. Eutectic Superalloys for Turbine Blade Applications," NAS 3–19711 Final Report, April, 1978, NASA– Lewis Center, Cleveland, Ohio, pp. 4–1 to 4–20.
6. F.D. Lemkey, "Development of a Second Generation Ductile/Ductile γ/γ'–α Eutectic Alloys," *Proc. of the Third International Symposium,* Seven Springs, Pennsylvania, (September 1976), pp. 321–330.
7. O.H. Kriege and J.M. Baris, "The Chemical Partitioning of Elements in Gamma Prime Separated from Precipitation-Hardened, High-Temperature Nickel-Base Alloys," *Trans. ASM,* vol. 62, (1969), pp. 195–200.

Stevens, Vander Voort, McCall, eds. Microstructural Science, Volume 8

FATIGUE FRACTURE SURFACE FEATURES: FRACTOGRAPHY AND MECHANISMS OF FORMATION

M.–W. Lui* and I. Le May**

ABSTRACT

Fractographic observations made on the fracture surfaces of fatigue specimens of AISI 4140 steel in the quenched, and quenched and tempered conditions, are reported. It is shown that the fracture path and the fracture surface features are affected by material condition and microstructure, in particular carbide morphology. Mechanisms to explain the formation of some of the surface features observed are proposed and discussed.

INTRODUCTION

Fatigue is the most common mechanism leading to service failure, and the stepwise propagation of a fatigue crack under stress cycles leads to the production of distinctive fracture surface features. These can, in turn, be distinguished using fractographic and microfractographic methods of examination, and positive identification of these features may allow a definite conclusion to be drawn concerning the initial cause of failure, i.e., that it was by fatigue.

The best known fatigue fracture surface features are striations, although these can be produced by other means, such as stress corrosion cracking[1] ; however, they are extremely useful in analyzing load spectra, and, in particular, overloading which may have occurred (see e.g.,[2]). Other features to be found on fatigue fracture surfaces include slip band cracks and tire tracks.

The formation of these various surface features and their resulting morphology are affected by the crystal structure and the microstructure of the material, this despite the overall macroscopic similarity between crack propagation in different materials ranging from polymers to high strength steels, which can result in striation formation in all of them. Thus, fractographic fractures can provide some information concerning the mechanisms leading to their formation and the part played by microstructural features, such as second phase particles, in modifying their mechanisms of formation and of fatigue crack propagation.

In this paper, some features observed on the fatigue fracture surfaces of a high strength, low alloy steel are reviewed, the effects of heat treatment and resulting microstructure examined, and the mechanisms of their formation discussed in the light of observations.

OBSERVATIONS

During the course of a study of fatigue fracture in AISI 4140 steel, quenched and tempered to different degrees, a large number of fracture surfaces were examined in detail using microfractographic techniques. Table I details the processing

* Meyer Aluminum Ltd., Kowloon, Hong Kong.
** University of Saskatchewan, Saskatoon, Canada S7N OWO.

Table I
Processing Conditions and Mechanical Properties of AISI 4140
Fatigue Specimens

Preliminary Heat Treatment: Solution treated at 850°C for 1.5h in argon, oil quenched, immersed in liquid nitrogen for 0.5h.

Subsequent Heat Treatment	Hardness, HRC	Yield Strength, MPa	UTS, MPa	Fatigue Limit, MPa
None	60	1643	2408	607
1 h at 200°C & WQ	58	1753	2133	772
1 h at 300°C & WQ	53	1650	1862	717
1 h at 400°C & WQ	49	1487	1612	690
1 h at 550°C & WQ	43	1194	1304	676
1 h at 650°C & WQ	31	948	1029	607
1 h at 700°C	25	652	729	538

conditions and resulting mechanical properties. All tests were made in air at approximately 22°C; they were made in bending with zero-to-tension loading (R=0), and at 33Hz.

Fractographic features observed during preliminary studies of 300°C and 650°C-tempered material were described in a preliminary report[3]; thus, these observations will be summarized here, with emphasis on the previously unpublished observations of fractographic features, on observations relating to crack path, and on mechanistic aspects.

As-Quenched Material

The fatigue cracked area was small before finial fracture occurred, and on it were seen regions of both intercrystalline and transcrystalline fracture. Fine striations were observed on both, although to a much lesser degree on the intercrystalline regions. Carbides were observed on the intercrystalline fracture surfaces (prior austenite grain boundaries) and there was evidence of plastic deformation around them. Typical features are illustrated in Fig. 1.

200°C-Tempered Material

The fracture appearance of this was similar to that of the as-quenched material, but with reduced intercrystalline cracking. Slip band cracks were observed occasionally in localized areas, and it is speculated that they may have formed in regions of retained austenite because of the extensive crack tip deformation required in their formation. Tire tracks were observed on occasion at high cyclic stress level, and Fig. 2 shows this feature, together with striations and dimples.

300°C-Tempered Material

As illustrated previously[3], most of the fracture surface was transcrystalline in this material, with a small amount of intercrystalline fracture only in most cases. At high stress level, some pulling out of carbides and cracking at prior austenite grain boundaries was apparent in intercrystalline-cracked regions, as shown in Fig. 3.

400°C-Tempered Material

Here, fatigue fracture surfaces were primarily transcrystalline in nature with only localized intercrystalline fracture. Striations were less well defined than for 300°C-tempered material and were irregularly shaped. Slip band cracking was extensive, and tire tracks were frequently seen when the stress level was high. Typical microfractographs are shown in Fig. 4, and the small region of intercrystalline fracture surrounded by transcrystalline fatigue fracture shown in Fig. 4b may be noted. It illustrates the dependence of intercrystalline crack propagation on a favorable grain boundary orientation with respect to the propagating crack.

550°C-Tempered Material

In this, as in material tempered at higher temperatures, all fatigue fracture took place in a transcrystalline manner. Tire tracks were frequently seen, as were slip band cracks.

650°C-Tempered Material

This material was discussed previously[3], but we may note here that the severity of tire tracks and slip band cracks increased still further, with increase in ductility of the steel.

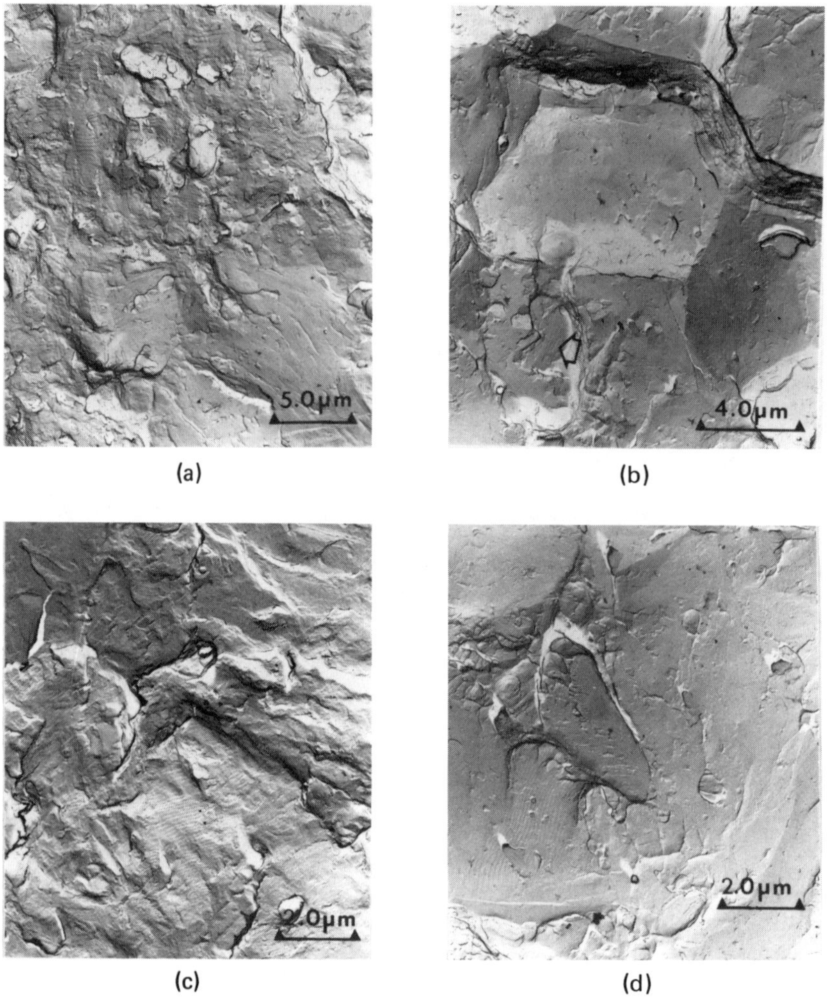

Fig. 1. Replica electron microfractographs from as-quenched material fatigue fracture surfaces: (a) At low stress (N_f = 1.7 x 10^5 cycles) showing transcrystalline fracture with striations and some intercrystalline fracture; (b) same specimen as (a), showing fracture along prior austenite grains – note arrowed intercrystalline striations; (c) at high stress (N_f = 7.9 x 10^3 cycles) showing striations; (d) same specimen as (c) showing intercrystalline fatigue fracture.

700°C-Tempered Material

In this condition the steel was completely recrystallized with Fe_3C spheroids in a ferrite matrix, those carbides lying along grain boundaries being of much larger size.

Slip band cracks were a dominant feature as shown in Fig. 5a, while tire tracks were long and frequent, as seen in Fig. 5b. At high stress, dimple rupture was seen in some regions, similar in appearance to a tensile fracture.

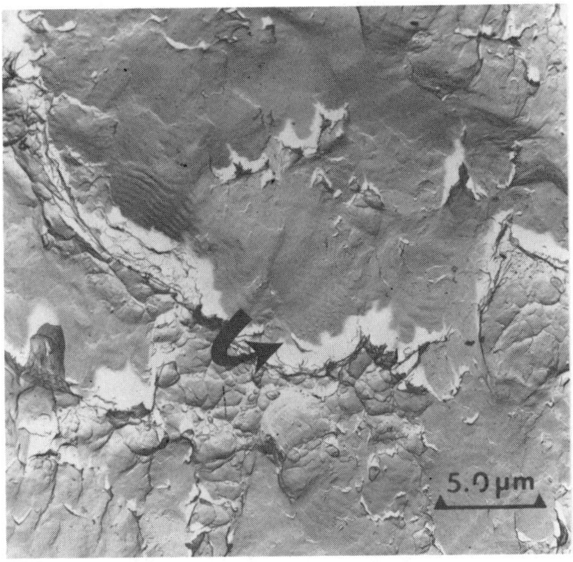

Fig. 2. Fatigue fractured surface of 200°C-tempered specimen (N_f = 7.8 x 10^3 cycles) showing striations, tire tracks (arrowed) and dimples.

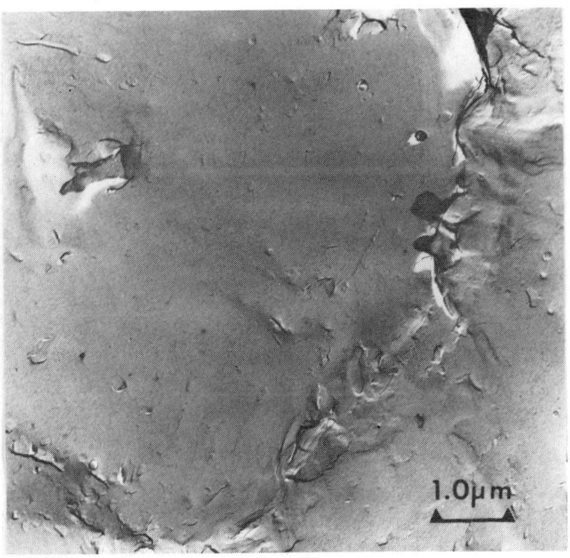

Fig. 3. Fatigue fractured surface of 300°C-tempered specimen (N_f = 1.3 x 10^4 cycles) showing cracked boundaries. Note that some carbides appear to have been pulled out.

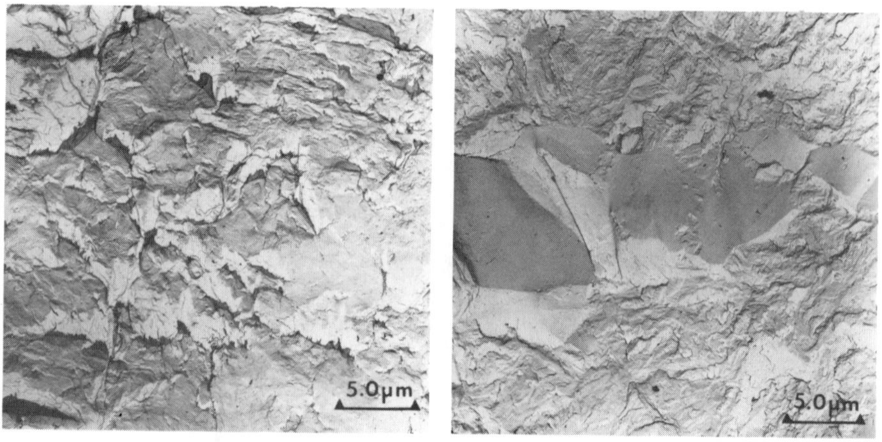

(a) (b)

Fig. 4. Fatigue fracture surfaces of 400°C-tempered specimens: (a) Extensive slip band cracking at high stress level on transgranularly fractured surface ($N_f = 2.6 \times 10^4$ cycles); (b) Intergranular fracture virtually surrounded by transgranular fracture ($N_f = 1.6 \times 10^5$ cycles).

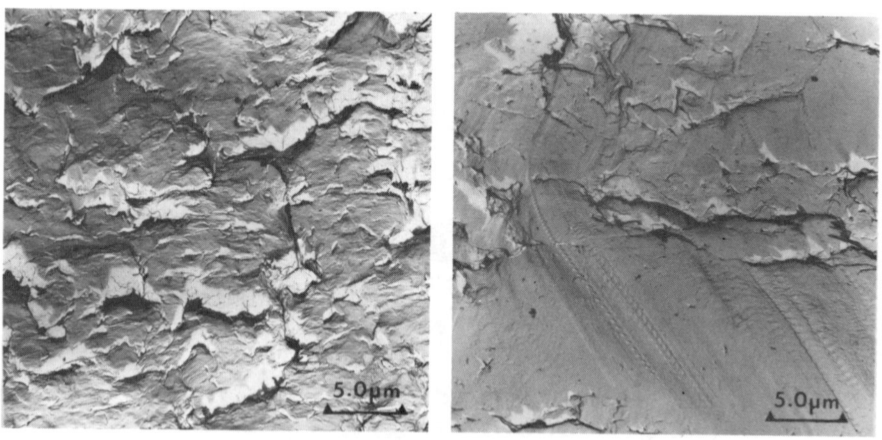

(a) (b)

Fig. 5. Fatigue fracture surfaces of 700°C-tempered specimens: (a) Dominant slip band cracking ($N_f = 6.1 \times 10^5$ cycles); (b) Tire tracks ($N_f = 2.3 \times 10^4$ cycles).

Fracture Profile Examinations

Sections taken through the fatigue fracture surfaces can show directly the crack path and the effect of inclusions on this. Although electron microfractography, as just presented, gives definite indications, e.g., as regards intercrystalline or transcrystalline fracture, artifacts associated with replication can sometimes be misleading. Thus, fracture surfaces of specimens tempered at 200°C, 550°C and 700°C were coated with electrodeposited nickel, sectioned perpendicular to the surface, and prepared for optical metallographic examination.

In all cases the initial crack started along a direction at ~45° to the specimen surface, i.e., along a maximum shear stress plane, subsequently moving to a plane of maximum tensile stress. Figure 6a shows the crack following prior austenite grain boundaries where these are suitably oriented, this being in 200°C-tempered material, and the effect of inclusions is shown in Fig. 6b, where cracking has taken place along the inclusion after it was reached. No intercrystalline fracture was seen in 550°C or 700°C-tempered material, in conformity with the fractographic observations. Dimple rupture was seen clearly in some of the high stress 700°C-tempered specimens, with large carbides appearing on the fracture surface, some being apparently at the bottom of the dimples.

MECHANISTIC ASPECTS

Striation Formation

It was earlier assumed that, because of the lack of ductility enabling material at the crack tip to deform plastically to an appreciable extent, fatigue striations were normally absent in high strength steels[4]. However, as shown here and elsewhere by others, this is not the case.

The observations reported here have shown that the striation morphology depends on material properties and microstructure. In less-tempered material, where the crack tip plastic zone would be small, slip would be restricted to a small number of planes, and the resulting striations are well defined. By contrast, in soft material, where the plastic zone size would be relatively large and slip would occur on a large number of planes, the striations are poorly delineated or even absent.

To explain observed striation morphologies, many models have been proposed. One which can explain the range of morphologies observed was proposed by one of the present authors[5], and is a modification of the model of Bowles and Broek[6]: it is similar to that proposed independently by Neumann[7]. Consider an FCC metal, with crack opening as shown in Fig. 7, slip taking place on all intersecting $\{111\}$ planes, rather than on a simple pair of intersecting $\{111\}$ planes at the crack tip as suggested by Bowles and Broek[6]. The crack tip profile is assumed to depend on the ease of slip, the tip being more blunt for easy slip and sharper for more difficult slip as in higher strength material.

Upon unloading, reverse slip will take place on the old slip planes and on new ones. According to the relative ease of slip on these planes, many different striation morphologies can be achieved, two such being shown in Fig. 8. Hence, soft materials will generally have flatter and more poorly defined striations, in contrast to the sharp striations of higher strength materials. Modifying the model to the more complex slip geometry of BCC materials does not change the overall principles.

The model can account for the slip wrinkles observed between striations in low strength materials, and for the formation of slip band cracks as described below.

Slip Band Crack Formation

Reverse slip taking place along the same planes as operate in forward slip can lead to slip band crack formation. The presence of large carbides and the large extent

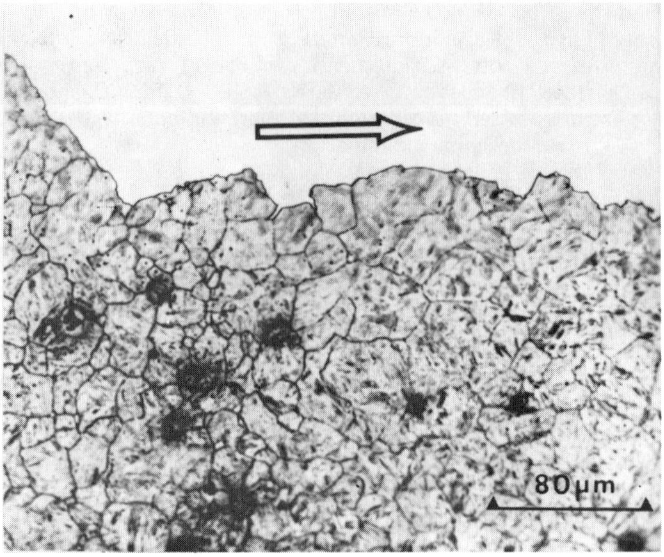

(a)

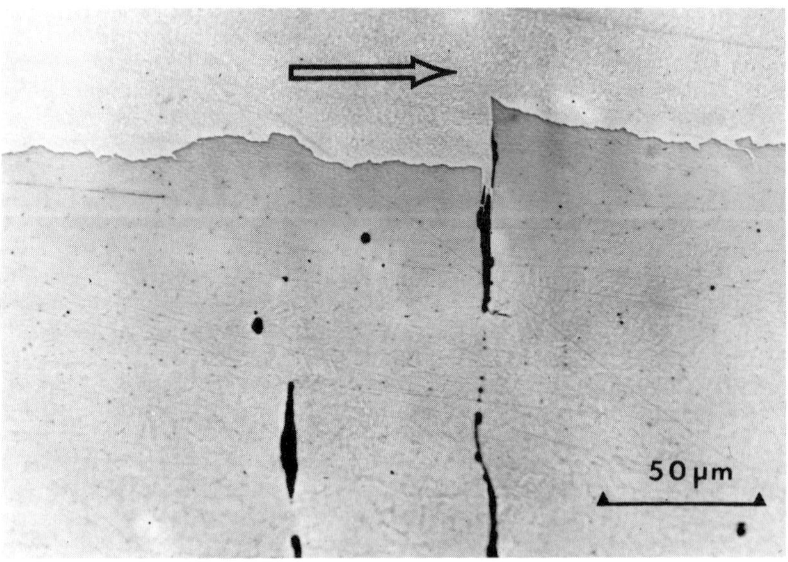

(b)

Fig. 6. Sections through fatigue fracture surfaces: (a) 200°C-tempered material, showing crack path along prior austenite grain boundaries; (b) Crack deflection on meeting an inclusion (550°C-tempered material). The arrows indicate the crack propagation direction.

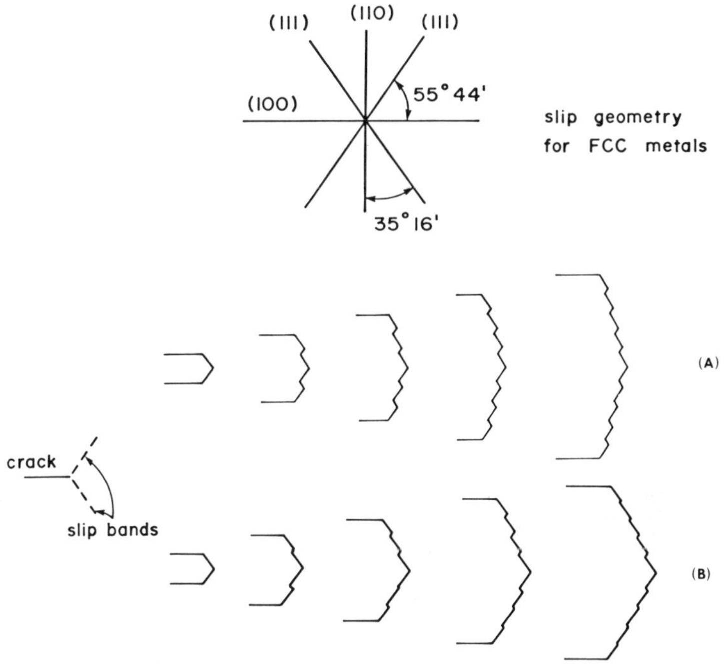

Fig. 7. Proposed model for fatigue crack growth. The slip planes shown correspond to FCC metals: (a) shows the crack opening for low strength material while (b) shows that assumed for high strength material.

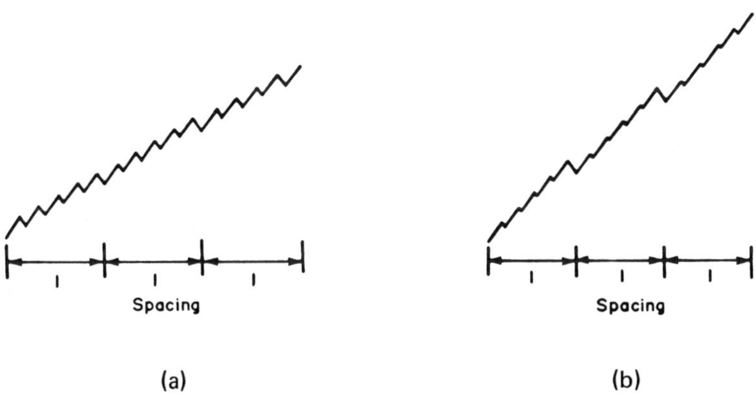

Fig. 8. Fatigue crack profile and striation morphologies: (a) for a low strength material; (b) for a higher strength material.

of forward slip in low strength material, i.e., in the steel tempered at higher temperatures, will cause difficulty in reverse slip taking place during unloading, so leading to fissures as shown in Fig. 9.

Intergranular Fatigue Crack Propagation

The mechanism for this has been discussed in some detail elsewhere[8], and it is believed that dislocations pile up within the crack tip plastic zone at carbides lying on prior austenite boundaries, allowing separation along the carbide/matrix interface of suitably oriented boundaries. This is illustrated in Fig. 10.

Tire Track Formation

As noted already, tire tracks were seen frequently in the material tempered at higher temperature, and it is believed they are related to carbide particles trapped within the propagating crack, Beachem[9,10] having discussed this previously. More detailed aspects of the formation of different types of tire track will be discussed at a later date by the present authors.

CONCLUDING REMARKS

The fractographic evidence presented in this paper has shown clearly that the detailed fatigue fracture surface features of a high strength, low alloy steel depend to a considerable degree on microstructure and mechanical properties.

The morphology of striations varied from well defined in the case of as-quenched material and material tempered at low temperature, to poorly defined with very extensive slip band cracking in the more heavily tempered material. The presence of striations on prior austenite grain boundaries and the intercrystalline crack propagation along these in as-quenched or less tempered material, as indicated by fractographic evidence, was confirmed by sectioning through the fracture surface. In order to explain these observations, mechanisms have been proposed for striation formation and intergranular fatigue fracture, the former being able to explain the formation of slip band cracks, while the latter relates to the presence of intergranular carbide layers.

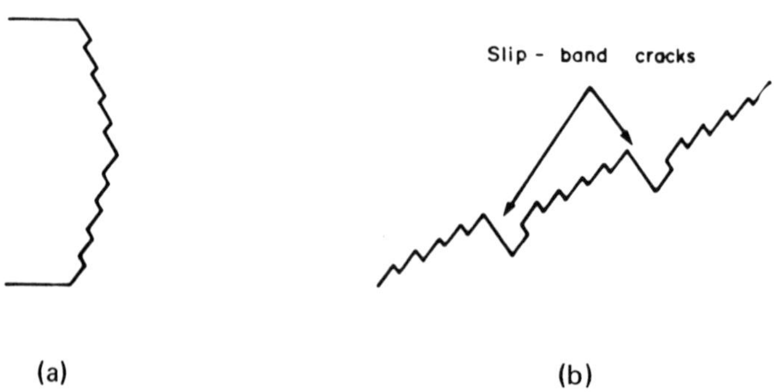

Fig. 9. Slip band crack formation: (a) fully opened fatigue crack; (b) striation profile with slip band cracks.

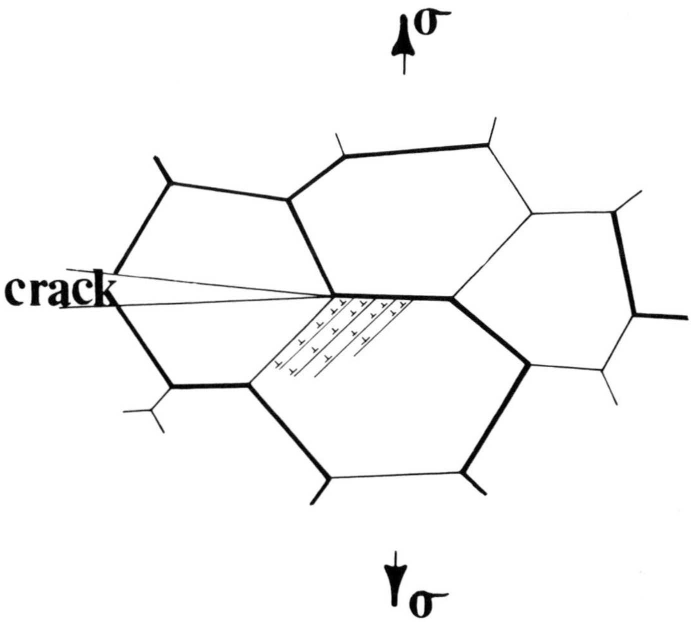

Fig. 10. Mechanism of intergranular fatigue crack growth. The heavy lines on some grain boundaries represent films of carbide.

The importance of carbide morphology is illustrated clearly by the observations reported. Grain boundary films resulting from auto-tempering during quenching can cause intercrystalline fatigue fracture, as noted above. This leads to rapid propagation once a crack is initiated, a point emphasized by measurements of the striation spacing. For example, in 200°C-tempered material, the spacing was about three times larger than in 650°C- tempered material, both fatigued at a stress close to their respective fatigue limits. This indicated that, as the strength is decreased by tempering, a larger proportion of fatigue life is spent in crack propagation.

It is interesting to note the effect of residual stresses produced during the quenching process. Because of removal of residual stresses by tempering at 200°C, and also because the dispersed carbide particles were too small to be effective stress concentrators, fatigue life was greater than for as-quenched material at the same stress level, and the fatigue limit was raised, as seen in Table I. Thus, it appears that residual stresses are more detrimental to crack initiation than to propagation. This is supported by the fact that, although the fatigue fracture area increased with tempering temperature, the difference was not sufficient to account for the reduction in fatigue strength between as-quenched and 200°C-tempered specimens. Also, the striation spacing was approximately identical in specimens at comparable stress levels.

In conclusion, it is again emphasized that fractographic features are important evidence in postulating mechanisms for fatigue crack growth, and that microstructural variations do affect them significantly.

ACKNOWLEDGEMENTS

The authors are grateful to the National Science Council of Canada (now the Natural Sciences and Engineering Research Council) for support of their studies of fatigue and fracture.

REFERENCES

1. I. LeMay, "Failure Mechanisms and Metallography: A Review" *Metallography in Failure Analysis,* J.L. McCall and P.M. French, eds., Plenum Press, New York, pp. 1–31 (1978).
2. E.C. Roberts, "Integration of Optical and Electron Fractography Methods in the Study of Fatigue", *Microstructural Science Volume 6,* J.E. Bennett, L.R. Cornwell and J.L. McCall, eds. Elsevier, New York, pp. 131–141 (1978).
3. M.—W. Lui and I. Le May, "A Fractographic Study of Fatigue in AISI 4140 Steel", *Proceedings of the 4th Annual Technical Meeting of the IMS,* J.L. McCall, ed., pp. 227–241 (1972).
4. W.J. Plumbridge and D.A. Ryder, "The Metallography of Fatigue", *Metall. Reviews,* No. 136, pp. 119–142 (1969).
5. M.—W. Lui, Ph.D. Thesis, University of Saskatchewan, Saskatoon (1973).
6. C.Q. Bowles and D. Broek, "On the Formation of Fatigue Striations", *Int. J. Fracture Mechanics,* Vol. 8, pp. 75–85 (1972).
7. P. Neumann, "The Geometry of Slip Processes at a Propagating Fatigue Crack — II", *Acta Met.,* Vol. 22, pp. 1167–1178 (1974).
8. M.—W. Lui and I. Le May, "Effects of Grain Boundary Carbides on the Fatigue Fracture of an AISI 4140 Steel", *Grain Boundaries in Engineering Materials,* J.L. Walter, J.H. Westbrook and D.A. Woodford, eds., Claitor's Publishing Division, Baton Rouge, La., pp. 397–407 (1975).
9. C.D. Beachem, "Microscopic Fatigue Fracture Surface Features in 2024–T3 Aluminum and the Influence of Crack Propagation Angle Upon Their Formation", *Trans. ASM,* Vol. 60, pp. 324–343 (1967).
10. C.D. Beachem, "Origin of Fatigue 'Tire Tracks' ", NRL Progress Report, pp. 22–23 (May 1966).

Stevens, Vander Voort, McCall, eds. Microstructural Science, Volume 8

MICROFRACTOGRAPHY OF THE FRACTURED SURFACES OF CAST ALUMINUM–MAGNESIUM–SILICON ALLOY AFTER DIFFERENT HEAT TREATMENTS

J. Pirs*

ABSTRACT

Aluminum alloys used for castings differ from wrought alloys in many respects. In the first place they differ in their high fluidity which permits the casting of articles with thin walls and complicated cross sections. The mechanical properties of most aluminum alloys depend heavily on chemical composition, casting methods and heat treatment. Alloys of type G–AlMg5Si (DIN 1725) are known for good corrosion resistance and mechanical properties at elevated temperatures. Under the designation of Hydronalium (Hy 511) they are used for the production of heads for diesel engines which are air cooled. This paper deals with results obtained during investigations of the fractured surfaces of tensile test specimens obtained from samples cast using procedures which are used for making cylinder heads. To obtain better mechanical properties different heat treatment procedures were adopted. In order to establish relationships between the structures and the properties, especially toughness and fracture mode, detailed microfractographic investigations of the fractured surfaces were performed.

The eutectic phase of the investigated aluminum – magnesium – silicon alloy had no correlation to the toughness of the alloy. In all cases the microfractographic investigations showed only brittle fracture and only the elongation values showed improvement.

INTRODUCTION

Aluminum alloys used for castings differ from wrought alloys in many respects. In the first place they differ by having high fluidity thus permitting the casting of articles with thin walls and complicated cross sections. High purity aluminum does not possess this property and moreover has low strength. Therefore, pure aluminum is not used much for intricate castings; rather it is used, for example, for parts of chemical apparatus, electrical devices (squirrel cages of electric motors, supports for brush holders, supporting rings of rotating converters), and in the food industry (boilers, kettles and various containers).

Aluminum has the remarkable property of preserving its basic physico – chemical properties in nearly all alloys. Thus, for instance, the specific gravity of aluminum is 2.70 and the specific gravities of its alloys are between 2.50 and 3.0 g/cm^3. The thermal conductivity of aluminum is 0.50 and that of its alloys is found to be 0.3 to 0.48 cal/cm.s.^{o} C.

The high corrosion resistance of aluminum is due to the formation of a dense protective aluminum oxide film on its surface. All industrial aluminum alloys with a high content of magnesium or silicon (but containing neither zinc nor copper) also display high corrosion resistance.

The mechanical properties of most aluminum alloys depend heavily on chemical composition, casting methods and heat treatment. However, the basic properties (normal and tangential elasticity moduli and Poisson's ratio) change very little.

*Technical Faculty Rijeka, 51ooo Rijeka, Narodnog Ustanka 58, Yugoslavia.

Some advantages of cast aluminum alloys are as follows:

- good castability (high fluidity) permitting the casting of parts of nearly any configurations, with a relatively small linear shrinkage and good machinability
- high specific strength, practically approaching that of straight carbon steels
- high corrosion resistance (except of alloys containing copper)
- high thermal and electrical conductivity.

All of these advantages permit cast aluminum alloys to compete successfully with many industrial alloys based on other elemental systems.

During this study an aluminum alloy with relatively high magnesium content, of type G–AlMg5Si (DIN 1725), known for good corrosion resistance and mechanical properties at elevated temperatures [1,2] was investigated. Under the commercial name of Hydronalium (Hy 511) this alloy is known to give satisfying results in the production of cylinder heads for air-cooled engines. [3]

The present paper deals with results obtained during microfractographic investigations of the fractured surfaces of tensile test specimens.

To obtain better and more satisfying results of the mechanical properties of the cylinder heads after casting, different heat treatment methods were adopted. In order to establish relationships between the structures and the properties, especially toughness and fracture mode, detailed microfracture mode investigations of the fractured surfaces were performed. [4-7]

Finally, the results of the microfractographic investigations of the fractured surfaces of tensile test specimens taken from melts where small additions of titanium had been given are reported.

MATERIALS AND METHODS OF INVESTIGATION

The materials which were investigated during this study were tensile test specimens obtained during the production of cylinder heads for a 6-stroke, air-cooled diesel engine. The alloy was melted in an induction heated coreless crucible electric furnace having a capacity of 600 kg. The melt was later degassed in another furnace and then the crystal grains were refined. After this procedure the melt stayed in the furnace for several hours. The cylinder heads were cast in metallic forms; the pouring temperature was 740°C and the temperature of the metallic form was 400°C. After the casting of the cylinder heads was completed they were subjected to a heat treatment. For each melt of cylinder heads the mechanical tests were performed.

The chemical composition of the investigated alloy was:

Mg	4.5 – 5.5 %
Mn	0.1 – 0.5 %
Si	0.6 – 1.5 %
Ti	0.2 %
Fe	max. 0.6 %
Cu	max. 0.6 %
Zn	max. 0.2 %

After heat treatment the following values of mechanical testings should be obtained:

Tensile strength	190 – 210 N/mm^2
Hardness	75 – 85 BHN
Elongation	min. 3 %

In Table I the adopted procedures of heat treatment of the castings are given.

Table I. Heat treatment procedures

Type	Procedure
No. 1	Solution heat treat at 530°C for 8.5 h, water quench, precipitation harden at 170°C for 18.5 h.
No. 2	Solution heat treat at 550°C for 8.5 h, water quench, precipitation harden at 150°C for 17.5 h.
No. 3	Solution heat treat at 550°C for 8.5 h, water quench, age for 30 hours at room temperature, precipitation harden at 150°C for 17.5 h.

Tensile testing of the specimens was performed on a Mohr Federhaff tensile test machine.

After tensile testing, the measurement of the eutectic phases was performed on each specimen. For this, Image Analyzing Techniques were applied. The specimens were examined under a Reichert microscope with an automatic XY stage and imaged into a Vidicon scanner. The Quantimet 720 Control Processor consisted of a shade corrector, a standard detector, a MS—3 computer and a multiple accumulator; all modules were under the control of the 720 programmer.

The preparation of samples consisted of an initial polishing on 600-grit paper; this was followed by diamond polishing to one micron. The objective lens magnification was chosen with regard to the size of the field of view, depth of focus across the field and magnitude of the precipitated phase as follows;

Objective = X 6.3
Scanner Lens = X 8
1 Picture Point = 1.35 microns
1 Picture Point (area) = 1.82 microns^2

The measurements recorded were the area, intercept and perimeter of the eutectic phase. From these measurements, the percentage area, the mean free path and the length of perimeter per square millimeter were calculated.

Scanning electron microscopy was used to provide high resolution micrographs to interpret the fracture mode. The fractured surfaces of tensile test specimens, with exception of tensile specimens with small additions of titanium, were first cleaned ultrasonically and later carbon was deposited on the surfaces. The tensile test specimens with the small amounts of titanium were only ultrasonically cleaned, with no carbon deposition, because energy dispersive X-ray analysis in the SEM was used to analyze the effect of the titanium on the fracture mode.

Scanning electron microscopy was performed using a Cambridge Stereoscan electron microscope. Another scanning electron microscope, a Leitz — AMR 1200 in conjunction with Kevex μX analytical spectrometer, was used to detect the titanium additions and their influence on the fracture mode.

RESULTS

In Table II the results of the chemical analyses, the mechanical tests and the image analyses of the investigated tensile test specimens after different heat treatment procedures are shown.

The primary aim of the present investigation was to obtain quantitative relationships between the grain size and mechanical characteristics of the investigated alloy, specially elongation values.

From the results in Table II we can see that several of the melts, Nos. 5, 127, 129, 266, 273 and 274, showed excessive values of silicon. Also several melts, Nos. 116,

Table II. Results of chemical analysis, mechanical testings and measurements of eutectic phase in investigated aluminum-magnesium alloy.

Melt No	Chemical composition in % Mg	Mn	Si	Cu	Ti	Results of tensile testings σ_B N/mm^2	δ_5 %	Hardness HB N/mm^2	Results of image analysis by Quantimet 720 Control Processor Area %	Intercept	Mean Free Path (λ)	Perimeter (pp)	Perimeter μ/mm^2	Heat Treatment
5	4,85	0,38	1,85	0,43	–	214	2,85	81,3	11,4	11449	54,7	42232	62,5	No. 1
8	5,22	0,34	1,42	0,50	–	191	1,57	81,7	17,8	12262	51,1	43270	64,1	
22-0	4,80	0,36	1,17	0,41	–	212	2,71	85,8	8,6	9433	66,4	34148	50,5	
44	5,10	0,34	1,27	0,45	–	211	1,43	87,2	10,0	8825	71,0	31602	46,8	
115	5,04	0,35	1,30	0,92	–	182	1,14	77,1	11,05	9214	68	32692	48,4	
116	5,80	0,35	1,32	0,51	–	200	1,42	84,4	11,5	9479	66,1	35337	52,3	
119	5,90	0,39	1,38	0,45	–	197	1,42	85,8	10,4	8630	22,6	32692	48,4	
124	4,90	0,27	1,32	0,48	–	203	1,43	92,8	10,05	8760	66,3	31600	49,6	
127	4,78	0,35	1,53	0,40	–	210	2,0	83,5	8,6	8820	61,4	42200	52,3	
129	5,24	0,34	1,53	0,40	–	186	1,43	79,6	10,0	8910	70,5	32325	47,3	
134	4,87	0,37	1,30	0,44	–	264	1,43	89,7	11,5	9125	67,0	34850	47,4	No. 2
135	4,84	0,37	1,39	0,38	–	245	1,14	97,7	12,0	9120	67	33740	49,5	
136	4,90	0,40	1,39	0,49	–	269	1,71	97,7	9,0	9430	69,8	31650	51,3	
137	4,87	0,38	1,40	0,32	–	185	1,40	72,4	10,0	8840	27,0	32541	46,3	
139	4,67	0,36	1,34	0,34	–	199	2,4	70,6	10,3	10020	62,5	34921	51,7	
140	4,86	0,37	1,31	0,41	–	235	2,0	92,3	11,6	12100	53,6	41231	61,5	
141	4,88	0,38	1,30	0,34	–	235	1,85	90,2	10,9	8787	73,0	33722	49,9	
147	4,45	0,35	1,46	0,38	–	207	1,0	89,7	10,3	9709	64,5	33874	50,1	
167	5,17	0,30	1,18	0,39	–	175	3,1	73,9	12,0	9414	69,0	32750	49,2	No. 3
266	5,20	0,32	1,59	0,32	–	231	2,28	91,7	11,8	12100	53,6	41234	62,0	
273	5,78	0,30	1,62	0,35	–	257	3,15	96,6	12,5	9320	65,0	33840	49,0	
274	5,05	0,30	1,51	0,32	–	240	2,28	92,3	11,6	12100	54,0	42100	61,5	

119 and 273, showed higher percentages of magnesium than prescribed. Melt 115 showed an excess percentage of copper, but all the other melts in Table II were within the prescribed limits of chemical composition.

The observed differences in the chemical composition did not have any great influence on the results of the measurements of the eutectic phases, which are also given in Table II. The grain sizes of the investigated alloys, obtained after different heat treatments, seem to be mostly equal in all the melts. A further conclusion which can be made is in connection with the measured elongation values, where the differences in elongation values can not be explained satisfactorily only on the basis of the values of the microstructural measurements and the quantitative assessments of the eutectic phases.

After the measurements given in Table II were completed, the fractured surfaces of the tensile test specimens were investigated in the scanning electron microscope. From many results obtained during this investigation only a few representative results are shown here. In Fig. 1 we see the results of microfractographic investigations of melts Nos. 129, 137 and 167. As can be seen from Fig. 1 different magnifications were adopted during this procedure. All results of the microfractographic investigations shown in Fig. 1 show clearly the fracture mode, which was in all cases brittle. It can be concluded from these results that the results of the microfractographic investigations do not allow any conclusion to be made about the fracture mode until scanning electron microscopy has been made. From the results in Table II, it can be concluded that the third procedure of heat treatment has given the most satisfactory results regarding the elongation values, but the fracture mode is still brittle in this case.

In the second phase of these investigations small amounts of titanium and nickel were given to the melt because the elongation values after different heat treatments did not conform to the prescribed values. The alloy additions were expected to yield finer grained materials with better elongation values.

Table III shows the results of chemical analysis, mechanical tests and image analyses of the melts.

The results of chemical analysis, Table III show that in all investigated melts the magnesium content was below the prescribed values and the manganese was higher than prescribed. Again it can be seen from Table III that all other elements were in the prescribed limits.

The results of mechanical tests showed improvements, especially elongation values corresponding to the results of metallographic structure determinations and the results of image analysis, where greater percentages of eutectic phases and shorter free paths allow the conclusion to be drawn, that the grain size of α-crystals was smaller.

Here again, as in the previous case, the fractured surfaces of tensile test specimens were investigated in the scanning electron microscope and from the many results obtained only some representative results are shown here. The results of scanning electron microscope investigations of the fractured surfaces of the tensile test specimens of the melts No. 2/1, 53/2 and 60/1 are shown in Fig. 2. Two magnifications were adopted during this procedure. The results of microfractographic investigations, Fig. 3, show the fracture mode is no more only brittle, but mixed brittle and ductile. It is evident that small additions of titanium helped to gain better mechanical values, being revealed by more ductile mode of the fractured surfaces.

Further research was devoted to the identification of titanium on the fractured surfaces of the tensile test specimens, thus to study its role on the formation of smaller grains of primary α-crystals. The prospect of performing chemical analysis on compact specimens at any desired level of resolution is particularly interesting, where commercial STEM – based systems are now available which virtually provide this capability.

From the X-ray analysis results, using a Kevex energy spectrometer, we wanted to obtain the concentrations of titanium present in small quantities at the intergranular voids or possibly the segregations of heavier elements.

Table III. Results of chemical analysis, mechanical testings and measurements of eutectic phase in investigated aluminum alloy, with addition of titanium.

Melt	Chemical composition in %							Results of tensile testings		Hardness	Results of image analysis by Quantimet 720 Control Processor				
	Mg	Mn	Si	Cu	Ni	Fe	Ti	σ_B N/mm^2	δ_5 %	HB N/mm^2	Area %	Intercept	Mean Free Path (λ)	Perim. (pp)	Perim. µ/mm^2
2/1	4,44	0,63	1,20	0,39	0,18	0,20	0,17	227	1,4	90,7	10.4	8630	22.6	32692	48.4
53/1	4,49	0,54	1,17	0,31	0,09	0,31	0,17	232	3,2	89,7	12.2	12480	58.8	34220	54.4
53/2	4,44	0,63	1,18	0,43	0,15	0,19	0,17	228	2,0	86,8	14.1	11800	56.8	36330	54.4
56/1	4,40	0,59	1,17	0,39	0,14	0,20	0,17	227	1,8	92,8	11.4	11449	54.7	42232	62.5
60/1	4,42	0,54	1,20	0,40	0,15	0,21	0,15	251	1,4	88,7	8.6	9433	66.4	34148	50.5
60/2	4,41	0,60	1,19	0,39	0,16	0,20	0,15	226	2,0	89,7	12.2	12013	56.7	43833	63.4

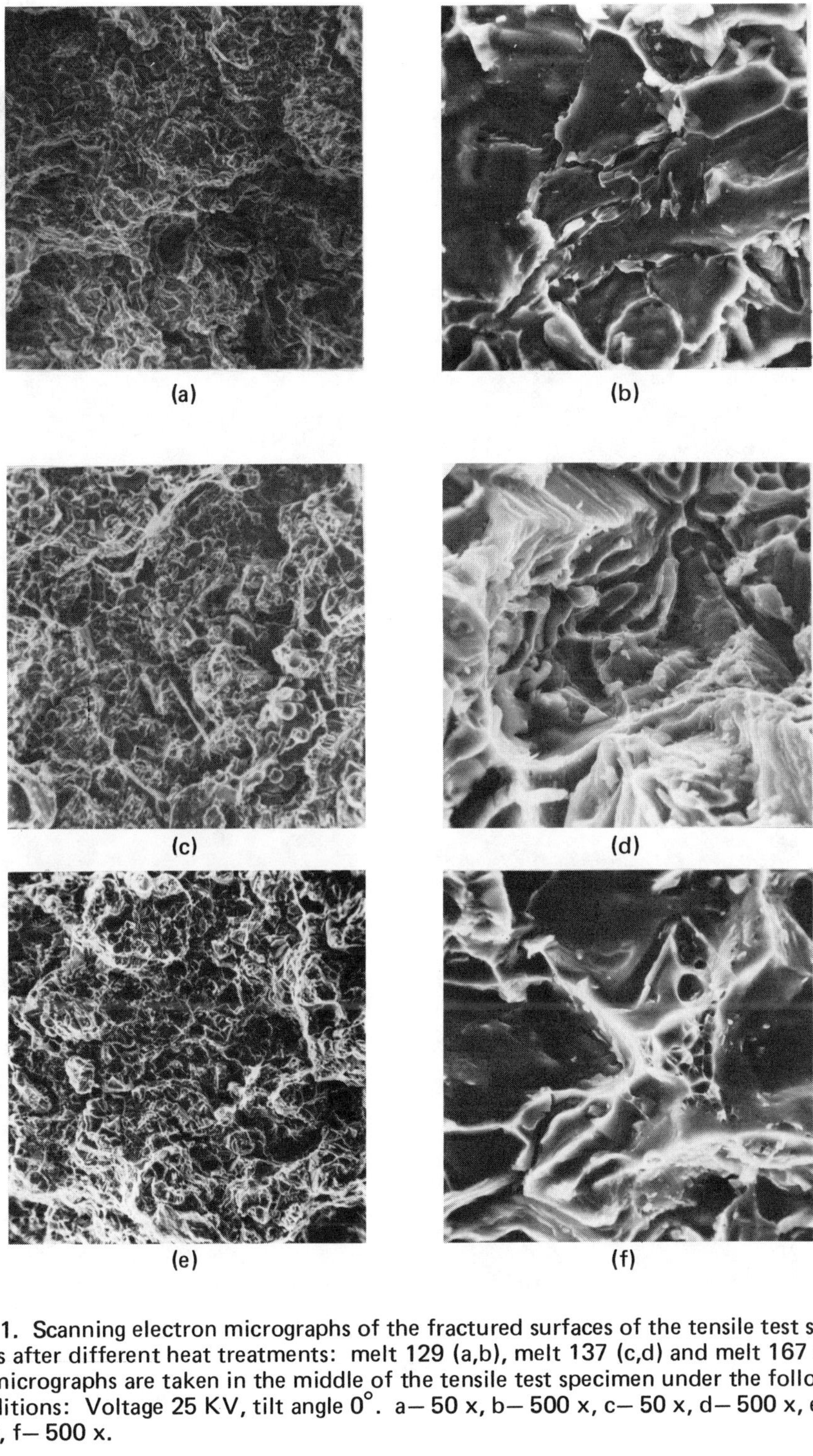

Fig. 1. Scanning electron micrographs of the fractured surfaces of the tensile test specimens after different heat treatments: melt 129 (a,b), melt 137 (c,d) and melt 167 (e,f). All micrographs are taken in the middle of the tensile test specimen under the following conditions: Voltage 25 KV, tilt angle 0°. a– 50 x, b– 500 x, c– 50 x, d– 500 x, e – 50 x, f– 500 x.

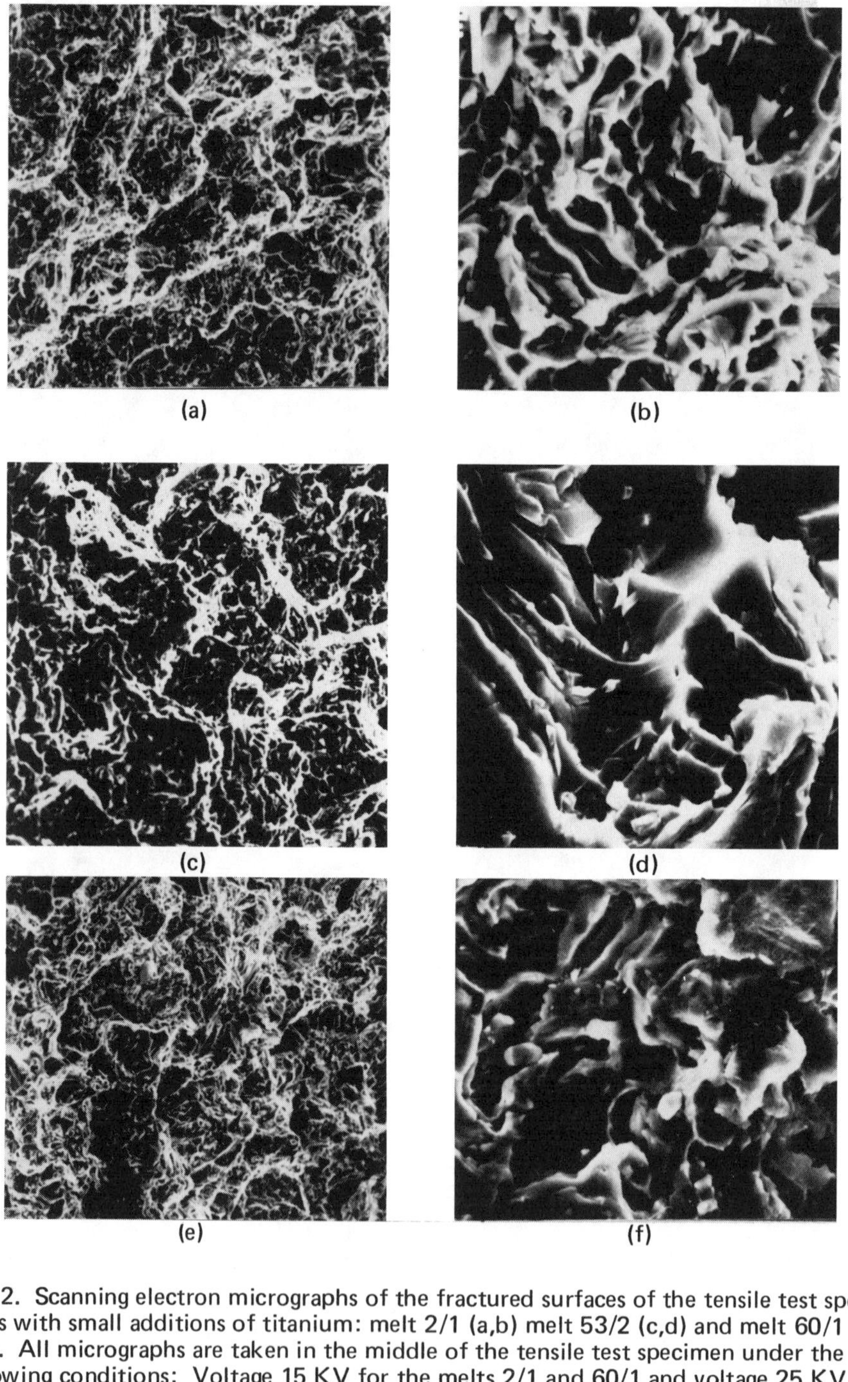

Fig. 2. Scanning electron micrographs of the fractured surfaces of the tensile test specimens with small additions of titanium: melt 2/1 (a,b) melt 53/2 (c,d) and melt 60/1 (e,f). All micrographs are taken in the middle of the tensile test specimen under the following conditions: Voltage 15 KV for the melts 2/1 and 60/1 and voltage 25 KV for the melt 53/2. Tilt angle in all cases 50°. a– 100 x, b– 1000 x, c– 100 x, d– 1000 x, e– 100 x, f– 1000 x.

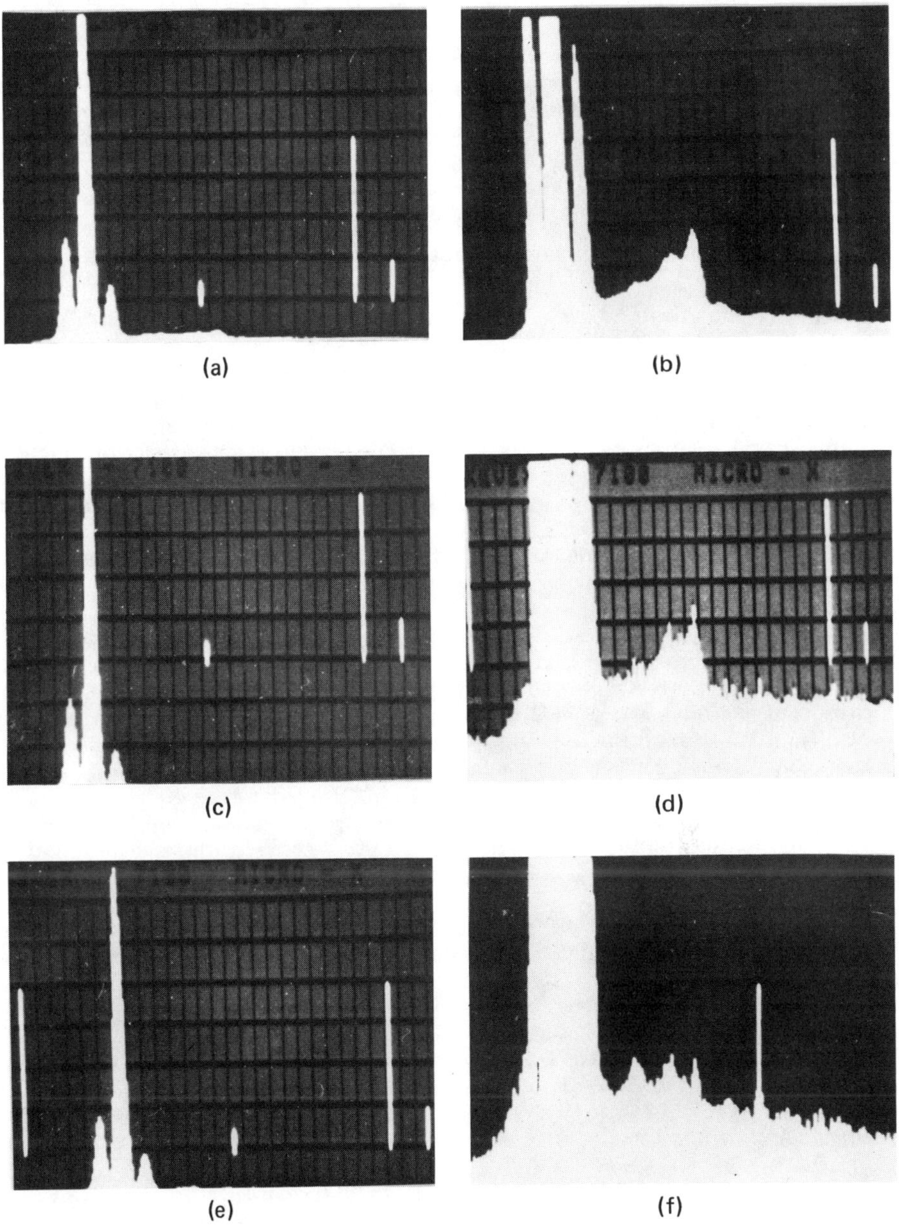

Fig. 3. Results of qualitative energy — dispersive X-ray spectrometric elemental analysis of the fractured surfaces shown in Fig. 2. a— X-ray spectrum from area of specimen in Fig. 2a, b— bulk analysis, c— X-ray spectrum from area of specimen in Fig. 2c, d— bulk analysis, e— X-ray spectrum from area of specimen in Fig. 2e, f— bulk analysis.

From all the results obtained during these investigations, only the results obtained on the fractured surfaces of the tensile test specimens Nos. 2/1, 53/2 and 60/1, as shown in Table II and in Fig. 2, will be presented here. In Fig. 3 we see the results of qualitative elemental analysis of the melts Nos. 2/1, 53/2 and 60/1. As can be seen from the results shown in Fig. 3, the same results being obtained also during investigation of the fractured surfaces of tensile test specimens of the other melts, no titanium was detected. On the fractured surface of the tensile test specimen of the melt No. 53/2 we detected segregations of calcium. We also detected concentrations of magnesium and silicon and this is shown in Figs. 4a and b. The detected concentrations of magnesium and silicon correspond to the chemical composition of Mg_2Si. As is known, the aluminum corner of the Al-Mg-Si phase diagram has a quasibinary section which divides the diagram into two secondary systems:

- Al – $AlMg_2Si$ with a ternary eutectic of α + Al_3Mg_2 + Mg_2Si
- Al – Mg_2Si – Si with a ternary eutectic of α + Si + Mg_2Si.

The binary eutectic, α + Mg_2Si, is also present in the aluminum corner of this ternary system.

With a low magnesium content the bulk of the binary eutectic, α + Mg_2Si, is found between the grains of the α-solid solution. With an excess of magnesium it will form large independent areas and the Mg_2Si phase becomes resolvable at 100 x magnification. The microstructure of this binary eutectic resembles Chinese script. Increasing amounts of the binary eutectic causes its microstructure to coarsen in the form of spider branching. The results of investigations of the fractured surface of tensile test specimen of the melt No. 53/2 shows the Mg_2Si phase having the described origin.

In order to determine the influence of titanium on the grain size of the investigated aluminum–magnesium–silicon alloy, Auger electron spectroscopy (AES) has been used.

Analogous to the standard electron microprobe the SAM can be used to obtain images, chemical maps and spot analyses of surfaces. A rather complicated inert ion sputter etching process was used to determine the extent of surface chemical variations. Auger spot analysis performed on the fractured surface of the tensile test specimen No. 53/2 showed the following peaks: aluminum, sulphur, carbon, calcium, nitrogen, oxygen, fluor, sodium and magnesium.

After ion sputtering (2 minutes, 3 KV, 15 mA) to a depth of 400 Å, besides aluminum only small concentrations of carbon, oxygen and argon were found. But after 5 minutes ion sputter etching to a depth of 1000 Å some locations were found to differ from each other in the concentrations of oxygen and titanium which were found.

It was essential for the results of the Auger analysis performed under the oxide layer that in the fundamental matrix consisting of aluminum segregations of titanium and oxidized places, where magnesium was expected, were found, It is interesting to note that the expected magnesium segregations were not identified.

On the sample No. 53/2 at the 1000 Å depth lineal analysis and Auger maps of the most interesting elements were performed. These elements were aluminum, oxygen and titanium. In Fig. 5 we see the results of Auger mapping and the results of lineal analysis of Al (67 eV), 0 (510 eV) and Ti (417 eV). The results of lineal analysis which are shown in Fig. 5d, were obtained in the middle of the Auger map, shown in Fig. 5a.

From the results of the Auger analysis it can be concluded that we obtained a uniform distribution of titanium in the melt, causing smaller primary crystals to be obtained. The distribution of other elements was expected and confirmed by the Auger analysis, but the results obtained with magnesium are confusing, because in the depth analysis we obtained practically no concentration of magnesium.

In the same manner we tested an oxide layer on the surface of the investigated tensile test specimen.

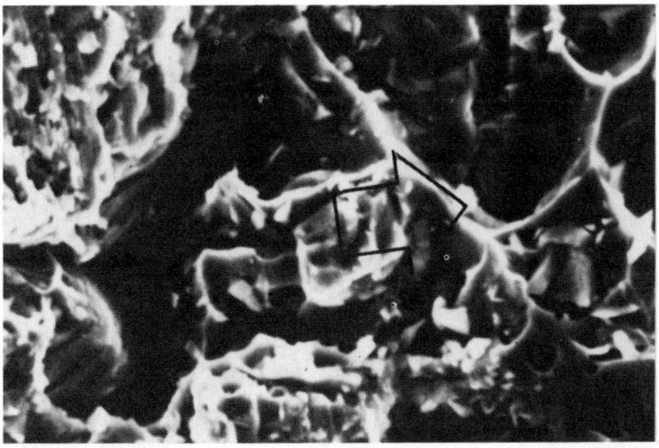

(a)

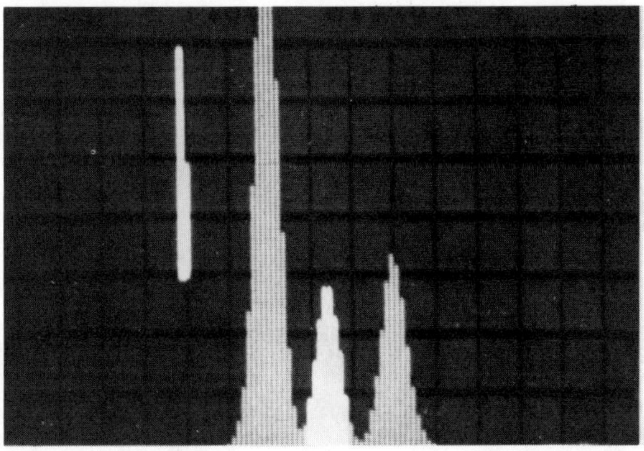

(b)

Fig. 4. Elemental analysis of inclusion on the fractured surface of the tensile test specimen. a— Scanning electron micrograph of the fractured surface of the specimen of melt 53/1 taken under the following conditions: Voltage 25 KV, tilt angle 50°. b— X-ray spectrum from the inclusion in the intergranular void (arrow) showing significant Mg and Si peak, corresponding to Mg_2Si.

DISCUSSION OF THE RESULTS

The results of the investigations of the fractured surfaces of tensile test specimens after different heat treatments show clearly that all the fractures are brittle. Of the mechanical property values, special importance was given to the elongation measurements.

All fractures obtained after different heat treatments were transcrystalline fractures being characterized by the presence of voids. These voids represent the final stage of the process of enlargement and coalescence of microvoids under plastic deformation. Small differences observed on some fractures, which showed smaller voids, have no significant influence on the general characterization. With the third procedure of heat treatment, we obtained improved values of elongation and this is interpreted as being the result of the longer annealing period at room temperature. This longer annealing

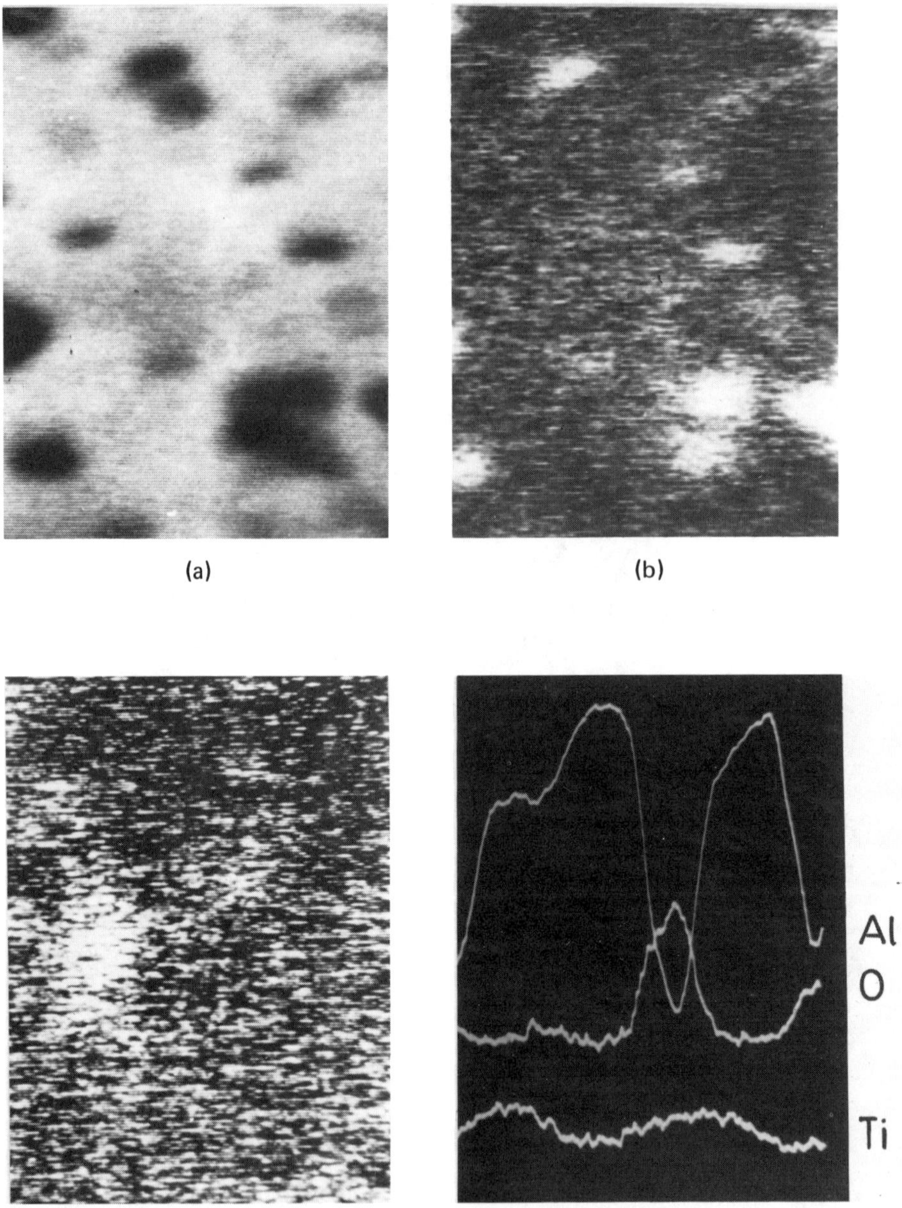

Fig. 5. Auger chemical maps. a– Aluminum Auger chemical map (Mag. 250 x)–67 eV, b– Oxygen Auger chemical map (Mag. 250 x)–510 eV, c– Titanium Auger chemical map (Mag. 150 x)–417 eV, d– Auger line analysis of aluminum, oxygen and titanium performed in the middle of Fig. 5 a, Aluminum – 67 eV, Oxygen – 510 eV, Titanium – 417 eV.

period did not allow the β-phase to precipitate. If the β (Al_3Mg_2) phase is precipitated from the solid solution it causes strong embrittlement. In such a way we obtained higher elongation values with the third procedure of heat treatment, but this change was not great enough to produce also the change from brittle to ductile fracture. Thus we were not able to detect the changes in elongation values through corresponding changes in fracture mode.

The differences in values of the mechanical tests after different heat treatments, especially the elongation values, may be due to the differences in the distributions of slip bands. The differences in distribution of slip bands may be investigated by transmission electron microscopy. Our further investigations will be directed in this field and the results obtained will be reported at a later date.

The second question is the distribution of alloying elements, especially titanium. This was not solved with the methods of classical metallography, but we wanted to investigate the fracture mode of different heat treated tensile test specimens together with the role of titanium and other alloying elements by using energy dispersive analysis.

With the method of energy dispersive analysis we detected Mg_2Si together with the segregations of calcium on the fractured surface. These elements may have caused embrittlement of the investigated alloy. With this method of analysis we wanted to obtain the distribution of small amounts of titanium in order to obtain better mechanical property values as a result of smaller primary crystals. We measured smaller primary crystals and with the scanning electron microscope detected a change from a brittle fracture mode to a ductile mode.

Finally, the detection of titanium and its distribution was performed by Auger spectroscopy.

CONCLUSIONS

This paper presents the results of investigations of the fractured surfaces of separately cast tensile bars which were obtained during the production of the cylinder heads of air-cooled, 6-stroke diesel engines. For this purpose the aluminum-magnesium alloy of type G—AlMg5Si, known under the commercial name of Hydronalium, was used. After the casting procedure was completed, cylinder heads were heat treated to obtain the prescribed values of mechanical properties. In order to obtain better values of elongation, these heat treatments were subject to small changes.

The principal aim of this investigation was to obtain an explanation for the different values of elongation after different heat treatment procedures. In order to do this, the chemical composition, results of mechanical tests, percentage of eutectic phase and microfractographic examinations of the fractured surfaces of tensile bars were compared.

The results of scanning electron microscope investigations showed only brittle fracture modes of the fractured surfaces of the test bars, regardless of the heat treatment procedure. From these results it can be concluded, that there was no definite relationship between the measured quantities of eutectic phases and values of mechanical tests.

Later we investigated the fractured surfaces of tensile test bars having small additions of titanium. Small additions of titanium obviously caused fine grained material to be obtained and caused the change in fracture mode from brittle to a mixed brittle-ductile mode. Besides these measurements, energy dispersive analysis of inclusions on the fractured surfaces were performed in order to investigate the influence of titanium and other elements on the microstructure.

In this paper it was again shown that the properties of materials are structure-sensitive. Structure in turn is determined by composition, heat treatment and processing. Thus it is necessary to characterize both composition and microstructure at the highest levels of resolution possible in order to understand materials behavior. Such characterization requires advanced and sophisticated methods of analysis using microscopic, diffraction and spectroscopic techniques. For this, electron microscopy is particularly versatile since we are now routinely synthesizing structure almost at atomic levels of resolution. The interaction between composition, heat treatment and properties is complex but this interaction must be understood if materials are to be improved or new materials to be designed.

Electron microscopy is playing an important role in research in materials science and engineering (e.g. failure analysis). The problem solving approach is not limited to a single technique and it is not implied that electron microscopy can solve all problems. The new analytical instruments of microdiffraction and X-ray and electron energy loss microanalytical capabilities are welcome additions to the materials scientists "bag of tools" as these methods offer large gains in spatial resolution compared to more conventional analytical methods.

ACKNOWLEDGEMENTS

This work presents a part of a research project which was financially supported by the Selfgoverning Interest Community for Scientific Work in Zagreb, under grant 1/13 and their support is gratefully acknowledged. Thanks are due to Marcelo Kura for his help during metallographic preparations of the samples. Dipl. engineer A. Zalar performed the Auger spectroscopy at the Institute for Vacuum Technique and Electronics in Ljubljana, and his help is gratefully acknowledged.

REFERENCES

1. R.W. Guard, "Mechanics of fine particles strengthening", *Strengthening Mechanisms in Solids,* ASM Cleveland, (1962).
2. H. Vosskuhler, "Untersuchungen an gegossenen Aluminum-Zylinderkopflegierungen", *Aluminium,* Vol. 28, pp. 99–106, (1952).
3. H. List, "Die Behandlung konstruktiver Probleme neuzeitlicher Dieselmotoren," *MTZ Motortechnische Zeitschrift,* Vol. 3, No. 3, pp. 100–108, (1972).
4. H.C. Burghard and D.L. Davidson, "Fracture Mechanisms and Fracture Surface Topography", *Proceedings of the First International Conference on Fracture,* Vol. 2, Sendai University, Tokyo, pp. 571–596, (1965).
5. R.M.N. Pelloux and J.C. McMillan, "The Analysis of Fracture Surfaces by Electron Microscopy", *Proceedings of the First International Conference on Fracture,* Vol. 2, Sendai University, Tokyo, pp. 547–569, (1965).
6. D.R. Miller and F.M.C. Besag, "Metallographic Studies of ductile fracture in high purity aluminum and lead", *Proceedings of the First International Conference on Fracture,* Vol. 2, Sendai University, Tokyo, (1965).
7. C. Crussard, J. Plateau et al., *Fracture,* J. Wiley & Sons Inc., New York – London, p. 524, (1959).

METALLOGRAPHIC METHODS

Chairpersons: J.A. Nelson and J.M. Hoegfeldt

Stevens, Vander Voort, McCall, eds. Microstructural Science, Volume 8

APPLICATION OF METALLOGRAPHIC TECHNIQUES TO THE STUDY OF METALLIC WEAR MECHANISMS

Peter J. Blau*

ABSTRACT

Optical and scanning electron microscopy were used to study features on and below unlubricated worn surfaces of several copper alloys. Each imaging technique and magnification range contributed different details towards an understanding of the deformation of these materials by sliding. Cross-sectioning of wear surfaces for metallography necessitated special edge preservation methods including copper electroplating and directionally controlled abrasion. Taper sectioning was used to observe the structure of debris pile-ups in wear tracks.

BACKGROUND

Studies of the relationships between microstructure and wear present numerous challenges for the metallographer. In many ways wear research is concerned with the same mechanisms responsible for the ease or difficulty in mechanically polishing samples, because mechanical polishing is essentially controlled abrasive wear. In a similar way, electropolishing and chemical polishing involve controlled corrosion.

Nearly every metallographic tool, both light-optical and electron-optical, can be utilized in tribological studies. Much of the early work in studying wear mechanisms used low magnification light microscopy for wear surfaces, because depth of field limited the use of higher powers. Later on, cross sections of worn surfaces were prepared and metallographers encountered the difficulties of edge preservation. For example, Samuels[1] addresses many of these problems in his excellent text on mechanical polishing. This paper briefly discusses the use of several complementary methods of wear surface examination with particular focus on edge preservation in studies of the unlubricated wear of some alloys of copper.

WEAR TEST METHOD

Two device types were used to produce wear in the current study. The first was a pin-on-disk type using a 52100 steel bearing for the standard rider pin. The disk was mounted vertically so debris could be collected beneath it. Sliding velocity was 5 cm/sec at a load of 250 g. The second device was a block-on-ring type (LFW-1 , given in ASTM D 2714-68) using the same velocity and a load of 20.4 Kg. The standard rings were 440 C stainless steel. Both systems were enclosed in test chambers purged with dried argon.

RESULTS AND DISCUSSION

Figure 1 shows a disk wear track for a Cu-10wt%Zn alloy at relatively low magnification. A myriad of fine details ranging from deep grooves to elongated flat plateaus to tiny chevrons can be seen. With the aid of the SEM another level of detail can be observed. Figure 2 shows an accumulation of metallic debris on a Cu-15wt%Zn disk wear track. Clear evidence of layering can be seen on the edges of the flakes, and considerable range in debris sizes is apparent. The fine chevron patterns which could be easily observed by light optics were not seen when using the SEM to examine the same areas.

---

*Department of Metallurgical Engineering, The Ohio State University, Columbus, Ohio 43210 USA.

Fig. 1. Photomicrograph of a disk wear track for a Cu-10wt%Zn alloy. Arrow (upper right) indicates sliding direction. Marker: 10 microns.

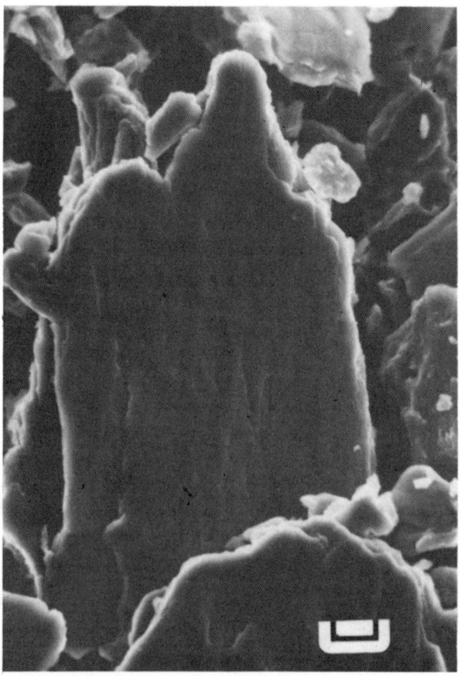

Fig. 2. SEM photomicrographs of a debris pile-up on a disk wear track in Cu-15wt%Zn. Sliding direction was upwards. Marker: 5 microns.

Observation of as-worn surfaces from above is only part of any broadly-based approach to understanding wear mechanisms. To more fully understand the role of microstructure in affecting wear properties, detailed studies of cross sections of worn samples must be performed. It is here that the proper use of metallographic procedures can help clarify the mechanisms of subsurface deformation.

As Samuels[1] has shown for several ferrous and non-ferrous alloys, the machining and polishing processes used in sample preparation can greatly affect the depth of damage and can create misleading artifacts in the polished section. As part of a recent study[2], it was found that by fly-cutting rather than by shaper-planing brass samples, microstructural damage was reduced from about 85 microns to about 10 microns. Furthermore, experiments with diamond wafering of large grained high purity copper rods produced evidence of damage to a depth of 25-30 microns depending upon the grain orientation. Therefore, in attempting to study the subsurface deformation due to sliding there exists a situation where two zones of material damage intersect: the first zone produced by the wear test and a second zone due to sample preparation. The polishing technique must be controlled to eliminate the artifacts of the second zone so as to observe the features of the first.

A recent article by Torrence[3] has presented a clear discussion of many of the problems associated with the preparation of wear surfaces, together with suggestions on how to solve some of them. In addition, Ahn, et al[4] have suggested additional techniques to minimize damage to edge protective coatings. In contrast to the method recommended by Waterhouse[5] for polishing away from the sample into the mount, Ahn et al recommend polishing toward the sample when edge protective plating is present. Abrading in this direction tends to keep the plating in compression rather than pulling it away from its substrate.

A taper section was made of a disk wear track in a Cu-7.5%Al alloy which showed similar accumulations of flakes as in the previous scanning electron micrograph (Fig. 2). The taper effect helped to reveal the structure of a debris particle pile-up as well as to show the deformed metal below it (Fig. 3).

Figure 4 shows the gap in between a copper electrodeposit and an OFHC copper wear block surface. This gap originally may have been filled by a tightly adherent mechanical compact of fine debris particles and/or transferred material which might have eroded during polishing. Frequent ultrasonic cleaning of the specimen during polishing was necessary to help alleviate abrasive scratching by lost clumps of debris. Another explanation of the gap involves loss of a portion of the highly deformed layer[6]. Note that electropolishing could not be used because the highly deformed near-surface zone is removed much faster than the bulk and microstructural information could be lost. In addition, use of etchants or coarse wear scar cleaning prior to plating could also disrupt the structure of the highly deformed zone.

With pin-on-disk tests of the same copper alloys, gaps in the plating were not nearly as serious a problem in sample preparation. The main reason for this was that the debris was more flake-like and metallic (other than compacted oxide-metal mixtures) facilitating the attachment of the electroplate. Figure 5 shows the cross-sectional view of a wear track on a copper disk (wear direction out of the plane of the section). Because gap development was less severe, a 20 second chemical polishing treatment could be used (57 ml orthophosphoric acid, 20 ml nitric acid, and 16 ml acetic acid at 50-60°C.) Not only is the highly deformed surface region visible, but possible evidence for crystallographic texturing of the less plastically deformed grains below it is also visible. This texturing evidence is observed in the form of a change in the etch pitting character of the grains as a function of distance from the wear surface. As distance below the surface decreased, increasing plastic strain due to sliding forces could have produced a crystallographic rotation sufficient to alter the etching response of these grains.

Fig. 3. Polished taper section of a wear product pile-up on a disk wear track in Cu-7.5wt%Al. P=Cu plating, B= heavily deformed region at the base of the wear track. Markers: Vertical 2.5 microns, horizontal 10 microns. Etchant: 2 g. potassium chromate, 1 g. sodium chloride, 8 ml. sulphuric acid, 100 ml. dist. water.

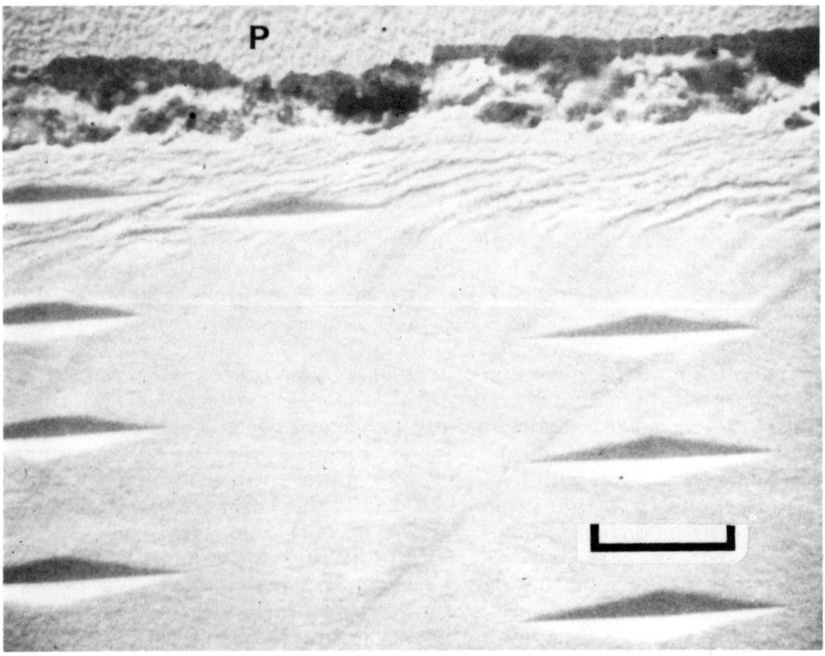

Fig. 4. SEM photomicrograph of a cross section of a block wear scar showing response to left to right sliding. Marker: 30 microns. Etchant: 20% ammonium persulphate.

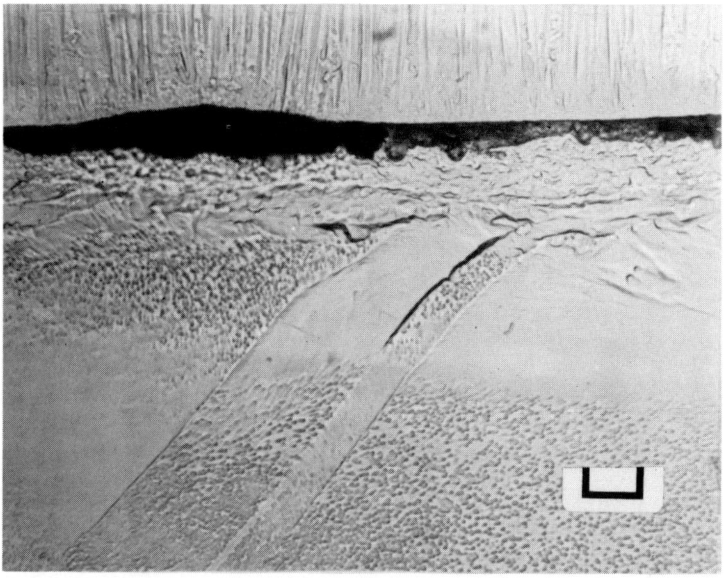

Fig. 5. Photomicrograph of a cross section of a pure Cu disk track with the sliding direction normal to the plane of polish. Chemical polished about 20 seconds with 57 ml phosphoric acid, 15 ml. nitric acid, 20 ml. acetic acid at 55°C. Marker: 10 microns.

SUMMARY

This paper has described some of the problems involved with the metallographic preparation of wear surface cross sections. Several methods of examination were used to emphasize the need to combine complementary techniques to acquire a more complete picture of the microstructures of wear. Certainly, metallurgical studies in tribology represent some of the greatest challenges to the metallographer.

ACKNOWLEDGEMENTS

The author would like to thank D.A. Rigney and R.V. Farrar of The Ohio State University for their critical reviews of this paper. The wear studies cited were funded by a grant from the National Science Foundation.

REFERENCES

1. L.E. Samuels; *Metallographic Polishing by Mechanical Methods,* Pitman and Sons Ltd., 194 pp, (1967).
2. P.J. Blau; PhD dissertation research, The Ohio State University, 1976-79.
3. A.A. Torrence; The Metallography of Worn Surfaces and Some Theories of Wear, *Wear*, Vol. 50, pp. 169-182, (1978).
4. T.M. Ahn, P. Blau, K.-L. Hsu, D.A. Rigney and J. Schell; Metallographic Techniques for Wear Test Specimens, submitted to *Wear*, (1979).
5. D.A. Rigney; private discussion, (1979).

Stevens, Vander Voort, McCall, eds. Microstructural Science, Volume 8

DIAMOND MACHINING OF ALUMINUM — MACROMETALLOGRAPHY

S.A. Levy*

ABSTRACT

Nonmetallic inclusions in aluminum alloys are nonuniformly distributed. To increase the probability of encountering such inclusions, a large sample size is desirable. This paper describes the use of diamond machining to answer this need. Ingot slices up to 18 inches in dia. have been machined, with a finish approaching that of a metallographic polish. The structure can first be scanned at low magnification, and any suspicious areas can be studied to 500X. The development of the technique and equipment is presented with emphasis on their use as a quality control tool for inclusion location.

INTRODUCTION

Two important aspects of quality control for aluminum ingot production are low hydrogen and inclusion content. As discussed in previous publications,[1,2] the state of technology with respect to hydrogen content is much more advanced because of the existence of instruments such as Telegas and Nitrogen Carrier Gas Fusion. The problem with respect to inclusion assessment is much more difficult and has resisted adequate practical solution. Hydrogen tends to be fairly uniformly distributed in a molten aluminum stream, and sampling becomes a minor concern. Particulate, especially aluminum oxide, is randomly distributed, and the larger the sample, the greater the probability that a sample is representative of the melt.

Buxmann *et al* [3] refer to the use of diamond machining as an answer to this need, but give little procedural details. The current paper describes efforts to establish a capability for inspecting large aluminum samples for the presence of structural discontinuities, such as porosity and inclusions. The goal was to develop a technique which would (1) highlight the presence of discontinuities when a large sample was viewed with low magnification [1 to 30X], (2) permit suspicious indications to be studied with the metallurgical microscope to 500X, and (3) avoid the introduction of artifacts. It was highly encouraging that diamond machining would meet these requirements, given actual experience in attempting to supply aluminum to a major copying machine manufacturer. The acceptance test for this application is the diamond machining of extruded aluminum drums. Inferior metal quality is readily apparent in that inclusions are dislodged during machining, which scratch an otherwise mirror-like finish. To prevent artifacts from forming, proper tool selection and vibration-free machining[4] must be achieved.

*Reynolds Metals Company, Metallurgical Research Division, Richmond, Virginia 23261 USA.

PROCEDURE

Early tests with a conventional lathe indicated that cutting, with a minimum of smearing, could be obtained with a "Milo 144" diamond tool.* As seen in Fig. 1, constituent particles were clearly visible, but a "phonograph ring" pattern of machining marks was present. Surface roughness traces indicated that these rings had a highly regular spacing, which tended to indicate that lathe vibration was the cause. Tests were conducted by renting time on a special lathe sold by Bryant Symons in England. These tests were very encouraging in that a mirror-like finish was obtained, except where an inclusion was encountered. A trail of debris was generally found to follow an inclusion, leaving a "comet tail" of dull surface. These appeared to be ideal conditions for locating inclusions.

Unfortunately, such specialized lathes cost about $45,000. It became necessary to build a lathe, embodying the features of the commercial lathe, to demonstrate the utility of the technique for metal quality problems. The lathe which was built had the following features: (1) a heavy base mounted on pneumatic pads, (2) a clutch mechanism such that machining could be done either with the motor running or with the motor off and power supplied by the kinetic energy of a flywheel, (3) a high-precision spindle, accurate to a few millionths of an inch, and (4) a pneumatically driven cross-slide with an adjustable feed rate of 1 to 15 inches per minute.

The lathe which was constructed was suitable only for facing. Figure 2 shows the main elements of the lathe. An electric motor (1) drives a heavy duty "Pope" spindle (2). The sample (3) is bolted to a very large steel flywheel (4), weighing 120 lbs. The flywheel can rotate up to 1500 RPM, thereby allowing cuts to be made using only the kinetic energy of the flywheel. The hydraulic cross-feed (5) is adjustable

Fig. 1. Sample machined with a diamond tool and conventional lathe. Blocky features at the center of the photomicrograph are machining chips. 100X

*Milo Manufacturing, Union, New Jersey.

Fig. 2. Special lathe constructed for diamond machining.

in speed, but for the majority of the work, 12 in. per min was used. The speed and feeds are relatively unimportant compared to the angle between the tool and the work piece. The latter must be adjusted through trial and error, using a small vernier adjustment on the tool post. Adjustment was based upon observation of the amount of constituent smearing when viewing the sample at 200X.

The final cut was made with a surface removal of about 0.0005 in. The tool moved from the center of the sample to the edge with an air stream blowing the chips away from the newly formed surface. A number of lubricants have been employed, including "Spray Tap"* and kerosene. After machining, the samples were cleaned with a vapor degreasing unit. Inspection began with a low power (10 to 30X) binocular microscope in a dark room. Oblique illumination was employed, and indications were checked with a soft brush to insure that they were not dust particles. All true indications were circled and checked with the metallurgical microscope.

Figure 3 illustrates the type of surface achieved with this procedure. The sample is an ingot slice of alloy 2011 at 100X. Swirly oxides are seen near the surface. Constituent particles are also readily visible. Figure 4 shows how porosity and constituent particles can readily be seen in the machined samples. To increase the chances of encountering defects, the current procedure is to inspect the sample ultrasonically to determine the depth of the defects. Once uncovered by machining, the defect can be removed and analyzed by x-ray diffraction or scanning electron microscopy. Oxides, borides, porosity, and salt inclusions have been located and identified. While not the ultimate answer for the problem of assessing cast metal quality, this technique has been quite useful in locating porosity and inclusions in aluminum castings.

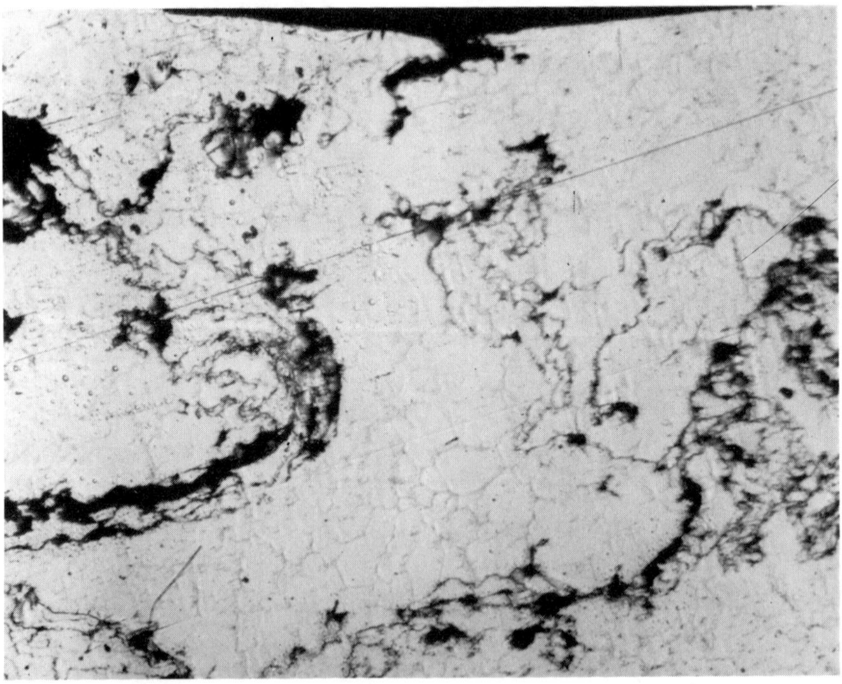

Fig. 3. Swirly oxides visible near the surface of a 2011 alloy ingot. 100X.

*Chem Tech Inc., Wilmington, Delaware.

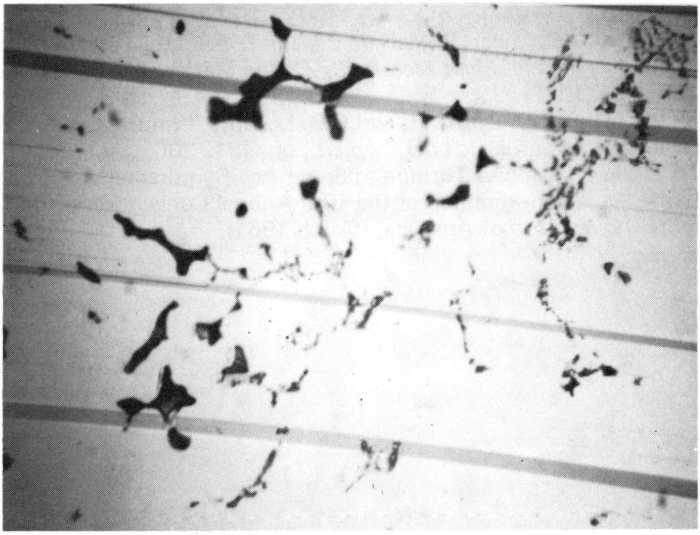

(a)

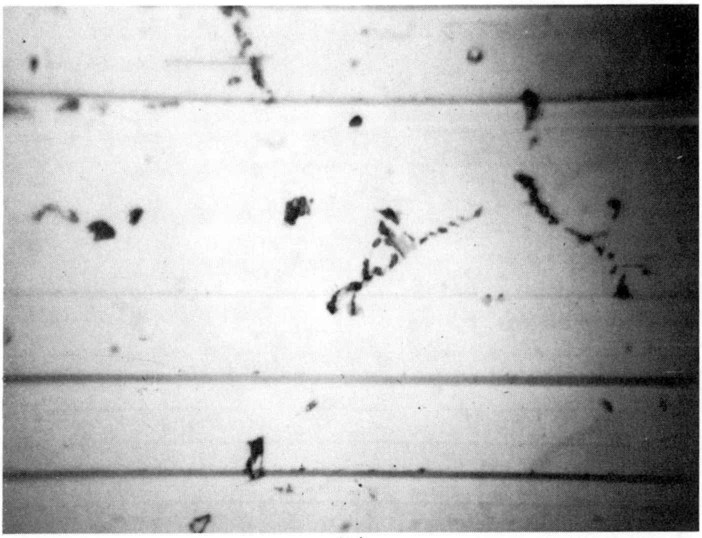

(b)

Fig. 4. Constituent particles and patches of porosity visible in diamond machined samples of 5083. Samples directly from lathe. (a) 320X, (b) 450X.

REFERENCES

1. J.C. Miller, S.A. Levy, and P. McNamara, "Recent Developments in the Evaluation of Molten Metal Quality," *Light Metals 1976*, The Metallurgical Society of AIME, pp. 229–246 (1976).
2. S.A. Levy, J.C. Miller, P. McNamara, and D.A. Feitig, "Molten Metal Quality—— Particulate Tests," *Light Metals 1977*, The Metallurgical Society of AIME, pp. 149–170 (1977).
3. K. Buxmann, J.D. Bornand, and G.B. Leconte, "Impact of Purification Methods on Inclusions and Melt Loss," *op.cit.*, pp. 191–206.
4. J.V. Milo, "Diamond Turning to Space Age Requirements with Specially Designed Lathe," a paper presented at the 19th Annual Convention of the Industrial Diamond Association of America, (March 1964).

Stevens, Vander Voort, McCall, eds. Microstructural Science, Volume 8

METALLOGRAPHY FOR GLASS–METAL SEAL QUALITY DETERMINATION

L. Zakraysek *

ABSTRACT

A new approach is suggested for the testing and inspection of glass-to-metal seals in high reliability electronic components. This approach is based upon the use of a metallographic section after sealing and an evaluation of the metal and glass microstructures in the vicinity of the glass-metal interface. The procedure is intended as a control mechanism for the pre-oxidation process used in seal-making. The objective of the metallographic analysis is to determine whether or not predetermined criteria have been met and maintained through the sealing process.

INTRODUCTION

Silicon integrated circuits presently require the use of hermetic packaging with glass-to-metal seals[1] for reliable operation. The packages must be capable of maintaining a clean and dry internal environment over a long period of time. Packages such as those shown in Fig. 1 are most often tested by helium leak detection to 10^{-8} torr or by dye penetrant through a 4–12 h exposure to the dye. Recent work[2] has shown that a package may meet the requirements of this kind of testing but that the same package will not remain hermetically tight after torque loading or after some fabrication processes. Sealed subassemblies will generally pass the least severe (Method 1011.2 Cond. A) thermal shock requirement of MIL–STD–883, but the same lots may exhibit a 6%, or higher, failure rate as a result of manufacturing stresses imposed by cover welding, de-lidding and thermal bakes. These failure rates can be reduced to under 1% through the use of metallographic acceptance criteria for defining package quality.

EXPERIMENTAL PROCEDURE

Standard procedures are used for the metallographic preparation of glass-to-metal seals. A sealed component can be sectioned either transverse or longitudinally. At the seal interface the glass provides a hard surface which helps to prevent rounding or other artifacts.

The inspection criteria of most interest include effects on the glass, the glass-metal interface and the metal members.

After sealing and sectioning, the glass can be examined for the volume fraction of voids due to gas evolution, and for the size and distribution of gas-formed voids. The glass-metal interface can be inspected for oxide precipitates in the glass, foreign material at the interface, and residual intergranular oxide in the metal at the interface. The metal member outside the seal area can be evaluated for the degree of intergranular degradation due to prior processing.

*General Electric Company, Syracuse, New York USA.

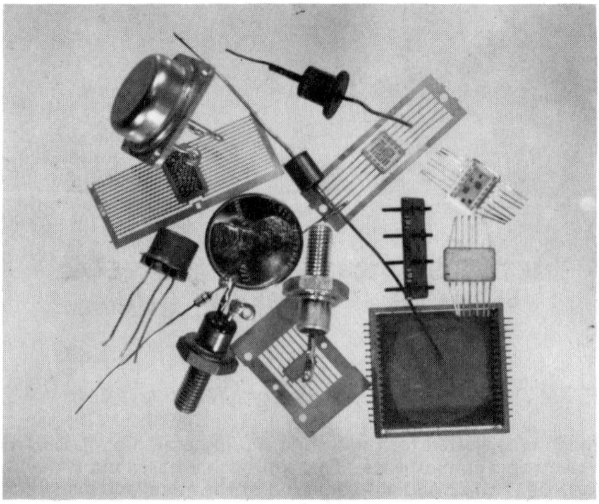

Fig. 1. Typical circuit packages

The metals involved in hard glass seal making are high-nickel alloys which normally require active etchants. In the case of glass-metal seals, the surface of the metals has been thermally affected by pre-oxidation and other heat treatments. Therefore, metallographic etching for the purpose of accentuating the glass-metal interface area is relatively simple. One of the most effective etchants is a 5% nitric acid in alcohol (nital) mixture, although other, more severe etchants can be used.

Packages were fabricated by the use of standard glass-sealing procedures. Representative samples were tested for hermetic integrity by helium and by dye penetrant leak testing methods. Metallographic sections were taken of hermetic and nonhermetic samples, and evaluation criteria were developed for the metallographic approach. From these comparisons, a metallographic test procedure was prepared, and the new test method was applied to the procurement of commercially available packages.

RESULTS AND DISCUSSION

Matched glass-seal making depends on the use of three thermal treatments for the metal member: decarburizing, pre-oxidizing and sealing. The present work has dealt specifically with pre-oxidation, where an oxide film, prior to sealing, promotes glass wetting and provides for the development of a chemical-mechanical bond at the glass-metal interface.

After the sealing process, the metal is fully annealed resulting in a coarse, equiaxed grain in the bulk material. Pre-oxidation for sealing results in the formation of an oxide scale and an intergranular oxide on all metal surfaces. When glass is bonded to a pre-oxidized metal surface, several characteristic microstructural features are consistently formed, and are visible in the metallographic section. Therefore, this test method can be used to determine the quality of a matched glass-to-metal seal on the basis of the following distinguishing features: (a) metal microstructure after pre-oxidation, (b) intergranular oxide layer formed during pre-oxidation, (c) layer of oxide scale formed during pre-oxidation, (d) metal microstructure after sealing, (e) residual intergranular oxide after sealing, (f) oxide precipitate from dissolved scale, and (g) gassiness in the glass. In the sealed sample, the residual intergranular oxide (e) is measured as an average depth of penetration from the glass-metal interface into the metal member. The measurement records the depth of the oxidation-affected zone after sealing is completed. This value is proportional to the total oxide prior to sealing, and residual quantities in a controlled range result in the highest seal quality. For practical purposes, an evaluation for gassiness, precipitated oxide, and residual intergranular oxide are sufficient for the quality determination.

Glass

During sealing, gasses are evolved in differing amounts as a function of glass composition, makeup, cleanliness and so forth. The resultant bubble quantity, size and distribution are readily discernible in the metallographic section. Shown in Fig. 2 is a sealed sample that exhibits the major features in regard to the formation of gas bubbles in the glass. The total volume is determined by a technique described by Rostoken and Dvorak[3], and the largest gas void is determined by direct measurement, as is the alignment of bubbles at the glass-metal interface. During this study, the acceptable criteria for these characteristics were determined to be: (a) 33% max. gas volume, (b) 0.75 mm max. bubble size, and (c) 20% of the seal length for stringers.

Interface

When glass is flowed on a pre-oxidized metal surface, the glass dissolves the oxide scale and bonds to the metal under it. As a result of wetting and bonding action, a characteristic product can be detected in a metallographic section, as shown in Fig. 3. The oxide scale appears as a precipitate in the glass, and the metal member contains the remnants of the total oxide that was present prior to sealing. The extent of this residual intergranular oxide is proportional to the degree of prior oxidation. This relationship can be used for the specification of a required oxide quantity. During the present study, a residual oxide in the range of 2.0 – 6.5 μm was determined to be acceptable. Oxide precipitates in the glass, as seen in this photograph, indicate the presence of excess oxide scale prior to sealing, which is an unacceptable condition.

CONCLUSIONS

Standard metallographic procedures are effective for the evaluation of glass-to-metal sealed components. The formation of a glass-metal seal results in several characteristic features for which acceptance criteria have been developed.

Inspection for residual intergranular oxide, oxide precipitates, and gassiness is suggested as a viable procedure for process control to assure hermetic integrity in high reliability electronic components.

It was found that commercial suppliers can produce assemblies that satisfy microstructural criteria and that the use of rigid metallographic requirements can improve component quality.

ACKNOWLEDGEMENTS

The author is grateful to John McCormick and the Rome Air Development Center for their support of this work, also for the assistance of George Trojanowski, Donald Blackwood and Mary Guerin.

REFERENCES

1. M.P. Borom, *The Glass Industry,* pp. 12–27 (March 1978).
2. J. McCormick and L. Zakraysek, "Alternate Method for the Evaluation of Fused Glass-to-Metal Seals", Final Technical Report RADC–TR–79–98, Rome Air Development Center, Griffiss AFB, NY 13441, (May 1979).
3. W. Rostoken and J.R. Dvorak, *Interpretations of Metallographic Structures,* p. 197, Academic Press, NY, (1965).

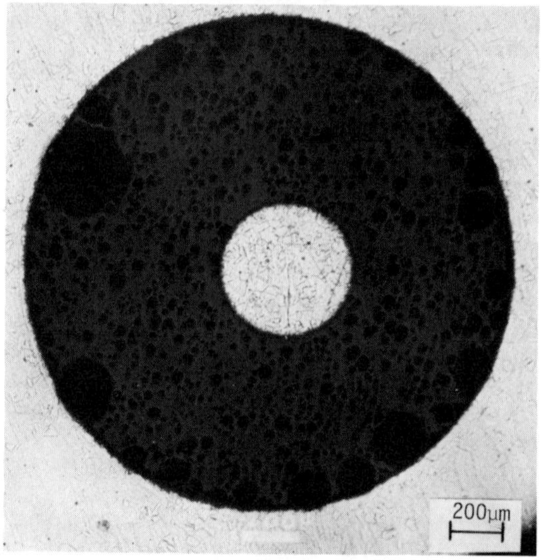

Fig. 2. Gassiness

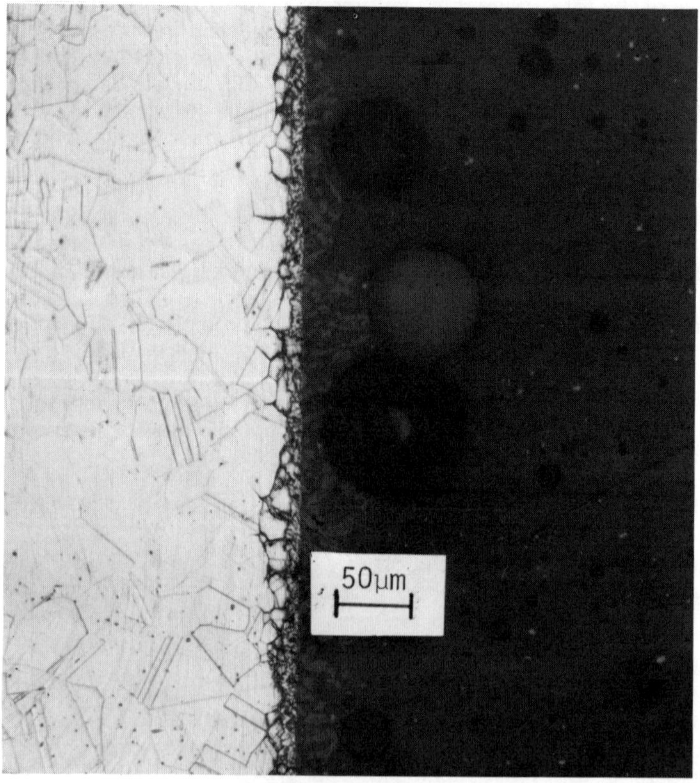

Fig. 3. Glass-metal interface

Stevens, Vander Voort, McCall, eds. Microstructural Science, Volume 8

STRUCTURAL CHARACTERIZATION OF THE SURFACE DAMAGE OF STELLITE–6B ALLOY DUE TO EROSION BY SOLID PARTICLES

Gordon A. Sargent*, M.K. Keshavan* and H. Conrad*

ABSTRACT

The topographic features and subsurface damage produced by the erosion of an annealed Stellite–6B alloy have been studied by the techniques of scanning electron microscopy (SEM), transmission electron microscopy (TEM), and X-ray diffraction (XRD). Specimens of the alloy were eroded by angular alumina particles of 240 and 400 grit (mean size 30 and 10 μm respectively) at velocities of 28 and 68 m/s for several angles of impingement between 10° and 90°. The SEM analysis indicated that the mechanism of erosion at 10° was micro-machining and that at 90° was battering. The depth of the damaged surface layer was determined for a x-ray diffraction line broadening analysis technique and found to be of the order 10–16 μm, increasing with angle of impingement. Dislocation cell sizes (~70 Å) and densities ($10^{12} cm^{-2}$) from the line broadening analysis were found to be in reasonable accord with those determined by TEM.

INTRODUCTION

The erosion of materials has recently been recognized to be a serious technological problem, especially in the transportation or processing of coal in particulate form. The phenomenon of erosion has been the subject of many investigations over the past ten years[1–10]. The emphasis of earlier work has been primarily to develop empirical relationships by which the erosion behavior of a specific material could be predicted as a function of particle velocity and angle of impingement on the surface. The importance of a material's structural condition and the relationship between plastic flow processes and erosion have been pointed out in several studies[4–6].

To date very little work has been done to characterize the structural damage produced at and below the surface of a material subjected to erosion. Ives and Ruff [14–15] and Wright and Edington[11–12] have studied the topographic and subsurface structural features associated with erosion damage using the techniques of SEM and TEM. In the present paper it will be shown that important structural information regarding the damage produced by erosion can be obtained from x-ray line broadening analysis. Data such as depth of damage, dislocation densities and cell sizes determined by this technique are compared with similar data obtained by more direct microstructure studies using TEM.

EXPERIMENTAL PROCEDURE

The material used in this study was annealed Stellite–6B (obtained from the Stellite Division of Cabot Corporation). The chemical composition of the material is given in Table 1.

*Department of Metallurgical Engineering and Materials Science, University of Kentucky, Lexington, Kentucky 40506 USA and associated with the Institute for Mining and Minerals Research, Kentucky Center for Energy Research Laboratory, Lexington, Kentucky 40583 USA.

Table 1
Nominal chemical composition of Stellite–6B (wt.%)

	Cr	Fe	Mo	Ni	Mn
Stellite–6B	28.0–38.0	3.0	1.5	3.0	2.0

	Co	Si	C	W
Stellite–6B	Bal.	2.0	0.9–1.4	3.5–5.5

Specimens for the erosion tests were in the form of 2.5 cm squares, which were cleaned in acetone in an ultrasonic cleaner prior to testing. The loss of material due to erosion was determined by weighing the specimens both prior to and after a test. The erosion was carried in a sand blaster type of erosion tester. Angular alumina particles of 240 and 400 grit (mean size 30 and 10μm respectively) were used for the experiments. The particles were accelerated to velocities of 28 and 68 m/s by means of pressurized air. The particle velocity was measured by the two-disc method of Ruff and Ives[16].

The SEM and x-ray diffraction studies were carried out on the 2.5 cm square specimens which were used for the weight loss measurements. For the TEM studies thin slices were cut from the Stellite–6B sheets using a diamond wafering machine. These specimens were then eroded under the conditions described above and small discs 3 mm in diameter were punched from the eroded slices to fit the TEM specimen holder. The eroded side of the disc was protected by lacquer and the specimen was thinned from the opposite side by jet electropolishing.

RESULTS AND DISCUSSION

Topography of the Eroded Surface

Figure 1 shows the effect of angle of impingement on the erosion of annealed Stellite–6B at a velocity of 68 m/s using alumina particles of 240 and 400 grit. To be noted is that increasing the particle size increases the amount of erosion at all angles. SEM studies of the surface (Fig. 2) show that the mechanism of erosion at low angles of impingement is different from that at high angles. It can be seen from Fig. 2a that at small angles (10°) the particles produce long tracks on the surface, which is characteristic of a cutting or micro-machining mechanism. At 90°, Fig. 2b, the surface damage is more characteristic of that resulting from continuous battering, with material being extruded up and around each individual particle. It is expected that the extruded material is subsequently removed from the surface when struck by another impinging particle.

Figure 3 shows an energy dispersive x-ray (EDX) analysis of the Stellite–6B surface before and after erosion with 240 grit alumina particles. It can be seen that after erosion the surface contains aluminum, which presumably is from the alumina particles either sticking to the surface or deposited as a thin layer onto the surface.

Subsurface Damage

X-ray diffraction line broadening measurements and TEM were used as techniques to characterize the subsurface damage. Figure 4 shows the (111) and (222) diffraction profiles of Stellite–6B prior to, and after, erosion at 90° with 240 grit alumina particles at a velocity of 68 m/s. It can be seen that after erosion there is a

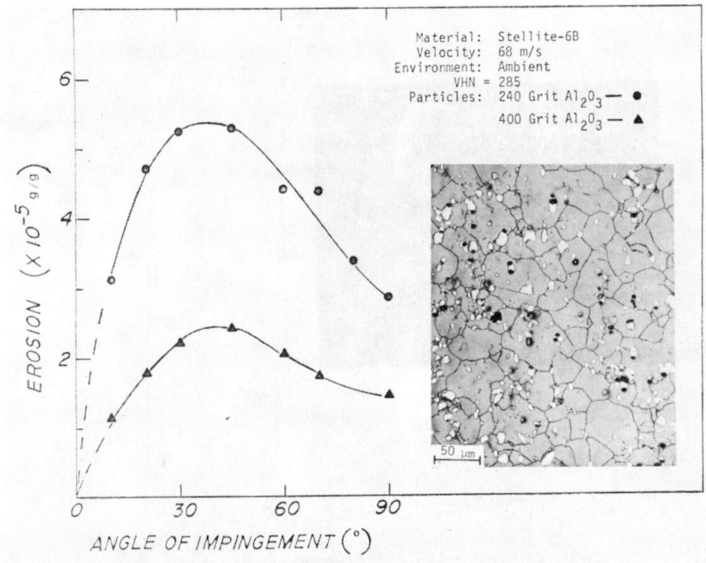

Fig. 1. Effect of angle of impingement on the erosion of annealed Stellite–6B (microstructure shown in insert) using 240 and 400 grit alumina particles.

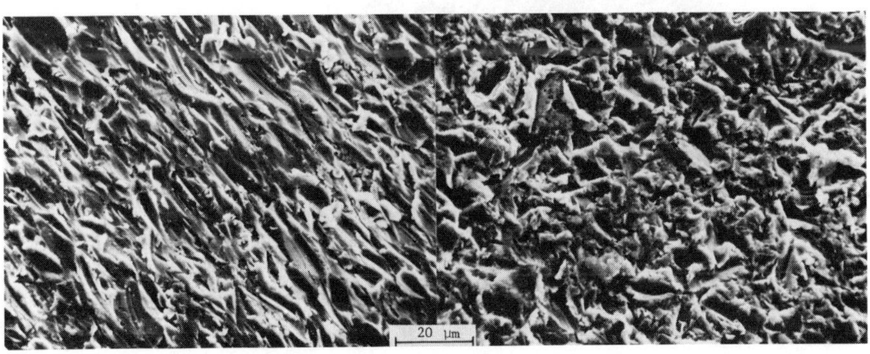

Fig. 2. Scanning electron micrographs of eroded Stellite–6B surfaces using 240 grit angular alumina particles at 68 m/s for (a) 10° and (b) 90° angle of impingements.

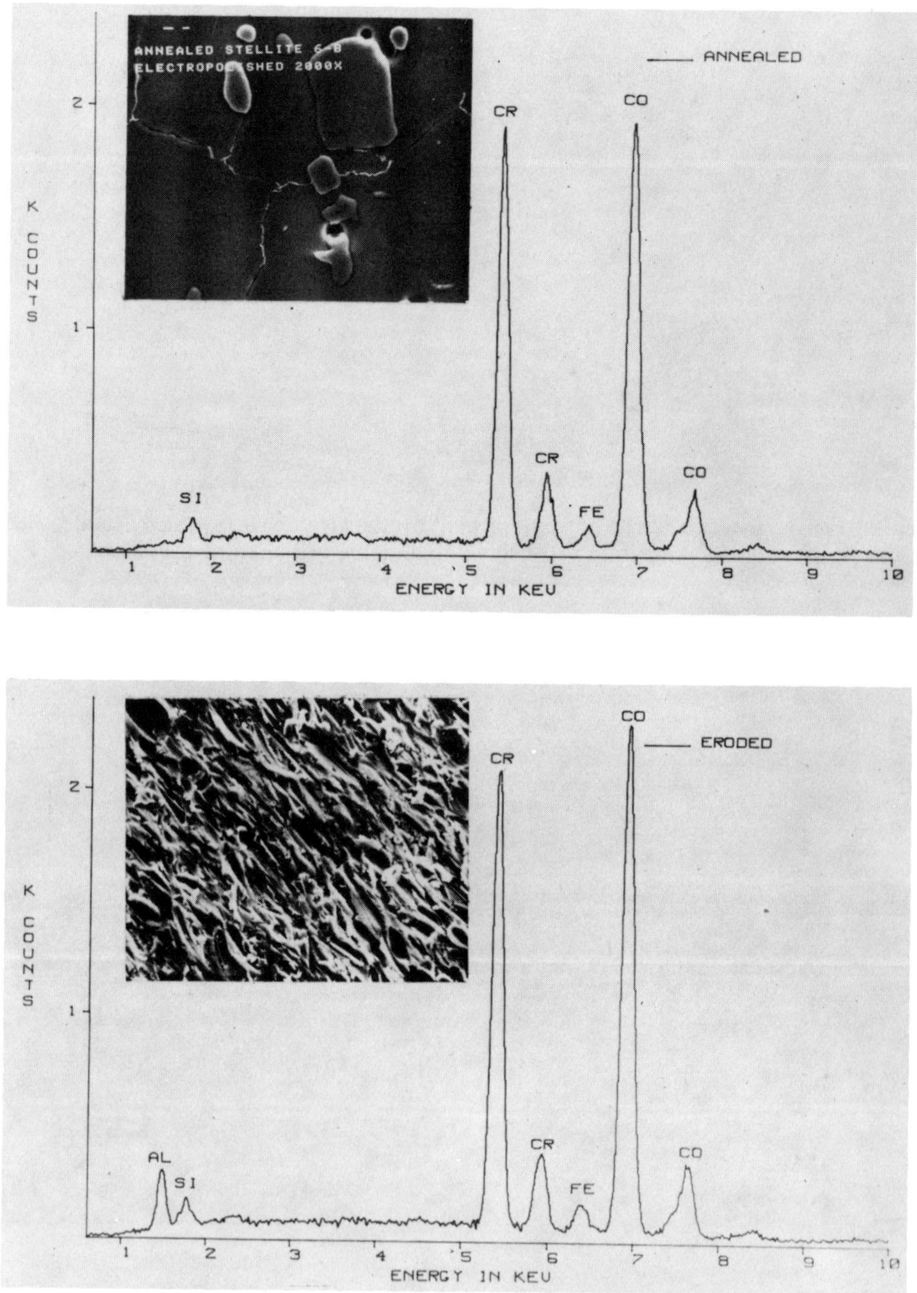

Fig. 3. EDX analysis of the initial annealed surface and following erosion with 240 grit alumina particles at 68 m/s and 10° angle of impingement.

large decrease in intensity and increase in the breadth of diffraction profiles. From an analysis of the diffraction profiles using either the Warren—Averbach[18] (double-order) or Ganesan-DeAngelis[13] (single-order) methods it is possible to determine the strain in the material due to the deformation caused by the erosion and to estimate the dislocation cell size and density.

To ensure that the line broadening shown in Fig. 4 was totally due to strain and not due to increased surface roughness caused by the erosion, both the annealed and eroded samples were given a stress-relieving treatment at 700°C for 20 minutes in vacuum. It can be seen in Fig. 5 that the diffraction profiles after the above treatment are identical; thus, surface roughness can be disregarded as contributing to the line broadening.

Figures 6–8 show the line broadening obtained after erosion with particles of different sizes, at different velocities and for different angles of impact. The peak intensity of the profiles decreases and the half-breath increases as the particle size, velocity and the angle of impingement increases.

Depth of Damage

To determine the depth of the damaged layer due to particle impingement, the diffraction profiles of the annealed and eroded samples were first obtained. The eroded surface was then electropolished to various depths and the corresponding diffraction profile was again measured. Figure 9 shows the diffraction profile obtained at a depth of 4μm below the surface of a specimen which had been eroded at an angle of 10° with 240 grit alumina particles at a velocity of 68 m/s. Figure 10 shows the variation below the surface as a function of erosion conditions of the half-breadth of the diffraction profile β_D corrected for instrumental broadening given by

$$\beta_D = (\beta^2 \text{ (hkl) eroded } - \beta^2 \text{ (hkl) annealed})^{1/2}$$

It can be seen that the damaged layer extends to about 16μm below the surface for specimens eroded by 240 grit alumina particles at 68 m/s and 90° angle of impingement. There is little change in the maximum depth of the damaged layer upon reducing the velocity to 28 m/s. However, the maximum depth of the damaged layer decreases with angle of impingement. Similar results were noted by Ives and Ruff [15] in Cu.

Severity of the Surface Damage

The severity of damage can be studied using x-ray diffraction line broadening and TEM analyses. It has been shown by the recent TEM work of Ives and Ruff[15] and Wright and Edington[12] that damage due to multiparticle impacts in copper and Stellite respectively results in the formation of a cellular type of dislocation structure, which is typical of metals which have been heavily plastically deformed. It is very difficult to measure the dislocation density from TEM photographs. Dislocation cell size and density can be readily determined from x-ray line broadening analysis.

There are several methods available to characterize the crystallite size or dislocation cell by x-ray diffraction[13], but the simplest method is the determination of the effective particle dimensions $D_{\text{half-breadth}}$ from the half-breadth of the diffraction profile by the Scherrer formula

$$D_{\text{half-breadth}} = \frac{\kappa\gamma}{\beta_D \cos\theta} \qquad (1)$$

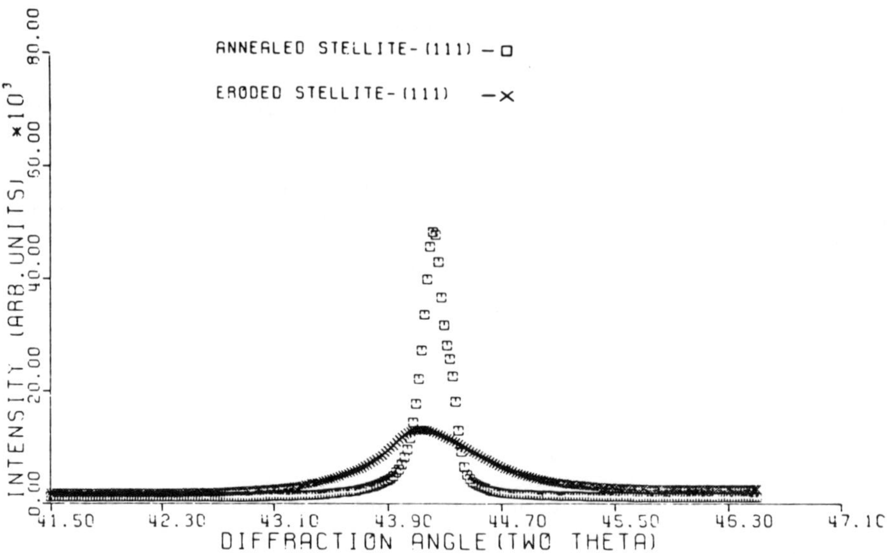

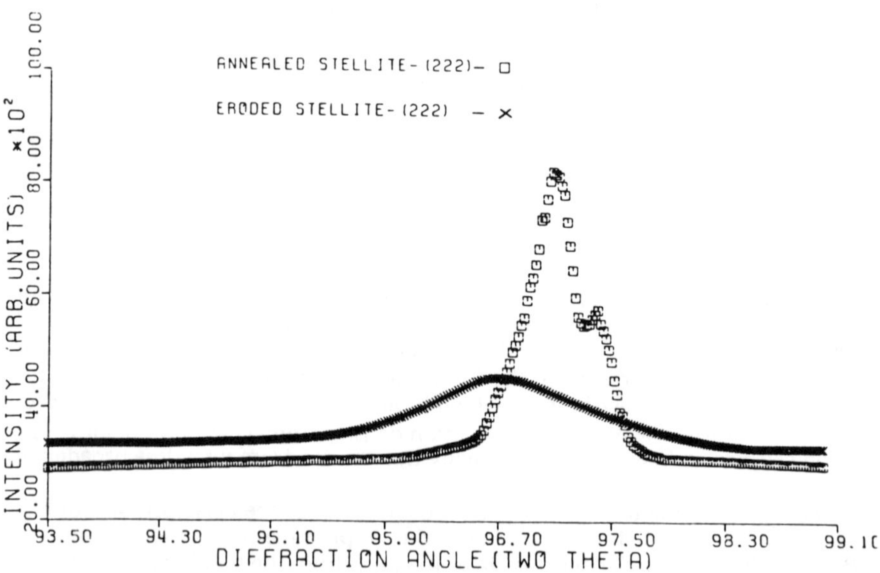

Fig. 4. (111) and (222) diffraction profiles of Stellite–6B before and after erosion using 240 grit alumina particles at 68 m/s and at an angle of impingement of 90°.

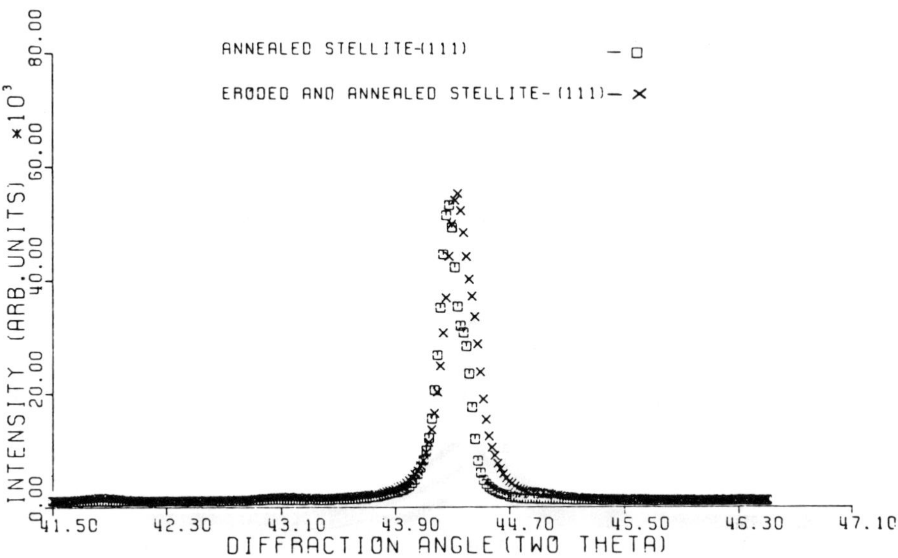

Fig. 5. (111) diffraction profile of the initial annealed surface and an eroded plus annealed surface.

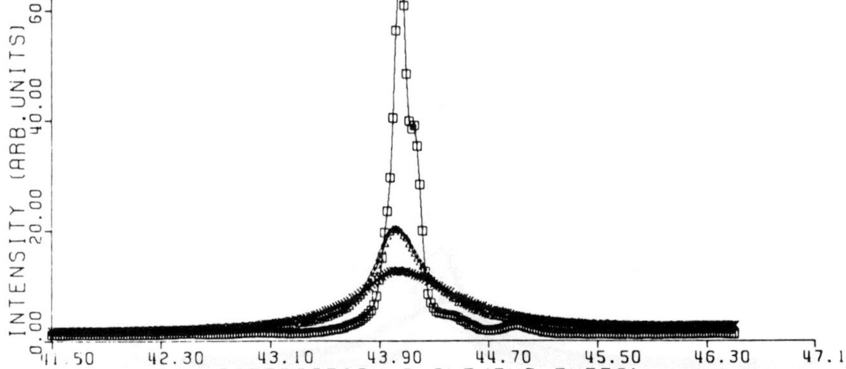

Fig. 6. (111) diffraction profiles of the initial annealed surface and following erosion with 240 and 400 grit alumina particles at 68 m/s and at an angle of impingement of 90°.

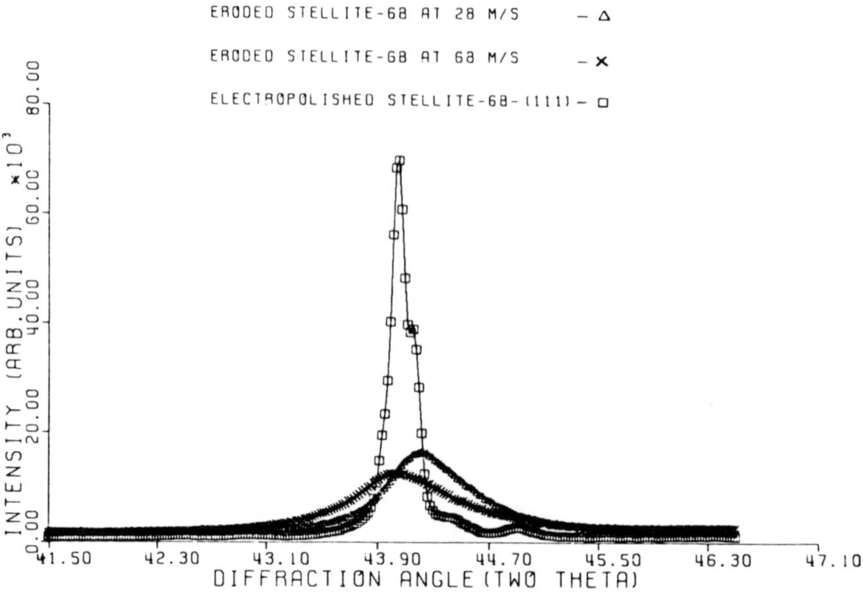

Fig. 7. Effect of the velocity of 240 grit alumina particles on (111) diffraction profiles when impacted at an angle of impingement of 90°.

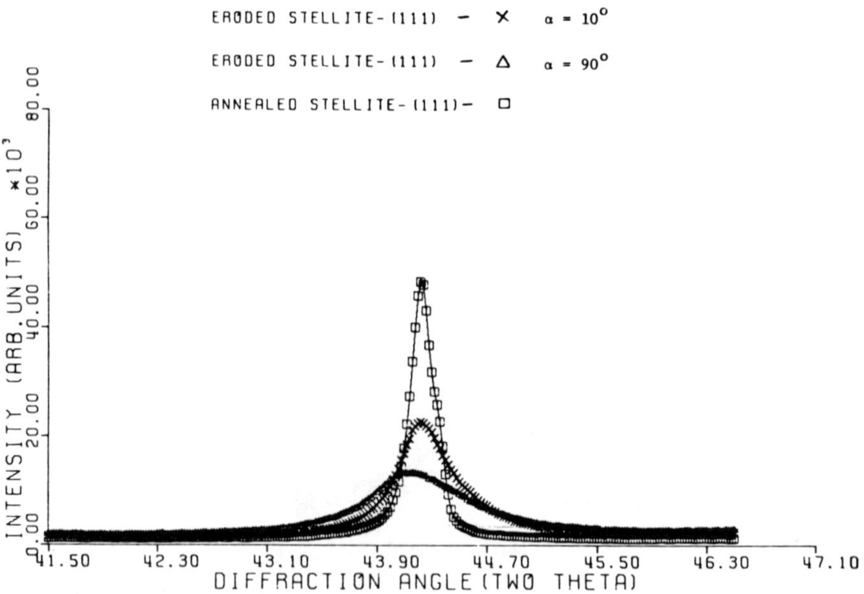

Fig. 8. (111) diffraction profiles of the initial annealed surface and surfaces following erosion with 240 grit alumina particles at 68 m/s and angles of impingement of 10° and 90°.

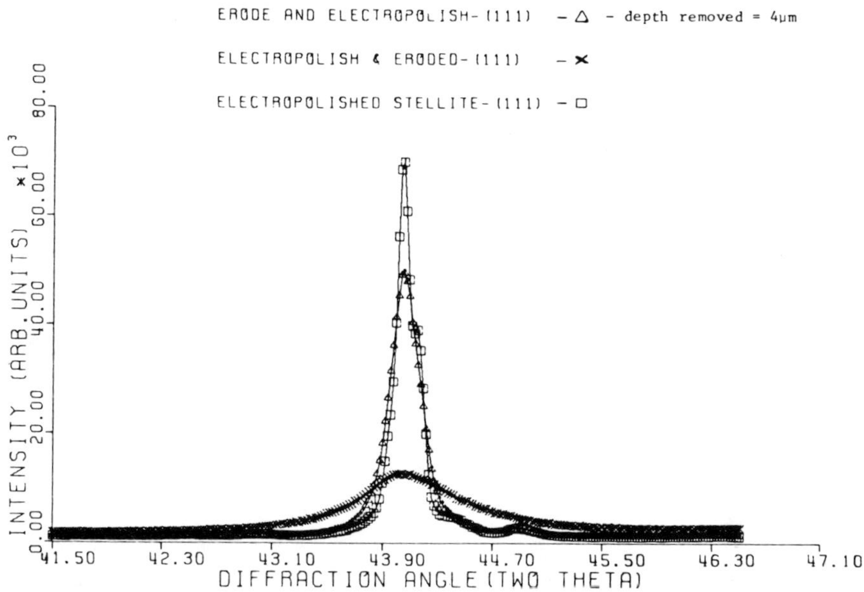

Fig. 9. (111) diffraction profile of eroded surface of Stellite–6B electropolished to remove 4μm from the surface eroded at 68 m/s using 240 grit alumina particles at an angle of impingement of 10°.

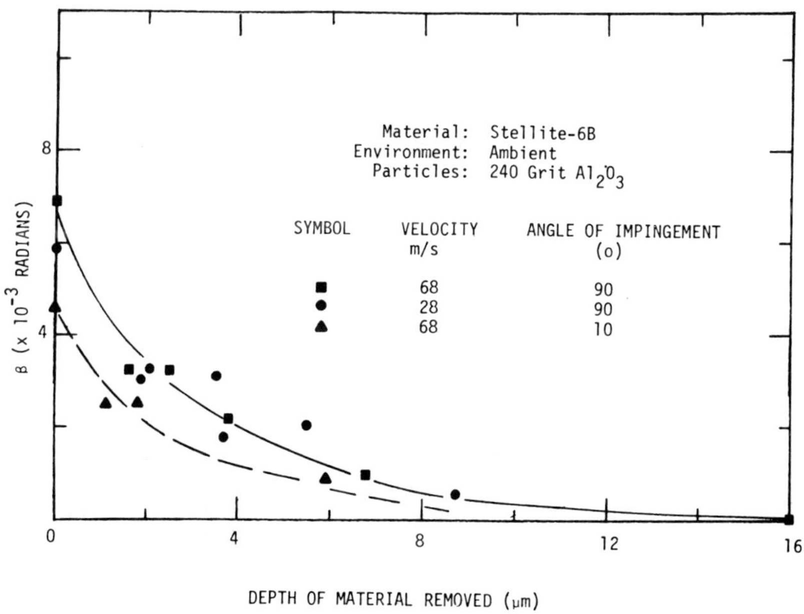

Fig. 10. Variation of the half-breadth (corrected for instrumental broadening) of the diffraction profiles as a function of depth of material removed.

where β_D is the half-breadth of the diffraction profile corrected for broadening, λ is the wavelength, θ is the diffraction angle and κ is a constant $\cong 0.9$. The dislocation cell size l_c calculated from half-breadth β_D may also be calculated from single profile analysis[13] or from double analysis[18] which are very accurate. Details of the methods for these are given elsewhere[17]. Values of the effective particle/dislocation cell sizes calculated from the line broadening analysis are listed in Table 2.

Table 2

Effective dislocation cell size at the surface of Stellite–6B eroded with 240 grit alumina particles at 68 m/s and 90° angle of impingement

	Single Profile Analysis (111)	Single Profile Analysis (222)	Double Order Analysis	Transmission Electron Micrographs
Dislocation Cell Size (Å)	72	70	73	100

It can be seen that there is a good correlation between the values obtained by the different x-ray analysis techniques. Figure 11 shows a TEM of the structure of Stellite–6B prior to, and after erosion. Cell sizes measured from such micrographs of eroded specimens were in the range 50–250 Å with a mean value of 100 Å, which is in reasonable agreement with the values obtained from the x-ray diffraction studies. The dislocation density values can be estimated using the relationship

$$\rho \cong \frac{1}{(\text{Dislocation Cell Size})^2} \tag{2}$$

This yields $\rho \cong 10^{12}$ cm^{-2}. In view of this high dislocation density, one can understand why particle size and velocity do not have a large effect on the degree of damage (see Figs. 6–9).

SUMMARY AND CONCLUSIONS

It is shown that x-ray line broadening analysis can be used to characterize the subsurface damage caused by multi-particle erosion of metals. The maximum depth of the damage layer was calculated from eroded surfaces of annealed Stellite–6B from the profile half-width. The total depth of damage was found to be relatively independent of particle velocity in the range of 28–68 m/s. However, it decreased with decrease in the angle of impingement. A depth of 16μm was obtained for erosion by 240 grit Al_2O_3 particles at 90° angle of impingement. The dislocation cell size at the surface of eroded specimen was calculated from x-ray line broadening analyses found to be in reasonable agreement with TEM measurement. An effective dislocation cell size of 70 Å and a very high density of dislocations (10^{12} cm^{-2}) was obtained. It is concluded that the dislocation density does not vary appreciably with velocity and particle size in the range considered here.

ACKNOWLEDGEMENT

The authors are grateful to Stellite Division, Cabot Corporation for supplying the Stellite samples and to Dr. P. Ganesan and Dr. Al Bland for TEM and x-ray work.

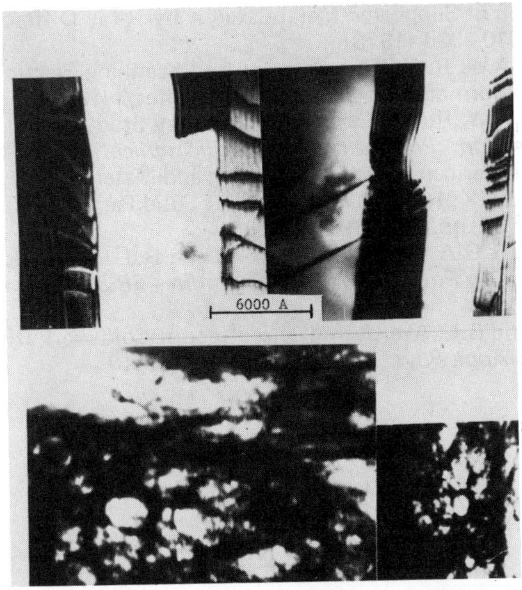

Fig. 11. Transmission electron micrographs of Stellite–6B before and after erosion.

REFERENCES

1. J.G.A. Bitter, "A Study of Erosion Phenomenon; Parts I and II," *Wear* 6, pp. 5–21 and pp. 169–190 (1963).
2. I. Finne, J. Wolak and Y. Kabil, "Erosion of Metals by Solid Particles," *Journal of Metals.* 2(3), pp. 682–700 (1967).
3. W. Head and M.E. Harr, "The Development of a Model to Predict the Erosion of Materials by Natural Contaminants," *Wear* 15, pp. 1–46 (1970).
4. W.H. Jennings, W.J. Head and C.R. Manning, Jr., "A Mechanistic Model for the Prediction of Ductile Erosion," *Wear* 40, pp. 93–112 (1976).
5. J.H. Neilson and A. Gilchrist, "Erosion by a Stream of Solid Particles," *Wear* 11, pp. 111–122 (1968).
6. G.L. Sheldon, "Similarities and Differences in the Erosion Behavior of Materials," *J. Basic Eng. (Trans ASME Ser. D)*, 92, pp. 619 (1970).
7. C.E. Smeltzer, M.E. Gulden and W.A. Compton, "Mechanisms of Metal Removal by Impacting Dust Particles," *J. Basic Eng. (Trans ASME Ser. D)*, 92, pp. 639–654 (1970).
8. G.L. Sheldon and A. Kanhere, "An Investigation of Impingement Erosion Using Single Particles," *Wear* 21, pp. 195–209 (1972).
9. G.P. Tilly, "A Two Stage Mechanism of Ductile Erosion," *Wear* 23, pp. 87–96 (1963).
10. G.P. Tilly and W. Sage, "The Interaction of Particle and Material Behavior in Erosion Processes," *Wear* 16, pp. 447–465 (1970).
11. J.W. Edington and I.G. Wright, "Study of Particle Erosion Damage in Haynes Stellite–6B –– Scanning Electron Microscopy of Eroded Surfaces," *Wear* 48, pp. 131–144 (1978).
12. J.W. Edington and I.G. Wright, "Study of Particle Erosion Damage in Haynes Stellite–6B –– Transmission Electron Microscopy," *Wear* 48, pp. 145–155 (1978).

13. P. Ganesan, H.K. Kuo, A. Saavedra and R.J. DeAngelis, "Particle Size Distribution Function of Supported Metal Catalyst by X-ray Diffraction," *J. of Catalysis,* 52, pp. 310–320 (1978).
14. L.K. Ives and A.W. Ruff, "Transmission and Scanning Electron Microscopy Studies of Deformation at Erosion Impact Sites," *Wear* 46, pp. 149–162 (1978).
15. L.K. Ives and A.W. Ruff, "Electron Microscopy Study of Erosion Damage in Copper," *Erosion: Prevention and Useful Applications,* ASTM STP 664, W.F. Adler, Ed., American Society for Testing and Materials, pp. 5–35 (1979).
16. A.W. Ruff and L.K. Ives, "Measurement of Solid Particle Velocity in Erosive Wear," *Wear* 35, pp. 195–199 (1975).
17. M.K. Keshavan, G.A. Sargent, P. Ganesan and R.J. DeAngelis, *Dislocation Cell Size Distribution Function of Eroded Stellite–6B by X-ray Diffraction,* to be published.
18. B.E. Warren and B.L. Averbach, "The Effect of Cold-work Distortion on X-ray Patterns," *J. Appl. Phys.* 21, pp. 595–599 (1950).

INTERNATIONAL METALLOGRAPHIC EXHIBIT

1979 WINNERS

Jacquet—Lucas Award
(Grand Prize)

M.J. Bridges and S.J. Dekanich, Union Carbide Corporation-Nuclear, Oak Ridge Gaseous Diffusion Plant, Oak Ridge, TN. USA

Corrosion Control in Aluminum-Silicon Brazed Joints Through Microstructural Modification

Optical Microscopy—Iron, Steel, Stainless Steel, Nickel, and Nickel Alloys (Class 1)

FIRST IN CLASS: C.W. Houck and K.F. Russell, Oak Ridge National Laboratory, Oak Ridge, TN, and R.T. King, Sprague Electric (formerly of ORNL) *Metallographic Examination of Type 310 Stainless Steel Air Distributor Plate from an Atmospheric Fluidized Bed Combustor (AFBC)*

SECOND IN CLASS: V.K. Sikka, S.A. David, C.W. Houck, and E.H. Lee, Oak Ridge National Laboratory, Oak Ridge, TN, *Ingot Grain Structure Causes Significant Variation in Creep-Rupture Properties of Cast Type 316 Stainless Steel*

THIRD IN CLASS: H. Chang, A. Carefoot, D. Buczek, S. Hahn, S. Sastri, B. Alexander, Gillette Company, Boston, MA, *A Comparison of Cosmic Metallurgy and Laboratory Metallurgy in Fe-Ni-P Alloys*

HONORABLE MENTION: Isolde Graf, Fhg Institute Fur Angewandte Materialforschung (IFAM), Bremen, West Germany, *Hard Solder Joint of NiCrSi Alloy*

HONORABLE MENTION: S.J. Dekanich, M.J. Bridges, Union Carbide Corporation-Nuclear, Oak Ridge Gaseous Diffusion Plant, Oak Ridge, TN, *Analysis of a Fractured Steering Knuckle Bolt*

HONORABLE MENTION: Capt. Jack M. Hyzak, Air Force Materials Laboratory, Wright-Patterson AFB, Dayton, OH, and R.D. Brodecki and R.J. Bacon, Systems Research Laboratories, Dayton, OH, *Intense Planar Slip Deformation—Stage 1 Cracking*

Optical Microscopy—Metals and Alloys Not Listed in Class 1 (Class 2)

FIRST IN CLASS: T.H. Sanders, Jr., Georgia Institute of Technology, Atlanta, GA, and R.H. Borland and R.H. Stevens, Aluminum Company of America, Alcoa Center, PA, *The Effect of Forging Temperature and Alloy Purity of the Microstructure of Al-5 wt% Mo*

SECOND IN CLASS: J.B. Bergin, S.L. DiGiallonardo, D.R. Lesuer, J.L. Robbins, Lawrence Livermore Laboratory, Livermore, CA, *Superplasticity of Delta-Phase Plutonium*

THIRD IN CLASS: L. Owen, J.C. Rigler, and W.J. Sabo, Gould Inc. Gould Laboratories, Cleveland, OH, *Electrolytic Copper Foil*

HONORABLE MENTION: W.C. Coons, Lockheed Missiles and Space Company, Inc. San Jose, CA and A. Heynen, Lockheed Missiles and Space Company, Inc. Palo Alto, CA, *Arc-Cast Wrought TZM Molybdenum*

HONORABLE MENTION: A.M. Garde, R.H. Lee, and W.K. Soppet, Argonne National Laboratory, Argonne, IL, *Effect of Cooling Rate on the Hydrogen Embrittlement of Transformed β-Phase Zircaloy*

HONORABLE MENTION: Y.R. Mahajan, Air Force Materials Laboratory, Wright-Patterson AFB, Dayton, OH, *Fatigue Crack Growth in Two Phase Alpha-Beta Titanium Alloy*

Optical Microscopy — Petrographic, Ceramographic, and Cermet Materials (Class 3)

FIRST IN CLASS: J.D. Holder, W.H. Farmer and G.W. Clark, Oak Ridge National Laboratory, Oak Ridge, TN, *Analytical Metallography Reveals Clustering in Cr2O3-Mo-Re Melts*

HONORABLE MENTION: C.D. Greskovich, C.R. Rodd, A.S. Holik, and J. Grande, General Electric Company, Schenectady, NY, *High Performance Sintered Silicon Nitride*

Electron Microscopy — Transmission (Class 4)

FIRST IN CLASS: D.N. Braski, Oak Ridge National Laboratory, Oak Ridge, TN, and H. Schroeder and H. Ullmaier, Kernforschungsanlage—Julich West Germany, *Helium Embrittlement of Austenitic Stainless Steel*

SECOND IN CLASS: D.N. Braski, R.W. Carpenter, and C.T. Liu, Oak Ridge National Laboratory, Oak Ridge, TN, *The Microstructure of Ordered Alloys (Co, Fe)3 V*

THIRD IN CLASS: M. Sarikaya, Lawrence Berkeley Laboratory, Berkeley, CA, *Crystallography and Distribution of Retained Austenite in Lath Martensitic Steels*

HONORABLE MENTION: C.C. Lo, J.A. Augis, and M.R. Pinnel, Bell Telephone Laboratory, Columbus, OH, *Relationship Between Deposition Temperature, Grain Size and Hardness in Sputtered Gold Thin Films*

HONORABLE MENTION: G.V. Rao, Westinghouse Electric Corporation, Pittsburgh, PA, *Electron Diffraction and Contrast Effects from Some Second Phase Particles*

Electron Microscopy — Analytical (Class 5)

FIRST IN CLASS: S.A. David, D.N. Braski, C.P. Haltom, and R.C. Robertson, Oak Ridge National Laboratory, Oak Ridge, TN, *Stem Microanalysis of Austenitic Stainless Steel Weld Metal*

SECOND IN CLASS: M. Raghaven, Exxon Research and Engineering Company, Linden, NJ, *Catalytic Decomposition of Acetylene*

THIRD IN CLASS: M. Firth, McGill University, Montreal, Quebec, Canada, *Identification of Corrosion Product on Head of Bolt*

HONORABLE MENTION: S. Komatsu, S. Nakamura, M. Nakahashi, and I. Watanabe, Tokyo Shibaura Electric Company, Ltd., Kanagawa, Japan, *Mo2C-V4C3 Carbides Precipitated in Cr-Mo-V Steel*

Electron Microscopy — Scanning (Class 6)

FIRST IN CLASS: N.J. Kim and G. Thomas, Lawrence Berkeley Laboratory, Livermore, Berkeley, CA, and J.Y. Koo, Rutgers University, New Brunswick, NJ, *Fracture Behavior of Dual Phase Fe/Si/C Steels*

SECOND IN CLASS: B. Strope, Systems Research Laboratories, Dayton, OH, J. Leonard and D. Fritts, Wright-Patterson AFB, Dayton, OH, *Cadmium Electrodes*

THIRD IN CLASS: J.F. Capes and C.E. Cross, Rockwell International, Golden, CO, *Modification of Eutectic Fusion Zone Microstructure*

HONORABLE MENTION: F. Furosawo, I. Taguchi, and R. Matsumoto, Nippon Steel Corporation, Kawasaki, Japan, *Three Dimensional Observation of Intermetallic Compound and Micro Facet Pit in High Alloy Using the Potentiostatic/Coulometric Etching Method with Non-Aqueous Solution System Electrolyte*

HONORABLE MENTION: D. Eylon, University of Cincinnati, Wright-Patterson AFB, Dayton, OH, and R. Omlor, R. Bacon, and G. Teeters, Systems Research Laboratory, Inc., Dayton, OH. and F.H. Froes, Air Force Materials Laboratory, Wright-Patterson AFB, Dayton, OH, *Characterization of Titanium Alloy Powders*

HONORABLE MENTION: Capt. J.M. Hyzak, Air Force Materials Laboratory, Wright-Patterson AFB, Dayton, OH and G.L. Teeters and R.J. Bacon, Systems Research Laboratory, Dayton, OH, *Transition in Fatigue Initiation for P/M Superalloys*

Pretty Microstructures (Class 7)

FIRST IN CLASS: P.E. Danielson, Teledyne Wah Chang Albany, Albany, OR, *Hafnium Crystal Bar Heat Tinted at 1000°F Showing Mechanical Twinning*

SECOND IN CLASS: R.J. Bacon and J.C. Conley, Systems Research Laboratory, Dayton, OH, *A Study of Particle Size and Shape Analysis of MnO_2 Powder*

THIRD IN CLASS: T. Piotrowski, Engelhard Minerals and Chemicals Corporation, Edison, NJ, *Four Seasons*

HONORABLE MENTION: R.H. Lee, Argonne National Laboratory, Argonne, IL, *Substructure of Slow Cooled Zircaloy Containing Oxygen*

HONORABLE MENTION: V.P. Laptik, United Technologies Research, East Hartford, CT, *Unidirectionally Solidified Terbium-Dysprosium-Iron ($Tb_{0.27}Dy_{0.73}Fe_2$) Crystal*

HONORABLE MENTION: D. Eylon, University of Cincinnati, Wright-Patterson AFB, Dayton, OH, and R. Omlor, R. Bacon, and G. Teeters, Systems Research Laboratory, Inc., Dayton, OH, and F.H. Froes, Air Force Materials Laboratory, Wright-Patterson AFB, Dayton, OH, *Making Titanium Powders the Natural Way*

Unique, Unusual or Other Techniques (Class 8)

FIRST IN CLASS: A.T. Santhanam, L.E. Willertz, and P.M. Yuzawich, Westinghouse Research and Development Center, Pittsburgh, PA, *How Can You Distinguish Low Cycle from High Cycle Fatigue in Copper*

SECOND IN CLASS: S.C. Huang, General Electric Company, Schenectady, NY, M.E. Glicksman, Rensselaer Polytechnic Institute, Troy, NY, *Growth of a Dendrite*

THIRD IN CLASS: M.R. Pinnel, H.G. Tompkins, and J. Augis, Bell Telephone Laboratory, Columbus, OH, *Grain Boundary Diffusion and Surface Oxidation in a Gold-Copper Layered Structure*

HONORABLE MENTION: I. Graf, Fhg Institut fur Angewandte Materialforschung (IFAM), Bremen, West Germany, *Hard Solder Joint of NiCrSi Alloy*

HONORABLE MENTION: J.R. Hitchings, Foote Mineral Company, Exton, PA, and W.E. Gresham, Micron Incorporated, Wilmington, DE, *Plasma Etching of Graphite*

HONORABLE MENTION: G.V. Rao and R. Gopal, Westinghouse Electric Corporation, Pittsburgh, PA, *Acoustic Metallography*

Color Micrographs (from any class) (Class 9)

FIRST IN CLASS: J.D. Holder, W.H. Farmer, and J.W. Nave, G.W. Clark, Oak Ridge National Laboratory, Oak Ridge, TN, *Interference Colors Aid in Study of Cr_2O_3-Re Eutectic Composites*

SECOND IN CLASS: I. Taguchi, M. Nagumo, and R. Matsumoto, Nippon Steel Corporation, Kawasaki, Japan and T. Asaoko, Tokyo Denki University, Chiyode-Ku, Japan N. Morikawa, Tokyo University, Tokyo, Japan, *Tritium Color Autoradiograph of Pure Iron Strained Under Hydrogen Charging*

THIRD IN CLASS: S. Prochazka, C.R. Rodd, and A.S. Holik, General Electric Company, Schenectady, NY, *Sintered Aluminum Nitride Thin Section*

HONORABLE MENTION: E. Nakata and M. Imanaka, Weseda University, Tokyo, Japan, J. Iisaka, IBM Japan, Tokyo, Japan, *Application of Texture Analysis for Quantitative Metallography*

Student Entries (Class 10)

FIRST IN CLASS: M.J. Kozak, Case Western Reserve University, North Tonawanda, NY, *The Development of a Color Etching Technique to Study the As-Cast Solidification Structure of Low Carbon Steel*

SECOND IN CLASS: L.D. Brown and J.D. Smith, University of Texas, Austin, TX, *Fracture of Ductile Iron*

THIRD IN CLASS: S. Tauscher, Rensselaer Polytechnic Institute, Watervliet Arsenal, NY, *Incomplete Austenitizing of Thick-Walled Forgings*

HONORABLE MENTION: G. Kyle and T. Sharp, Seneca College of Applied Arts and Technology, Willowdale, Ontario, Canada, *Brazed Tubes of 4140 Steel Used in Lower Frame Joint of a Bicycle Frame*